MICROSENSORS

Edited by **Igor V. Minin** and **Oleg V. Minin**

Microsensors
Edited by Igor V. Minin and Oleg V. Minin

Published by InTech
Janeza Trdine 9, 51000 Rijeka, Croatia

Publishing Process Manager Sandra Bakic

Technical Editor Teodora Smiljanic

Cover Designer InTech Design Team

Additional hard copies can be obtained from orders@intechopen.com

Microsensors, Edited by Igor V. Minin and Oleg V. Minin
p. cm.
ISBN 978-953-307-170-1

For EU product safety concerns:
IN TECH d.o.o., Prolaz Marije Krucifikse Kozulić 3, 51000 Rijeka, Croatia,
info@intechopen.com or visit our website at intechopen.com.

Meet the editors

Oleg V. Minin received a B.A. in Physics from the Novosibirsk State University, a PhD in Physics from Tomsk State University in 1987 and a Doctor of science from NSTU in 2002. Currently he is a full Professor in the Department of Information Protection at Novosibirsk State Technical University (NSTU), Russia. From 1982 to 2001 he was Chief Research Scientist at the Institute of Applied Physics, Novosibirsk, Russia. Dr. Minin's research interests are in the areas of diffractive optics and antenna experiment (including explosive plasma antenna), millimeter wave and THz photonics and nanophotonics, information security, detection of hidden weapons as well as development of antiterrorism devices, experiment technologies, explosive physics. He is a member of SPIE, COST-284 and COST-ic0603 and he is the author of several books and book chapters in technical publications. For his work Dr. Minin was awarded the Commendation for Excellence in Technical Communications (LaserFocusWorld, 2003) and commendation by the Minister of Defense of Russia, 2000.

Igor V. Minin is a full Professor in the Department of Information Protection at Novosibirsk State Technical University (NSTU), Russia. Dr. Minin received a B.A. in Physics from the Novosibirsk State University, a PhD in Physics from Leningrad Electro-Technical University in 1986. and a Doctor of science from NSTU in 2002. Dr. Minin has over twenty years of international industrial and academic experience and has played key roles in a number of projects including 3D millimeter wave real-time imaging and antiterrorism applications. He is the author or coauthor of approximately 350 research articles, seven monographers (including Diffractive optics of millimeter waves (IOP Publisher, Boston-London, 2004), Basic Principles of Fresnel Antenna Arrays (Springer, 2008)), and has been awarded 24 patents and inventions. He is the author of several books and book chapters in technical publications and has been the Editor of several books including Microwave and Millimeter Wave Technologies Modern UWB antennas and equipment (InTech, Austria 2010) and Microwave and Millimeter Wave Technologies from Photonic Bandgap Devices to Antenna and Applications (InTech, Austria 2010).

Contents

Preface XI

Part 1 Magnetic Sensors 1

Chapter 1 **Magnetic Microsensors 3**
Căruntu George and Panait Cornel

Chapter 2 **Photoelectronic Magnetic Microsensor with a Digit Readout 49**
Hsing-Cheng Chang

Chapter 3 **Development of Resonant Magnetic Field Microsensors: Challenges and Future Applications 65**
Agustín L. Herrera-May, Luz A. Aguilera-Cortés, Pedro J. García-Ramírez, Nelly B. Mota-Carrillo, Wendy Y. Padrón-Hernández and Eduard Figueras

Part 2 Chemical Microsensors 85

Chapter 4 **A Heat Flux Microsensor for Direct Measurements in Plasma Surface Interactions 87**
Dussart Rémi, Thomann Anne-Lise and Semmar Nadjib

Chapter 5 **Microsensors for Microreaction and Lab-on-a-chip Applications 109**
Pawel Knapkiewicz and Rafal Walczak

Chapter 6 **Chemical Microsensors with Ordered Nanostructures 143**
Marina Vorozhtsova, Jana Drbohlavova and Jaromir Hubalek

Part 3 **Optical Microsensors 159**

Chapter 7 **Surface-Enhanced Raman Scattering Sensors based on Hybrid Nanoparticles 161**
Rafael Contreras-Cáceres, Benjamín Sierra-Martín and Antonio Fernández-Barbero

Chapter 8 **Optical Fiber Microsensor of Semidrop 185**
Esteban Molina-Flores, R. B. López-Flores, Daniel Molina-Flores, José A. Dávila-Píntle, Germán A. Muñoz-Hernández Carlos A. Gracios-Marín and Enrique Morales-Rodríguez

Chapter 9 **A Glass Capillary-based Microsensor for L-Glutamate in *in vitro* Uses 203**
Masao Sugawara and Atushi Shoji

Chapter 10 **Diffractive Optics Microsensors 217**
Igor V. Minin and Oleg V. Minin

Part 4 **Microsensors Application 235**

Chapter 11 **Strength Reliability of Micro Polycrystalline Silicon Structure 237**
Shigeru Hamada, Kenji Hasizume, Hiroyuki Nakaura and Yoshihide Sugimoto

Chapter 12 **MEMS Gyroscopes for Consumers and Industrial Applications 253**
Riccardo Antonello and Roberto Oboe

Chapter 13 **Planar Oxygen Sensors for Non Invasive Imaging in Experimental Biology 281**
Henning Tschiersch, Gregor Liebsch, Achim Stangelmayer, Ljudmilla Borisjuk and Hardy Rolletschek

Preface

Microsensors are appropriately categorized as "transducers", which are defined as devices that convert energy from one form to another. In the case of microsensors, the device typically converts a measured mechanical signal into an electrical signal. The critical physical dimensions of microsensors devices can vary from well below one micron on the lower end of the dimensional spectrum, all the way to several millimeters. Sensors and microsensors are: force and pressure microsensors, position and speed microsensors, acceleration microsensors, chemical microsensors, biosensors and temperature sensors.

The common trends in sensor technology today are: Miniaturization, Integration: sensor with signal processing circuits for linearising sensor output, etc., sensor with built-in actuator for automatic calibration, change of sensitivity etc., and Sensor arrays: one-function units (to improve reliability), multiple-function units.

The main advantages of microsensors, as it is well-known, are: lower manufacturing cost (mass-production, less materials), wider exploitation of IC technology (integration), wider applicability to sensor arrays, lower weight (greater portability).

Over the last years, advances in microsensors, computing, physics, chemistry, have enabled new and innovative tests that have allow to design a new devices to improve outcomes.

This book is planned to publish with an objective to provide a state-of-art reference book in the area of microsensors for engineers, scientists, applied physicists and post-graduate students. Also the aim of the book is the continuous and timely dissemination of new and innovative research and developments in microsensors.

This reference book is a collection of 13 chapters characterized in 4 parts: magnetic sensors, chemical, optical microsensors and applications.

This book provides an overview of resonant magnetic field microsensors based on MEMS, optical microsensors, the main design and fabrication problems of miniature sensors of physical, chemical and biochemical microsensors, chemical microsensors with ordered nanostructures, surface-enhanced Raman scattering microsensors based on hybrid nanoparticles, etc.

Several interesting applications area are also discusses in the book like MEMS gyroscopes for consumer and industrial applications, microsensors for non invasive imaging in experimental biology, a heat flux microsensor for direct measurements in plasma surface interactions and so on.

Igor V. Minin and Oleg V. Minin
Novosibirsk State Technical University,
Russia

Part 1

Magnetic Sensors

1

Magnetic Microsensors

Căruntu George and Panait Cornel
Faculty of Electronics, Electrotechnics and Computer Science, Constanta Maritime University,Constanta, Romania

1. Introduction

In the presence of a magnetic field, the Hall effect takes place in the active region of the transistors, however their magnetic sensitivity is insignificant.Moreover, the Hall effect may interfere with the action of a bipolar transistor in many ways which makes the analysis and optimization of devices much more difficult.
However, there are also magnetotransistors structures in which, under appropriate operating conditions the magnetic sensitivity increases to values useful in practical work. In this way integrated magnetic sensors useful for emphasizing and measuring mechanical and geometrical quantities can be obtained.

The double-collector bipolar magnetotransistors

1.1 The general characterization of the double-collector bipolar magnetotransistors

Figure 1.1 shows the cross section of a double collector *npn* vertical magnetotransistor operating on the current deflection principle [1]. This structure is compatible with the bipolar integrated circuit technology.

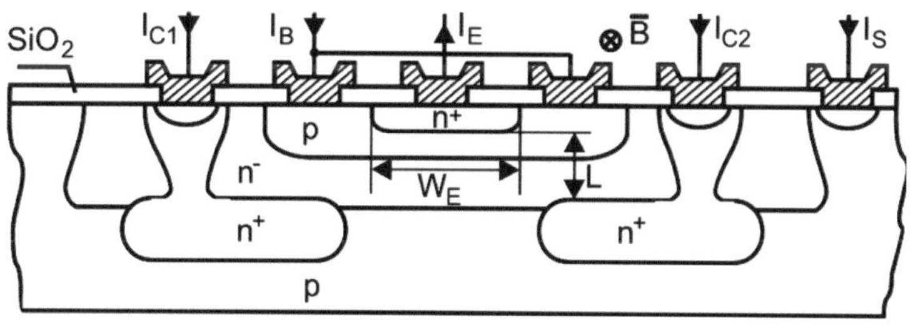

Fig. 1.1. The structure of a double-collector magnetotransistor

The most of the n type low-doped epitaxial layer serves as the collector region and is depleted of the charge carriers upon reverse biasing of the collector-base junction. The two collector contacts are realised by splitting the buried layer (n^+). L is the collector-emitter distance, and W_E is the width of the emitter. In the absence of the magnetic field the electron flow injected into the emitter, which crosses the base is symmetrical and the two collector currents are equal: $I_{C1} = I_{C2}$. In the presence of a magnetic field having the

induction $\overline{B}$ parallel with the device surface, the distribution of the emitter electron current becomes asymmetrical and causes an imbalance of the collector currents: $\Delta I_C = I_{C1} - I_{C2}$.
The analysed magnetotransistor operates in the Hall current mode and ΔI_C depends on the Hall transverse current. Assimilating the low-dopped epitaxial layer of the collector region with a short Hall plate, and based on the properties of dual Hall devices it results [2]:

$$\Delta I_C = \frac{I_H}{2} = \frac{1}{2}\mu_{Hn} \cdot \frac{L}{W_E} \cdot G \cdot I_C \cdot B \tag{1.1}$$

where μ_{HCh} is the carriers Hall mobility in the channel, G denotes the geometrical correction factor and $I_C = I_{C1}(0) + I_{C2}(0)$.

1.2 The sensor response and the sensitivity related to the bias current

The sensor response is expressed by:

$$h(B) = \frac{\Delta I_C}{(I_{C1} + I_{C2})_{B=0}} = \frac{1}{2}\mu_{Hn}\frac{L}{W_E} \cdot G \cdot B_\perp \tag{1.2}$$

and it is linear for induction values which satisfy the condition: $\mu_H^2 \cdot B_\perp^2 << 1$

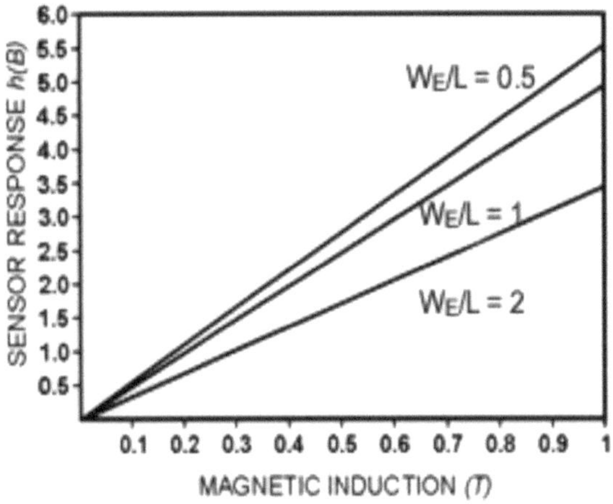

Fig. 1.2. The *h(B)* depending on *B* for three devices of different geometry

In figure 1.2 the geometry influence on h(B) values for three magnetotransistor structures can be seen ratios W_E / L $(W_E = 50\mu m)$.

MGT1: $W_E / L = 0.5$, $(L / W_E)G = 0.72$;

MGT2: $W_E / L = 1$, $(L / W_E)G = 0.68$;

MGT3: $W_E / L = 2$, $(L / W_E)G = 0.46$;

It is noticed that the response $h(B)$ is maximum for $W_E / L = 0.5$ structure. Decreasing the emitter-collector distance, $h(B)$ decreases with 37.5% for $W_E = 2L$, as compared to the maximum value. The sensor response decreases with 10.7%, comparative with $W_E / L = 0.5$ structure if the distance between emitter and collector doubles. For the same geometry $W_E / L = 0.5$, the response is depending on material features. In figure 1.3 $h(B)$ values of three sensors MGT1, MGT2, MGT3 are shown, realized on

Si ($\mu_{Hn} = 0.15 m^2 V^{-1} s^{-1}$),

InP ($\mu_{Hn} = 0.46 m^2 V^{-1} s^{-1}$)

GaAs ($\mu_{Hn} = 0.80 m^2 V^{-1} s^{-1}$).

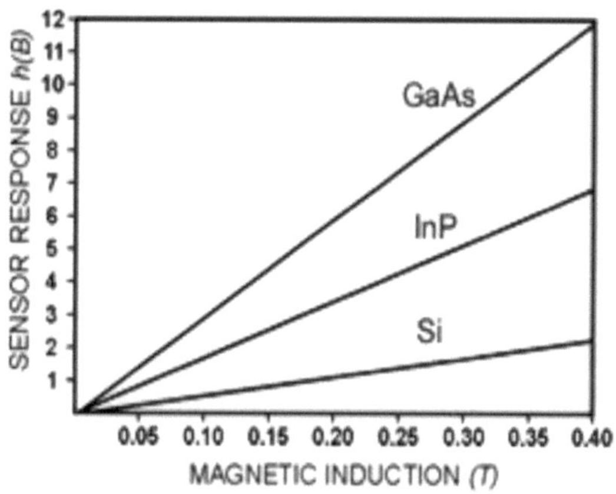

Fig. 1.3. The *h(B)* depending on *B* for three devices on different materials

A magnetotransistor may be regarded as a modulation transducer that converts the magnetic induction signal into an electric current signal.
This current signal or output signal is the variation of collector current, caused by induction $B_\perp$.
The absolute sensitivity of a magnetotransistor used as magnetic sensors is:

$$S_A = \left|\Delta I_C / B\right| = \frac{1}{2}\mu_{Hn} \cdot \frac{L}{W_E} G \cdot I_C \tag{1.3}$$

The magnetic sensitivity related to the devices current is defined as follows:

$$S_I = \frac{1}{I_C}\left|\frac{\Delta I_C}{B_\perp}\right| = \frac{1}{2}\mu_{Hn}\frac{L}{W_E} G \tag{1.4}$$

For a given induction $(B = 0,4T)$ and at given collector current $I_C = 1mA$, the sensitivity depends on the device geometry and the material properties. In table 1.1 the obtained values for five magnetotransistors structures are presented.

The analysis of the main characteristics of the double-collector magnetotransistor shows that the $W_E / L = 0.5$ structure is theoretically favourable to high performance regarding signal-to-noise ratio, as well as the offset equivalent magnetic induction. Also substituting the silicon technology by using other materials such as GaAs or InSb with high carriers mobility values assure higher characteristics of the sensors

	W_E / L	$\mu_{Hn} [m^2V^{-1}s^{-1}]$	$S_I [T^{-1}]$
MGT1	2	0,15 Si	0,035
MGT2	1	0,15 Si	0,05
MGT3	0,5	0,15 Si	0,055
MGT4	0,5	0,46 InP	0,168
MGT5	0,5	0,85 GaAs	0,292

Table 1.1. The numerical values of the supply-current-related sensitivity.

1.3 The offset equivalent magnetic induction

The difference between the two collector currents in the absence of the magnetic field is the offset collector current:

$$\Delta I_{Coff} = I_{C1}(0) - I_{C2}(0) \tag{1.5}$$

The causes consist of imperfections specific to the manufacturing process: the contact non-linearity, the non-uniformity of the thickness and of the epitaxial layer doping, the presence of some mechanical stresses combined with the piezo-resistive effect.
To describe the error due to the offset the magnetic induction is determined, which produces the imbalance $\Delta I_C = \Delta I_{Coff}$. The offset equivalent magnetic induction is expressed by considering the relation (4):

$$B_{off} = \frac{\Delta I_{Coff}}{S_I I_C} = \frac{2}{\mu_{Hn}} \cdot \frac{\Delta I_{Coff}}{I_C} \cdot \left(G \frac{L}{W_E} \right)^{-1} \tag{1.6}$$

Considering $\Delta I_{Coff} = 0.10 \mu A$ and assuming that the low magnetic field condition is achieved, in figure 1.4 the dependence of B_{off} on I_C for three magnetotransistors with the same geometry $W_E / L = 0.5$ realised from different materials is presented:

MGT1: Si with $\mu_{Hn} = 0.15 m^2V^{-1}s^{-1}$;

MGT2: InP with $\mu_{Hn} = 0.46 m^2V^{-1}s^{-1}$;

MGT3: GaAs with $\mu_{Hn} = 0.85 m^2V^{-1}s^{-1}$.

The geometry influence upon B_{off} is shown in figure 1.5 by simulating three magnetotransistors structures realised from silicon and having different W_E / L ratios.

$$\text{MGT1:}\quad W_E / L = 0{,}5; \quad GL / W_E = 0.73;$$

$$\text{MGT2:}\quad W_E / L = 1; \quad GL / W_E = 0.67;$$

$$\text{MGT3:}\quad W_E / L = 2; \quad GL / W_E = 0.46;$$

If the width of the emitter is maintained constant, B_{off} as the emitter-collector distance decreases. This kind of minimum values for the offset equivalent induction are obtained with the device which has $L = 2W_E$, and in the MGT3 device these values are 53.5% bigger.

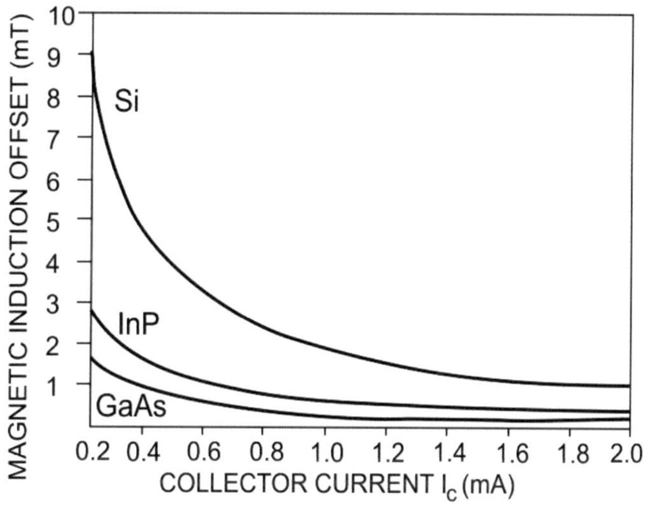

Fig. 1.4. The B_{off} depending on the collector current IC for three devices of different materials

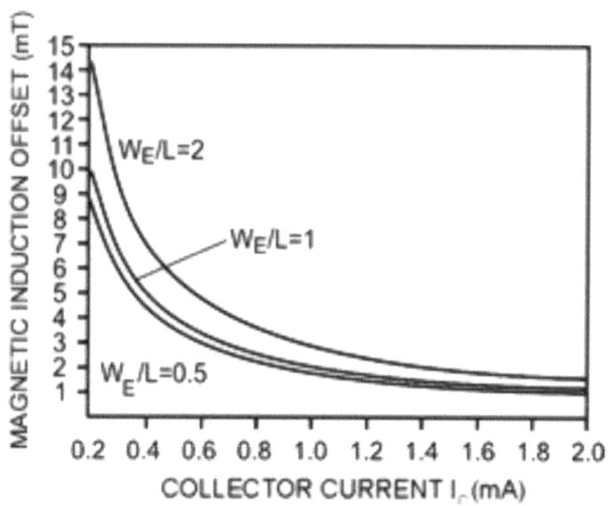

Fig. 1.5. The B_{off} depending on the collector current I_C for three devices of different geometry

1.4 Signal-to-noise ratio

The noise affecting the collector current of a magnetotransistor is shot noise and 1/f noise. Signal-to-noise ratio is defined by:

$$SNR(f) = \frac{\Delta I_C}{[S_{NI}(f) \cdot \Delta f]^{1/2}} \tag{1.7}$$

where Δf denotes a narrow frequency band around the frequency f, and $S_{NI}(f)$ denotes the noise current spectral density in the collector current.

In case of shot noise, the noise current spectral density at frequencies over 100 Hz is given by [3]:

$$S_{NI} = 2qI \tag{1.8}$$

where I is the device current.

In case of shot noise, in a narrow range f of frequency values, By substituting (1.1) and (1.8) into (1.7) it results that:

$$SNR(f) = \frac{1}{2\sqrt{2}} \mu_{Hn} (\frac{L}{W_E} G) \frac{I_C}{(q \cdot I \cdot \Delta f)^{1/2}} B_\perp \geq \frac{1}{2\sqrt{2}} \mu_{Hn} (\frac{L}{W_E} G) \frac{I_C^{1/2}}{(q\Delta f)^{1/2}} B_\perp \tag{1.9}$$

To emphasise the dependence of $SNR(f)$ on the device geometry there (figure 1.6) three magnetotrasistor structure realised on silicon ($\mu_{Hn} = 0.15 m^2 V^{-1} s^{-1}$)were simulated having different rations W_E / L ($W_E = 40 \mu m$; $\Delta f = 1$; $I_C = 1mA$).

MGT1: $W_E / L = 2$;

MGT2: $W_E / L = 1$;

MGT3: $W_E / L = 0.5$;

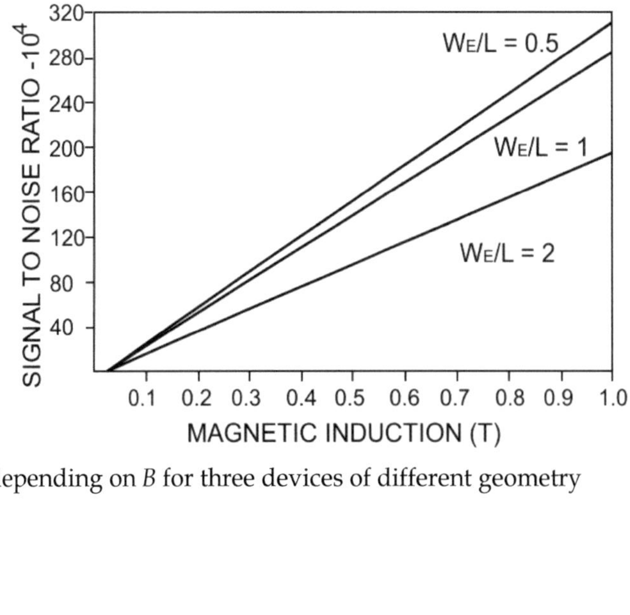

Fig. 1.6. *SNR(f)* depending on *B* for three devices of different geometry

The device were biased in the linear region at the collector current $I_C = 1mA$, the magnetic field has a low level ($\mu_H^2 B^2 << 1$).

It is noticed that the $SNR(f)$ is maximum for $W_E / L = 0.5$ and for smaller values of this ratio. For the same B magnetic induction, increasing the emitter width, $SNR(f)$ decreases with 37.2% for $W_E = 2L$ As compared to the maximum value. In case of $1/f$ noise, the noise current spectral density at the device output is given by [4]:

$$S_{NI}(f) = I^2 \alpha N^{-1} f^{-\beta} \tag{1.10}$$

where I is the device current, $N = n\delta LW_E$ is the total number of charge carriers in the device, α is a parameter called the Hooge parameter and $\beta \cong 1 \pm 0.1$ (typically). For semiconductors, it is reported that α values range from 10^{-9} to 10^{-7}. Substituting (1.1) and (1.10) into (1.7) it is obtained:

$$SNR(f) \cong \frac{(ndLW_E)^{1/2}}{2\alpha^{1/2}} \cdot \mu_{Hn} \left(\frac{f}{\Delta f}\right)^{1/2} \cdot \left(\frac{L}{W_E} G\right) \cdot B_\perp \tag{1.11}$$

To illustrate the $SNR(f)$ dependence on device geometry three split-collector magnetotransistor structures realised on Si were simulated (figure 1.7).

MGT1: $W_E / L = 0.5$;

MGT2: $W_E / L = 1$;

MGT3: $W_E / L = 2$.

It is considered that: $f = 4Hz$, $\Delta f = 1Hz$, $n = 4.5 \cdot 10^{21} m^{-3}$, $d = 4 \cdot 10^{-6} m$, $\alpha = 10^{-7}$, $q = 1.9 \cdot 10^{-6} C$, the devices being biased in the linear region and the magnetic field having a low level. For the same magnetic induction B, $SNR(f)$ is maximum in case of $L = 2W_E$.
The increasing of the emitter collector distance causes the decreasing of $SNR(f)$ with 35.2% for a square structure with 69.1% for $W_E = 2L$.

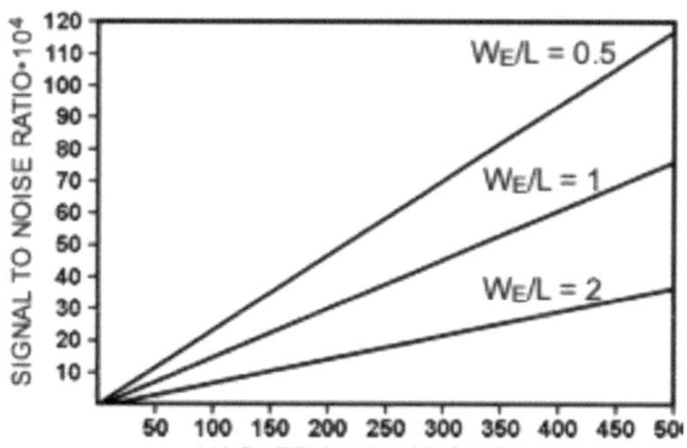

Fig. 1.7. *SNR(f)* depending on *B* for three devices of different geometry

1.5 The detection limit

A convenient way of describing the noise properties of a sensor is in terms of detection limit, defined as the value of the measurand corresponding to a unitary signal-to-noise ratio.
In case of shot noise, for double-drain magnetotransistors using (1.9) it results for detection limit it results that:

$$B_{DL} \leq \frac{2\sqrt{2}\left(q\Delta f\right)^{1/2}}{\mu_{Hn}\left(L/W_E\right)G} I_C^{-1/2} \quad (1.12)$$

To illustrate the B_{DL} dependence on device geometry (figure 1.8) three double-collector magnetotransistor structures on silicon $\left(\mu_{Hn} = 0.15 m^2 V^{-1} s^{-1}\right)$ were simulated having

MGT1: $W_E/L = 0.5$;

MGT2: $W_E/L = 1$;

MGT3: $W_E/L = 2$;

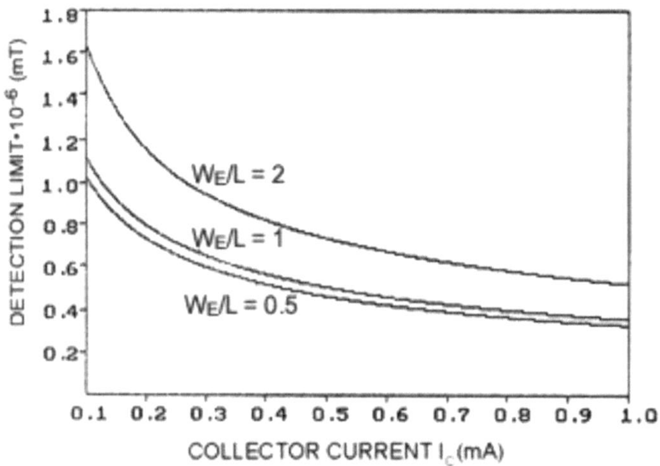

Fig. 1.8. B_{DL} depending on the collector total current for three devices of different geometry

It is noticed that the B_{DL} is minimum for $W_E/L = 0.5$ structure. For optimal structure B_{DL} decreases at materials of high carriers mobility.
In figure 1.9 the material influence on B_{DL} values for three double-collector magnetotransistor structures realised from *Si*, *GaSb* and *GaAs* can be seen having the same size: $L = 200\mu m$, $W_E = 100\mu m$.

MGT1: *Si* with $\mu_{Hn} = 0,15 m^2 V^{-1} s^{-1}$;

MGT2: GaSb with $\mu_{Hn} = 0,5 m^2 V^{-1} s^{-1}$;

MGT3: *GaAs* with $\mu_{Hn} = 0,8 m^2 V^{-1} s^{-1}$.

By comparing the results for the two types of Hall devices used as magnetic sensors a lower detection limit of almost 2-order in double-colletor magnetotransistors is recorded.

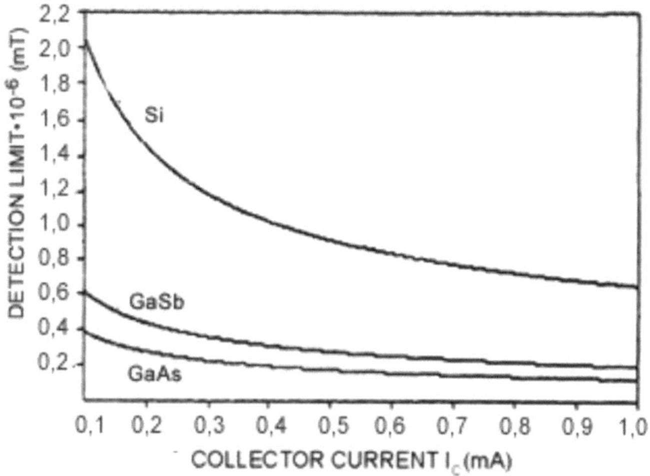

Fig. 1.9. B_{DL} depending on the drain current for three devices of different materials

1.6 The noise-equivalent magnetic induction

The noise current at the output of a magnetotransistor can be interpreted as a result of noise equivalent magnetic induction.

The mean square value of noise magnetic induction (NEMI) is defined by:

$$\langle B_N^2 \rangle = (\int_{f1}^{f2} S_{NI}(f) \cdot df)(S_I \cdot I_C)^{-2} \tag{1.13}$$

In case of shot noise, by substituting (1.1) and (1.8) into (1.13) it results that:

$$\begin{aligned} < B_N^2 > &= 2qI \cdot \Delta f \cdot 4\left(\frac{W_E}{L}\right)^2 \cdot \frac{1}{G^2 \mu_{Hn}^2} \cdot \frac{1}{I_C^2} \le \\ &\le 8q\left(\frac{W_E}{L}\right)^2 \cdot \frac{\Delta f}{G^2} \cdot \frac{1}{\mu_{Hn}^2} \cdot \frac{1}{I_C} \end{aligned} \tag{1.14}$$

Considering the condition of low value magnetic field fulfilled ($\mu_H^2 B^2 << 1$),a maximum value for $(L / W_E)G = 0.74$, if $W_E / L < 0.5$. [5] is obtained

In this case:

$$< B_N^2 >_{\min} \le 14.6q \frac{\Delta f}{I_C} \cdot \frac{1}{\mu_{Hn}^2} \tag{1.15}$$

In figure 1.10 NEMI values obtained by simulation of three magnetotransistors structures from different materials are shown*MGT*$_1$: Si with $\mu_{Hn} = 0.15 m^2 V^{-1} s^{-1}$

MGT$_2$: InP with $\mu_{Hn} = 0.46 m^2 V^{-1} s^{-1}$

MGT_3: GaAs with $\mu_{Hn} = 0.85 m^2 V^{-1} s^{-1}$

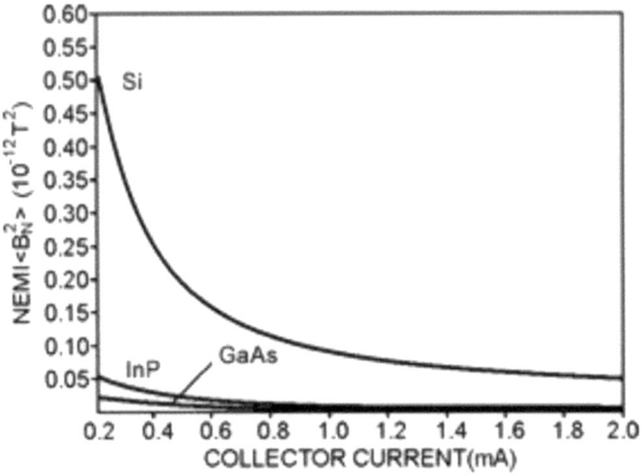

Fig. 1.10. NEMI depending on the collector current for three devices of different materials

To emphasize the dependence of NEMI on device geometry (figure 1.11) three double-collector magnetotransistors structures realised on silicon, $\mu_{Hn} = 0.15 m^2 V^{-1} s^{-1}$ were simulated, having different ratios $W_E/L\,(W_E = 50\mu m)$. The devices were based

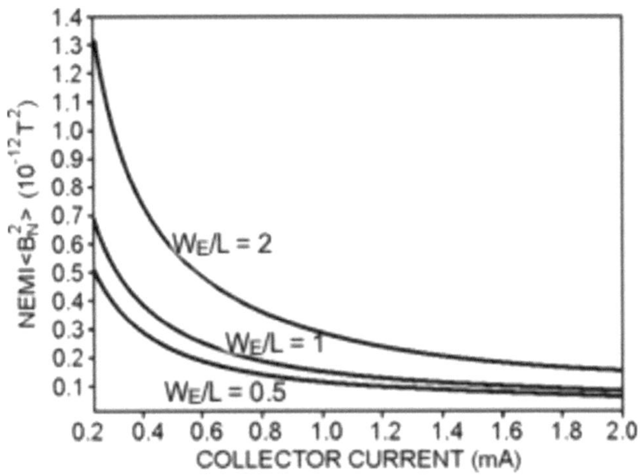

Fig. 1.11. NEMI depending on the collector current for three devices of different geometry

MGT_1 with $W_E/L = 0.5$ and $(LG/W_E)^2 = 0.576$

MGT_2 with $W_E/L = 1.0$ and $(LG/W_E)^2 = 0.409$

MGT_3 with $W_E/L = 02$ and $(LG/W_E)^2 = 0.212$

It is noticed that the NEMI is minimum for $W_E/L = 0.5$, and for smaller values of this ratio. The decreasing of the channel length causes the increase of NEMI with 40.8 % for a square structure $W_E = L$ and with 173 % for $W = 2L$.

Conclusions

The magnetotransistors have a lower magnetic sensitivity than the conventional Hall devices but allow very large signal-to-noise ratios, resulting in a high magnetic induction resolution. The analysis of the characteristics of two magnetotreansistors structures shows that the $W/L = 0.5$ ratio is theoretically favourable to high performance regarding signal-to-noise ratio, as well as the noise equivalent magnetic induction

Also substituting the silicon technology by using other materials such as GaAs or InSb with high carriers mobility values assure higher characteristics of the sensors

The uses of magnetotransistors as magnetic sensors allows for the achieving of some current-voltage conversion circuits, more efficient that conventional circuits with Hall plates.

The transducers with integrated microsensors have a high efficiency and the possibilities of using them ca be extended to some measuring systems of thickness, short distance movement, level, pressure, linear and revolution speeds.

1.7 System to monitor rolling and pitching angles

The efficient operation of the modern maritime ships requires the existence of some reliable command, watch and protection systems that permit transmission, processing and receiving of signals with great speed and reduced errors.

On most of the merchant ships the watch of the rolling and the pitching is done by conventional instruments as gravitational pendulum. The indication of the specific parameters is continuous, the adjustment operations are manual and the transmissions of the information obtained in the measurement process, at distance is not possible.

An automatic and efficient surveillance system ensures the permanent indication of the inclination degree of the ship , the optic and the sound warning in case of exceeding the maximum admissible angle and the simple transmission of the information at distance.

1.7.1 Installation for the measurement of the rolling and pitching thatuses magnetotransistors

The presentation of the transducers

The primary piece of information about the rolling and pitching angle is obtained with the help of the classical system used on ships, with the difference that at the free end of the pendulum,a permanent magnet with reduced dimensions is fixed provided with polar parts shaped like those used in the construction of the magnetoelectric measurement devices. Along the circle arc described by the free end of the pendulum , there are disposed at equal lengths, accordingly to the displacements of $1°$ for the rolling and of $1° \ 30'$ for the pitching, twenty magnetotransistors, ten on one side and ten on the other side of the equilibrium position.

Due to the high inertia moment, the pendulum maintains its vertical position, and actually during the rolling and pitching the graded scale, fixed on the wall, is the one that moves at the same time with the ship.

The transducer for the indication of the rolling is disposed in a vertical plane, transverse on the longitudinal axis of the ship, and the one for the pitching in a vertical plane that contains or is parallel with the longitudinal axis of the ship. In order to simplify the presentation will consider that the pendulum is the one that moves in with the graded scale. In figure 1.14 the principle diagram of the transducer is shown vertical bipolar magnetotransistor with double collector. In the absence of the magnetic field, the two collector currents are equal and the output of the comparator is in "DOWN" state (logical level „0"). In the presence of a field of induction *B*, parallel with the device surface, a lack of poise between the two collector currents ΔI_C is produced and at the input of comparator is applied the voltage:

$$\Delta V_C = \mu_H \left(L / W_E\right) G R_C I_C B \tag{1.18}$$

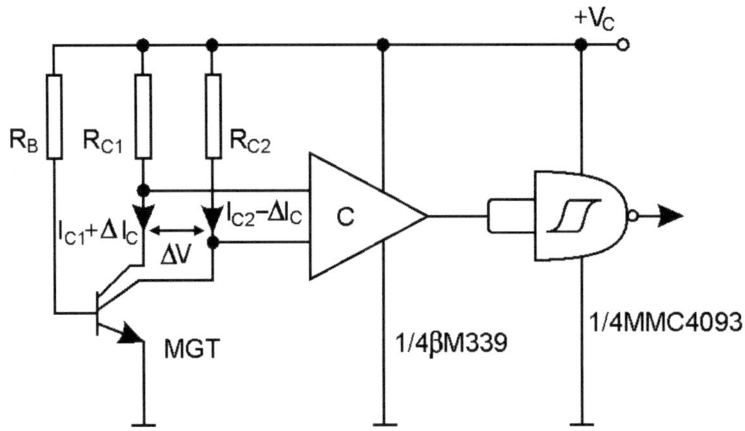

Fig. 1.14. The electric diagram of transducer

This voltage is applied to a comparator with hysteresis, which acts as a commutator. The existence of the two travel thresholds ensure the immunity of the circuit at noise monostable made with *MMC 4093* ensures the same duration for the transducers generated pulses. Applied to the comparator C, this voltage changes its state and the output goes on logical level "1".

The principle block diagram. The description of working

When the ship lists, the permanent magnet of the pendulum will scavenge in turn a number of magnetotransistors, and the signals from their outputs will determine the tipping of the comparators. We will thus obtain impulses which are applied through an "OR" circuit at the CBM input (figure 1.15). This commands the block for the interruption of the power supply (IPS), achieving the cancellation of the potentials in the thyristors anode for a time interval of milliseconds.

At the same time the impulses generated by the transducers are transmitted with the help of separator B1, B2, ..., B10 on the thyristors gates, determining their damping. Once the thyristors are damped, they maintain that state, therefore these are memorizing the last indicated value, until the power supply is cancelled. So if the rolling or the pitching have intermediate values ranging between the successive marks of the graded scale , the last complete measured value remains displayed.

For a rolling value noted with "K" , all the displays from one to "K" will work in "bright point" mode, when for the same "K" value of the rolling will be lighted, therefore the scheme allows the analogical display in bar mode.
Eliminating the diodes D1, D2, ..., D9, the display will be in "bright spot" mode when for the same value "K" of rolling only the "K" display is lighted.
If the inclination of the ship reaches a limit value L settled beforehand with the help of the „K" switch , then the output signal X_L (L=1,2,...,10) commands the bistable of T type which commutes, releasing the sound alarm device.

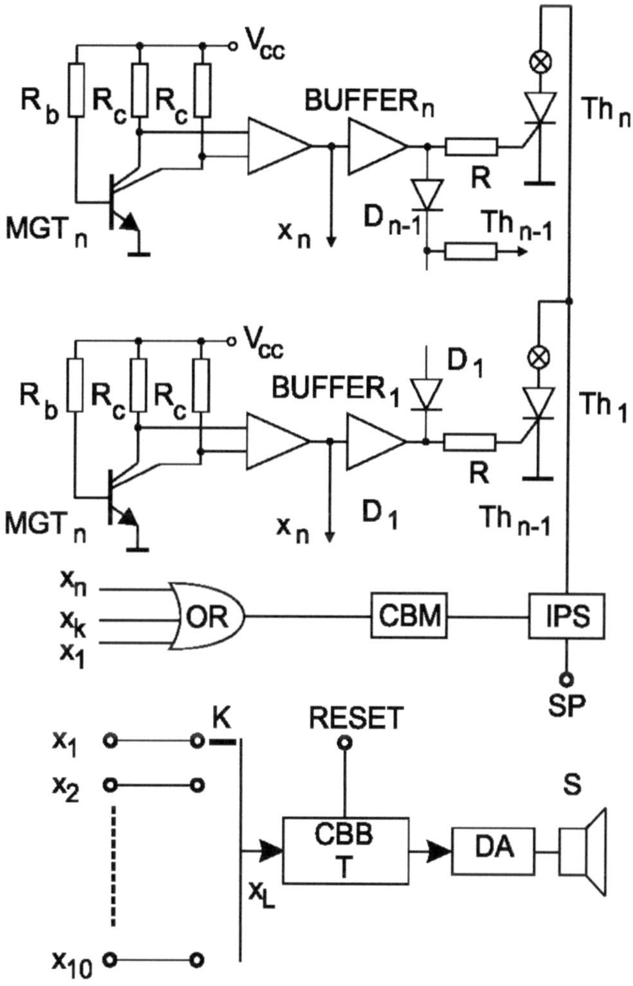

Fig. 1.15. The diagram of the installation for the measurement of the rolling and pitching

Supposing that the angle of the ship's list increases, the pendulum overtakes the „L" position and after it touches a maximum deviation it starts the return run in which it will pass again through the front of the magneto transistor. The impulse generated by this, will swing again the bistable and the sound alarm ceases.
An undesirable situation appears when the maximum inclination of the ship has precisely the pre-established "L" value or it exceeds very little this value. In this case, in the return

run of the pendulum a new "X_L" impulse, which will swing again the bistable, is no longer generated and therefore the acoustic alarm is maintained although the inclination angle has been reduced.
This drawback can be eliminated either by using the "X_{L-1}" output for the bistable command on the input or by replacing the bistable with one of R-S or J-K type commanded by "X_L" and "X_{L-1}" signals. The "X_L" signal establishes the placing in function of the alarming device and the X_{L-1}" signal the blockage of this device, which permits to obtain a safety hysteresis.
If in the scheme in figure 1.15 the diodes D1, D2, ..., D9 are eliminated, then in every moment a single thyristor will be in conduction, suitable at a certain angle.
The signals in the anodes of the other blocked thyristors will be at the logical level "1" and only in the anode of the commanded thyristor the signal will have the logical level "0". This signals are applied to a binary-decimal circuit, at the output of which is obtained the value of the angle in the binary code. A decoder seven segments which commands a display with seven segments and permits the digital display of the measured value.(figure 1.16).
The binary-decimal coder can be a matrix of diodes, a matrix of connected gates in a suitable way, or a specialized circuit like those used in numerical keyboards.

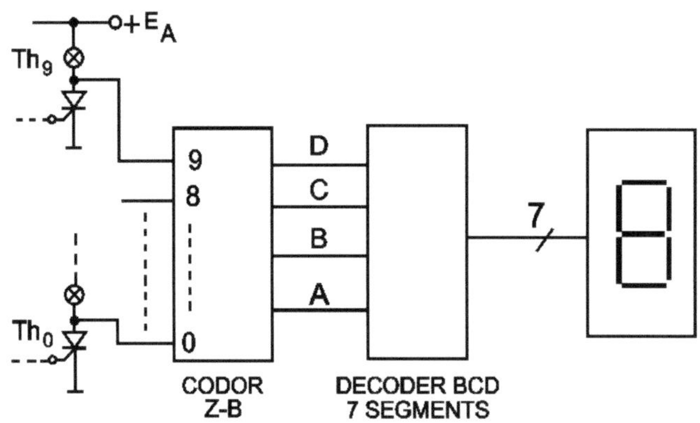

Fig. 1.16. The electrical diagram of the digital display

1.7.2 Installation for the measurement of the rolling and the pitching *that uses phototransistors*

The presentation of the transducer

The principle scheme of the photoelectric transducer is shown in figure 1.17. Classical system on board ships,is used but at the free end of the gravitational pendulum an electroluminiscent diode (LED) in visible or in infrared is set moves along an arc of circle on which the phototransistors are placed. The LED pendulum and the phototransistors are fixed in a box which protects them from the exterior light.
The power supply is ensured from a stabilized source of 12V.
Through one phototransistor, with off-load base / unconnected base and in the absence of the light the so-called "dark current" will flow between the emitter and the collector.

$$I_D = \beta \cdot I_{CB0} \tag{1.19}$$

where β denotes the amplification factor of the transistor and I_{CB0} is the current generated by the base-collector junction, in the absence of light.

When the base-collector junction is illuminated, through this an illumination current will appear (I_l) and this current is all the intense as the illumination is bigger, and the collector current becomes:

$$I_C = \beta\left(I_{CB0} + I_l\right) \tag{1.20}$$

Since the phototransistors are blocked in the absence of the illumination, the output voltage of the collector is practically equal with the value of the power supply $(+E)$. In the moment of the illumination the phototransistors are saturated and the collector voltage lowers to U_{CEsat} value. Therefore the signal given by the transducer is in shape of negative power impulses with the amplitude approximately equal to the value of the power supply

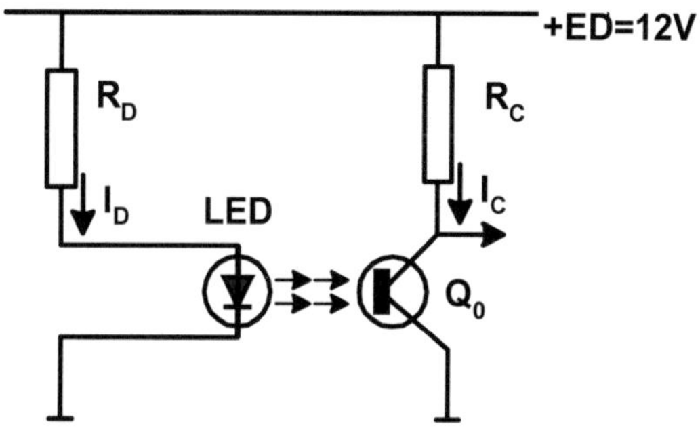

Fig. 1.17. The photoelectric transducer

The principle block diagram. The description of working

Because the measurement of the rolling and the pitching is in fact reduced to the measurement of an angle, the measurement system is made up of two identical measurement blocks, one for the rolling and the other for the pitching, with the difference that the measurement transducer will be placed differently: the one for the rolling on the transversal axis of the ship and the one for the pitching on the longitudinal ship's axis.

In figure 1.18 it is presented the block scheme of the measurement system of the inclination angle of the ship on the longitudinal (transversal) axis of the ship. In this scheme the following notations have been used: TM -measurement transducer of the rolling (pitching); CF - formatter circuit; CM-memory cell; IO - optic indicator; CR -resetting circuit; OC - commanded oscillator; AAF – audio frequency amplifier; SB - supply block; HA - warning block.

It is noticed that the measurement system is made up "2n+1" identical measurement and display chains:

- a chain corresponding to the equilibrium position (zero)
- "n" chains corresponding to „n" number of values of the port inclination angle (stern)
- "n" chains corresponding to „n" number of values of the starboard inclination angle (bow)

The measurement transducer (TM) supplies the necessary information about the inclination angle in a discrete way, meaning that the total angle can be measured has been divided into "2n" sectors ("n" port sectors "n" starboard sectors), and the transducer supplies one impulse for the output suitable for the inclination angle reached by the ship. The impulse suitable for this angle is processed in the formatter circuit (CF) and it is applied to the corresponding memory cell (MC), in which the information about the inclination angle of the ship is kept. This information is displayed by the optic indicator (IO) corresponding the same sector, until the ship, in her movement, reaches the same inclination angle, but moving in reverse direction. The measurement transducer generates at the same output a new impulse that wipes up the information from the memory cell and at the same time the display of the optic indicator. In this way, the measurement transducer supplies two impulses for every inclination angle of the ship, except for the maximum inclination angle, which is pre-established. In case of maximum inclination, the transducer produces a single impulse that is memorized and displayed until the ship, after she reaches the maximum inclination angle in the opposite direction, returns to the equilibrium position. In the passing moment through the equilibrium position, the impulse from the CF out put wipes up, with the help of a reset circuit (CR), the information from all the memory cells and therefore from the cells corresponding to the maximum angles for the two inclination ways of the ship. From now on the working of the system is continued from the equilibrium position.

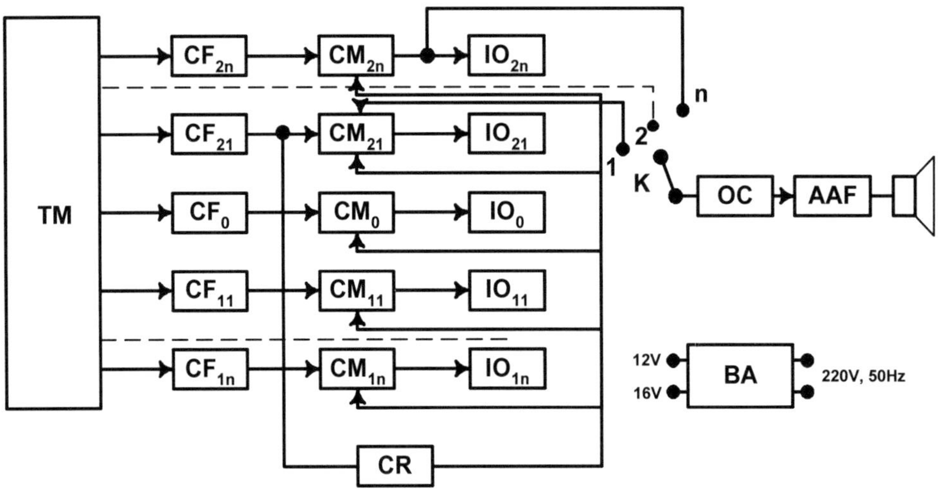

Fig. 1.18. The block diagram of the measurement system based on the phototransistors

Conclusions

The use of another transducer type for the measurement of the rolling and the pitching, namely the classic solution with the transducers in Gray code, would have reduced the number of the components. With such a transducer, using eight pairs of LED phototransistors a measurement accuracy of the angle, better with ±1°, could have been obtained. Besides, this system allows the long distance transmission of the information, but the solution would have required a sophisticated electronic part, the major difficulty being the precise construction achievement of the Gray transducer in eight bits.

The block scheme also contains a sound warning system, which is triggered at a maximum inclination angle pre-established with the help of "K" switch. Once this switch fixed on the desired position, the logical level from the output of the memory cell corresponding to the chosen angle, it will command and release the sound warning system.

This is made up of a command oscillator (OC), an audio frequency amplifier (AAF) and a warning horn (HA). The command signal of the sound warning system is taken over from the input of the optic indicator and not from its output, although it would have been easier to achieve this, out of safety operation reasons. The probability of the memory cell to go out of order is less than that of the optical indicator, moreover, if the optical indicator goes out of order that doesn't imply that the sound warning system gets inoperative.

2. Introduction

The possibility of modelling the channel depth by means of the external supply voltage and low value of area carrier density suggests the possibility of using the MOSFET channel as the active region of a Hall plate.
At the same time, the advantage of integrating on the same chip of a magnetic sensor and the signal processing circuit is outlined.
The Hall devices in MOS structure have some drawbacks: the carrier mobility in the channel is half of its value in the volume of the device; the increasing of *1/f* noise, and the instability of device surface.
The analysis made in this paper outlines the way in which the way of choosing the adequate choice device material and dimensions allows the improvement of CMOS technology sensors.

The double-drain magnetotransistor

2.1 The characterisation of the double-drain magnetotransistor

The double - drain MOS device (figure 2.1) is obtained from a MOSFET structure where its conventional drain region is replaced by two adjacent drain regions[7]. Consequently, the total channel current is shared between these two drain regions.

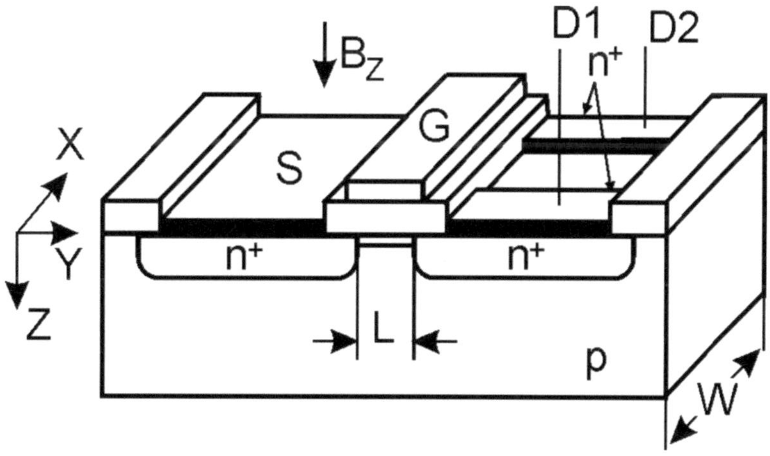

Fig. 2.1. Double-drain MOSFET magnetotransistor

The result of the bias is the linear region is the obtained of a continuous channel of approximately constant thickness, which can be assimilated with a Hall plate.
The deflection of current lines appears under the action of a magnetic field $B_{\perp}$, perpendicular to the device surface. The carrier deflection causes a discrepancy between two drain currents:

$$\Delta I_D = \left| I_{D1}(\bar{B}) - I_{D1}(0) \right| = \left| I_{D2}(\bar{B}) - I_{D2}(0) \right| \quad (2.1)$$

Since the output signal of the double-drain MOS magnetotransistors consists of the current variation between its terminals, this device operates in the Hall current mode. Using the features of dual Hall devices, and the Hall current expression it results [2]:

$$\Delta I_D = \frac{I_H}{2} = \frac{1}{2}\mu_{HCh} \cdot \frac{L}{W} \cdot G \cdot I_D \cdot B_\perp \quad (2.2)$$

The supply-current-related sensitivity of the devices is defined by:

$$S_I = \frac{1}{I_D} \cdot \left| \frac{\Delta I_D}{B_\perp} \right| = \frac{1}{2}\mu_{HCh} \cdot \frac{L}{W} G \quad (2.3)$$

where G denotes the geometrical correction factor and μ_{HCh} is the Hall mobility of the carriers in the channel.

For a given induction $(B = 0,4T)$ and at given drain current $I_D = 1mA$, the sensitivity depends of the device geometry and the material properties.

In table 2.1 the values for five magnetotransistors structures are presented.

Device	W/L		$S_I[T^{-1}]$
MGT1	2	0,07 Si	0,018
MGT2	1	0,07 Si	0,025
MGT3	0,5	0,07 Si	0,028
MGT4	0,5	0,23 InP	0,084
MGT5	0,5	0,42 GaAs	0,146

Table 2.1. The numerical values of the supply-current sensitivity

2.2 The sensor response

The sensor response is expressed by:

$$h(B) = \frac{\Delta I_D}{(I_{d1} + I_{D2})_{B=0}} = \frac{1}{2}\mu_{HCh} \frac{L}{W} \cdot G \cdot B_\perp \quad (2.4)$$

and it is linear for induction values which satisfy the condition: $\mu_H^2 \cdot B_\perp^2 << 1$.

In figure 2.2 the geometry influence on $h(B)$ values for three magnetotransistor structures can be seen, realised on silicon ($\mu_{HCh} = 0.07 m^2V^{-1}s^{-1}$) and having different ratios W / L

MGT1: $W / L = 0.5$, $(L / W)G = 0.72$;

MGT2: $W / L = 1$, $(L / W)G = 0.68$;

$$\text{MGT3: } W/L = 2, \ (L/W)G = 0.46;$$

It is noticed that the response $h(B)$ is maximum for $W/L = 0.5$ structure.
For the same geometry $W/L = 0.5$, the response depends on the material features.
Decreasing the channel length, $h(B)$ decreases with 37.5% for $W = 2L$, As compared to the maximum value.
The sensor response decreases with 10.7%, comparative with $W/L = 0.5$ structure if the channel length doubles.

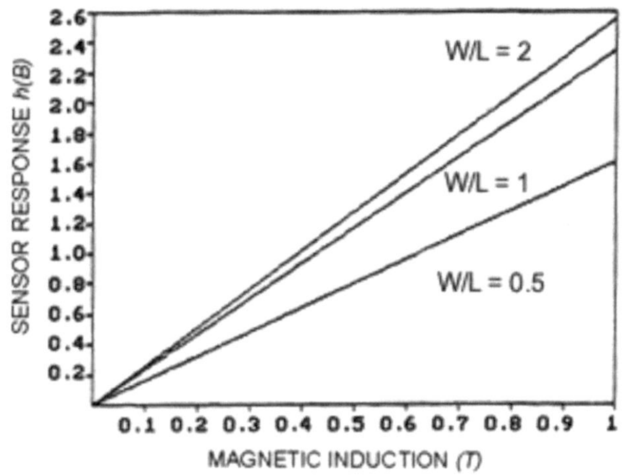

Fig. 2.2. The *h(B)* depending on *B* for three devices of different geometry.

In figure 2.3 are shown $h(B)$ the values of three sensors MGT1, MGT2, MGT3 realised on:

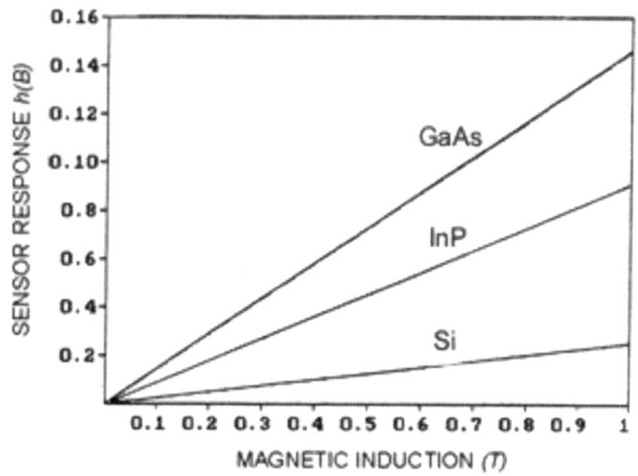

Fig. 2.3. The *h(B)* depending on *B* for three devices on different materials.

$$\text{Si}(\mu_{Hn} = 0.07 m^2 V^{-1} s^{-1});$$

$$\text{InP}(\,\mu_{Hn} = 0.23m^2V^{-1}s^{-1}\,);$$

$$\text{GaAs}\,(\,\mu_{Hn} = 0.40m^2V^{-1}s^{-1}\,).$$

2.3 The offset equivalent magnetic induction

The difference between the two drain currents in the absence of the magnetic field is the offset collector current:

$$\Delta I_{Doff} = I_{D1}(0) - I_{D2}(0) \tag{2.5}$$

The main causes of the offset in the case of Hall devices realised in the MOS integrated circuits technology consists of imperfections specific to the manufacturing process: the misalignment of contacts, the non-uniformity of both the material and channel depth, the presence of some mechanical stresses combined with the piezo-effect.

To describe the error due to the offset the magnetic induction, which produce the imbalance $\Delta I_D = \Delta ID_{off}$ is determined..

The offset equivalent magnetic induction is expressed by considering the relation (2.3):

$$B_{off} = \frac{\Delta I_{Doff}}{S_I I_D} = \frac{2}{\mu_{Hn}} \cdot \frac{\Delta I_{Doff}}{I_D} \cdot \left(G \frac{L}{W_E} \right)^{-1} \tag{2.6}$$

Considering $\Delta I_{Doff} = 0.10\mu A$ and assuming that the low magnetic field condition is achieved in figure 2.4 is presented the dependence of B_{off} on I_D for three magnetotransistors with the same geometry $W / L = 0.5$ realised from different materials:

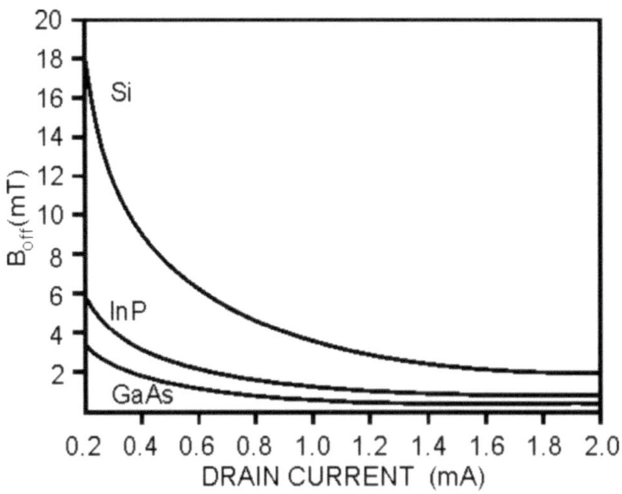

Fig. 2.4. The B_{off} depending on the drain current I_D for three devices of different materials.

$$\text{MGT1: Si, } \mu_{HCh} = 0.07m^2V^{-1}s^{-1};$$

$$\text{MGT2: InP, } \mu_{HCh} = 0.23 m^2V^{-1}s^{-1};$$

$$\text{MGT3: GaAs, } \mu_{HCh} = 0.43 m^2V^{-1}s^{-1}.$$

The geometry influence upon B_{off} is shown in figure 5 by simulating three magnetotransistors structures realised from silicon and having different $\frac{W}{L}$ ratios.

$$\text{MDD1: } \frac{W}{L} = 0,5; \quad G\frac{L}{W} = 0.73;$$

$$\text{MDD2: } \frac{W}{L} = 1; \quad G\frac{L}{W} = 0.67;$$

$$\text{MDD3: } \frac{W}{L} = 2; \quad G\frac{L}{W} = 0.46;$$

If the width of the channel is maintained constant, B_{off} increases as the channel length decreases. So that minimum values for the offset equivalent induction are obtained with the device which has $L = 2W$, and in the MDD3 device these values are 53.5% higher.

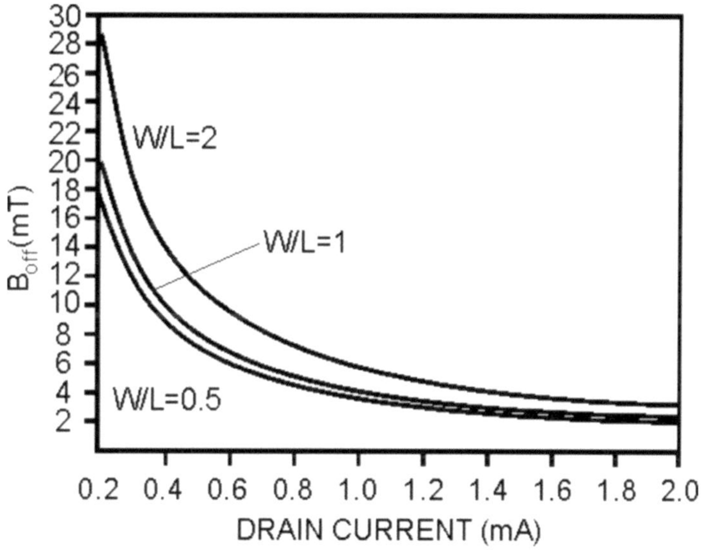

Fig. 2.5. The B_{off} depending on the drain current I_D for three devices of different geometry.

2.4 Signal – to – noise ratio

The noise affecting the drain current of a MOSFET magnetotransistors is shot noise and $1/f$ noise. Signal-to-noise is defined by [8]:

$$SNR(f) = \Delta I_D \cdot \left[S_{NI}(f) \cdot \Delta f \right]^{-1/2} \quad (2.7)$$

where Δf denotes a narrow frequency band around the frequency *f*, and $S_{NI}(f)$ is the noise current spectral density in the drain current. In case of shot noise by substituting (2.2) and (1.8) into (2.7) it results:

$$SNR(f) = \frac{1}{2\sqrt{2}}\mu_{HCh}\left(\frac{L}{W}G\right)\frac{I_D}{(q \cdot I \cdot \Delta f)^{1/2}}B_{\perp} \geq \frac{1}{2\sqrt{2}}\mu_{HCh}\left(\frac{L}{W}G\right)\frac{I_0^{1/2}}{(q\Delta f)^{1/2}} \cdot B_L \qquad (2.8)$$

In figure 2.6 is shown the *SNR(f)* dependence on magnetic induction of three MOS magnetotransistors structures of different materials ($W/L = 0.5$, $\Delta f = 1Hz$, $I_D = 1mA$)

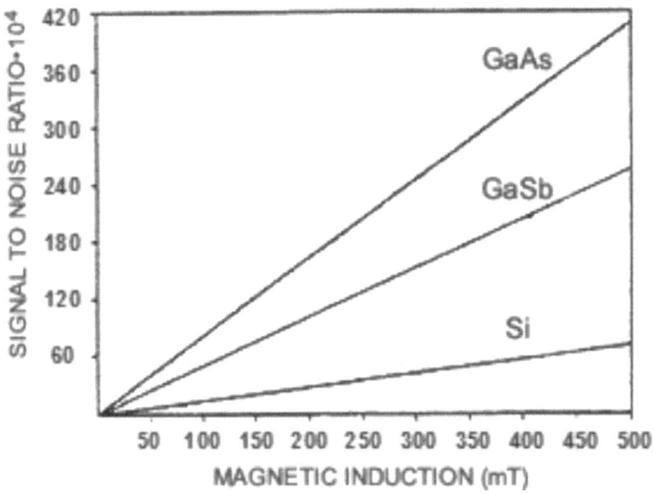

Fig. 2.6. SNR(f) depending on B for three devices of different materials.

$$\text{MGT1:}Si,\ \mu_{HCh} = 0.07m^2V^{-1}s^{-1}$$

$$\text{MGT2:GaSb},\ \mu_{HCh} = 0.25m^2V^{-1}s^{-1}$$

$$\text{MGT3:GaAs},\ \mu_{HCh} = 0.04m^2V^{-1}s^{-1}$$

A high value of carrier mobility causes the increasing of *SNR(f)*. So for $B = 0,5T$, *SNR(f)* increase with 60% for GaAs comparative with GaSb.

To emphasize the dependence *SNR(f)* on device geometry there (Fig. 2.7) three MOS magnetotransistors structures realised on silicon $\mu_{HCh} = 0.07m^2V^{-1}s^{-1}$ were simulated having different ratios L/W . ($W = 50\mu m$ $\Delta f = 1Hz$, $B = 0.2T$, $I_D = 1mA$).

$$\text{MGT1:}\frac{W}{L} = 2 \text{ and } \left(\frac{L}{W}G\right) = 0.212$$

$$\text{MGT2:}\frac{W}{L} = 1 \text{ and } \left(\frac{L}{W}G\right) = 0.409$$

$$\text{MGT3:} \frac{W}{L} = 0.5 \text{ and } \left(\frac{L}{W}G\right) = 0.576$$

It is noticed that the *SNR(f)* is maximum for $W/L = 0.5$, and for smaller values of this ratio. For the same *B* magnetic induction, increasing the channel, *SNR(f)* decreases with 44% width for *W=2L* As compared to the $W/L = 0.5$ structure.
In case of $1/f$ noise, by substituting (1.10) and (2.2) into (2.7) it is obtained:

$$SNR(f) \cong \frac{(ndLW_E)^{1/2}}{2\alpha^{1/2}} \cdot \mu_{Hn} \left(\frac{f}{\Delta f}\right)^{1/2} \cdot \left(\frac{L}{W_E}G\right) \cdot B_\perp \tag{2.9}$$

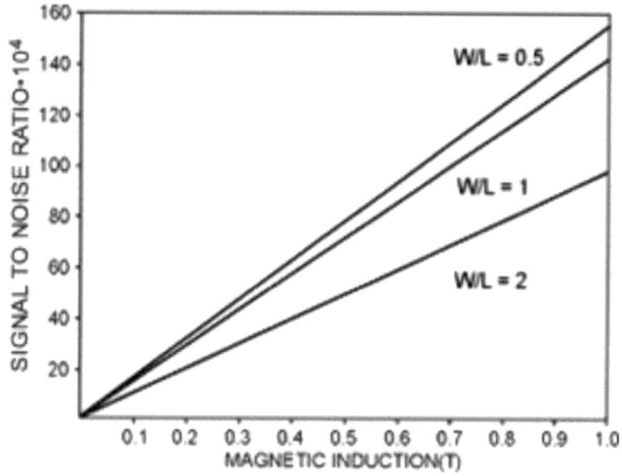

Fig. 2.7. SNR(f) depending on B for three devicesof different geometry.

To illustrate the $SNR(f)$ dependence on device geometry three split-drain magnetotransistor structures realised on Si were simulated (figure 2.8).

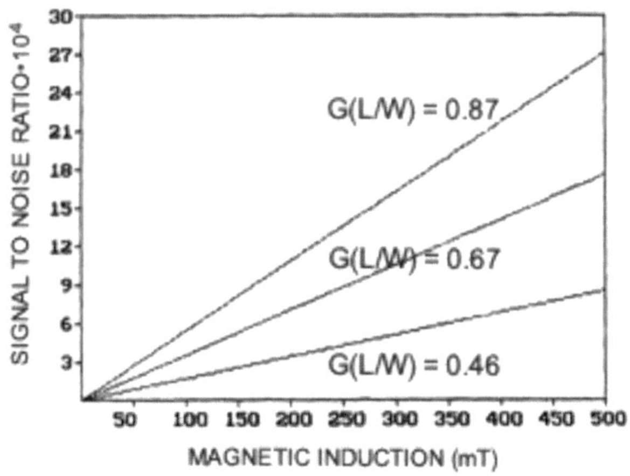

Fig. 2.8. $SNR(f)$ depending on B for three devices of different geometry.

$$G(L/W) = 0.46$$

$$MGT2: W = 50\mu m\,,\ L = 50\mu m\,,$$

$$G(L/W) = 0.67$$

$$MGT3; W = 50\mu m\,,\ L = 100\mu m\,,$$

It is considered that: $f = 4Hz$, $\Delta f = 1Hz$, $n = 4.5\cdot 10^{15} cm^{-3}$, $\alpha = 10^{-7}$, $\delta = 0.5\mu m$, $q = 1.9\cdot 10^{-6} C$ the devices being biased in the linear region and the magnetic field having a low level.
For the same magnetic induction B, $SNR(f)$ is maximum in case of $L = 2W_E$ .The increasing of the canal length causes the decreasing of $SNR(f)$ with 35.2% for a square structure and with 69.1% for $W = 2L$ .In figure 2.9 is presented the dependence of SNR on B for three magnetotransistors whit the same geometry $W / L = 0,5$, $L = 200\ \mu m$ realised from different:

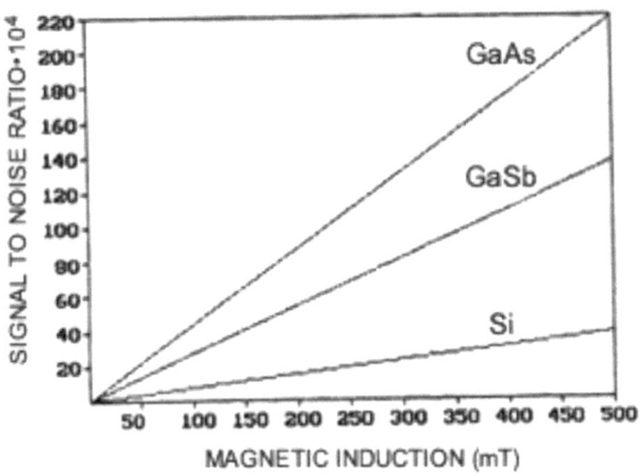

Fig. 2.9. $SNR(f)$ depending on B for three devices of different materials

$$MDD1(\text{Si},\ \mu_{HCh} = 0{,}07\, m^2V^{-1}s^{-1}),$$

$$MDD2(\text{GaSb},\ \mu_{HCh} = 0{,}25\, m^2V^{-1}s^{-1}),$$

$$MDD3(\text{GaAs},\ \mu_{HCh} = 0{,}42\, m^2V^{-1}s^{-1}),$$

A high value of carrier mobility cause the increasing of SNR So for $B = 0.5T, SNR(f)$ increase with 65% for $GaAs$ comparative $GaSb$

2.5 The detection limit of sensor in mos technology

A convenient way of describing the noise properties of a sensor is in terms of detection limit, defined as the value of the measurand corresponding to a unitary signal-to-noise ratio.
In case of shot noise, for double-drain magnetotransistors using (2.8) it results for detection limit:

$$B_{DL} \leq \frac{2\sqrt{2}(q\Delta f)^{1/2}}{\mu_{HCh}(L/W)G} I_D^{-1/2} \tag{2.10}$$

To illustrate the B_{DL} dependence on device geometry (figure 2.10) three double-drain magnetotransistors structures on silicon $\mu_{HCh} = 0.07m^2V^{-1}s^{-1}$ were simulated and having different ratios $(W = 100\mu m)$.

MGT1: $W/L = 0.5$;

MGT2: $W/L = 1$;

MGT3: $W/L = 2$

It is noticed that the B_{DL} is minimum for $W/L = 0.5$ structure. For optimal structure B_{DL} decreases at materials of high carriers' mobility.

In figure 2.11 the material influence on B_{DL} values for three double-drain magnetotransistor structures realised from *Si, GaSb* and *GaAs* can be seen having the same size: $L = 200\mu m$, $W = 100\mu m$.

MGT1: *Si* with $\mu_{HCh} = 0.07m^2V^{-1}s^{-1}$;

MGT2: *GaSb* with $\mu_{HCh} = 0.25m^2V^{-1}s^{-1}$;

MGT3: *GaAs* with $\mu_{HCh} = 0.42m^2V^{-1}s^{-1}$.

By comparing the results for the two types of Hall devices used as magnetic sensors it is recorded a lower detection limit of almost 2-order in double-drain magnetotransistors. A high value of carrier mobility causes the increasing of *SNR(f)*. So for $B = 0,5T$, *SNR(f)* increase with 60% for GaAs comparative with GaSb.

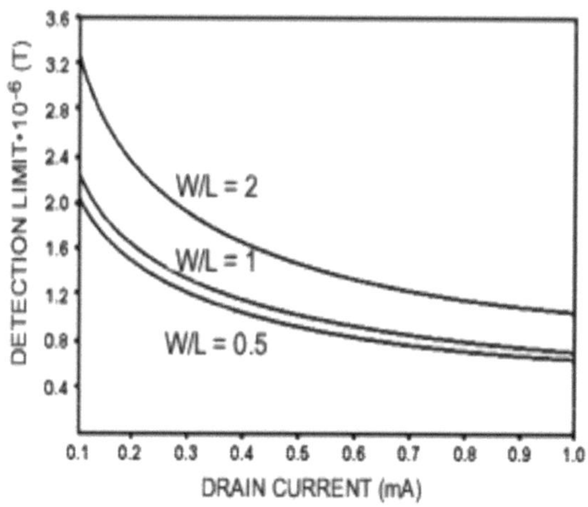

Fig. 2.10. B_{DL} depending on the drain-current for three devices of different geometry.

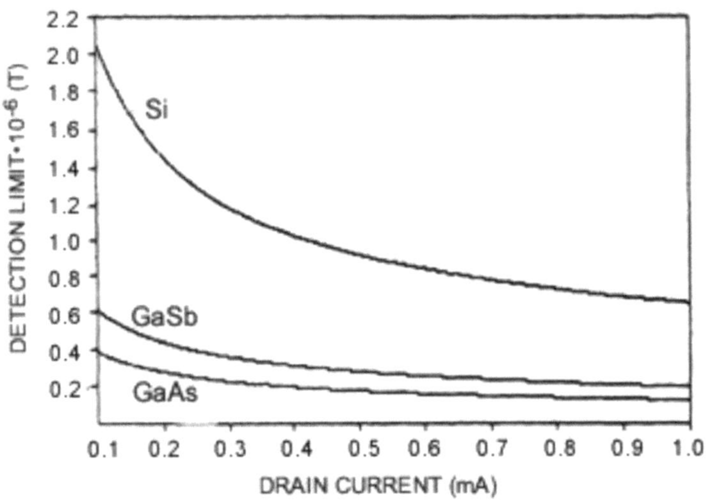

Fig. 2.11. B_{DL} depending on the drain current for three device of different materials.

2.6 The nemi for double-drain magnetotransistors

The noise current at the output of a magnetotransistors can be interpreted as a result of an equivalent magnetic induction.The mean square value of noise magnetic induction (*NEMI*) is defined by [8]:

$$\langle B_N^2 \rangle = \frac{\int_{f1}^{f2} S_{NI}(f) \cdot df}{(S_I \cdot I_D)^2} \tag{2.11}$$

Here S_{NI} is the noise current spectral density in the drain current, and (f_1, f_2) is the frequency range.

In case of shot noise, in a narrow frequency band around the frequency f by substituting (1.8) and (2.3) into (2.11) it results:

$$\langle B_N^2 \rangle = 2Iq\Delta f \cdot 4 \cdot \left(\frac{W}{L}\right)^2 \frac{1}{G^2 \mu_{HCh}^2} \cdot \frac{1}{I_D^2} \leq 8q\left(\frac{W}{L}\right)^2 \cdot \frac{\Delta f}{G^2} \cdot \frac{1}{\mu_{HCh}^2} \cdot \frac{1}{I_D} \tag{2.12}$$

Considering the condition of low value magnetic field fulfilled $\left(\mu_H^2 B^2 << 1\right)$, it is obtained a maximum value for $\frac{L}{W} G = 0{,}74$, if $\frac{W}{L} < 0{,}5$ [5]. In this case:

$$\langle B_N^2 \rangle_{\min} \leq 14{,}6q(\Delta f / I_D)\mu_{HCh}^{-2} \tag{2.13}$$

To emphasize the dependence of *NEMI* on device geometry there were simulated (figure 2. 12) three double-drain magnetotransistors structures realised on silicon, $\mu_{HCh} = 0{,}07 m^2 V^{-1} s^{-1}$, and having different ratios W / L ($W = 50\mu m$). The devices were based in the linear region and magnetic field has a low level $\left(\mu_H^2 B^2 << 1\right)$.

MGT1: $W / L = 0.5$ and $(L / W)G = 0.56$

$$\text{MGT2: } W/L = 1 \text{ and } (L/W)G = 0.409$$

$$\text{MGT2: } W/L = 2 \text{ and } (L/W)G = 0.212$$

It is noticed that the *NEMI* is minimum for $W/L = 0.5$, and for smaller values of this ratio. The decreasing of the channel length causes the increasing of $NEMI(f)$ with 40,8% for a square structure $W = L$, and with 173% for $W = 2L$.
In figure 2.13 NEMI values are shown obtained by simulation of three double-drain MOS magnetotransistors structures from different materials.

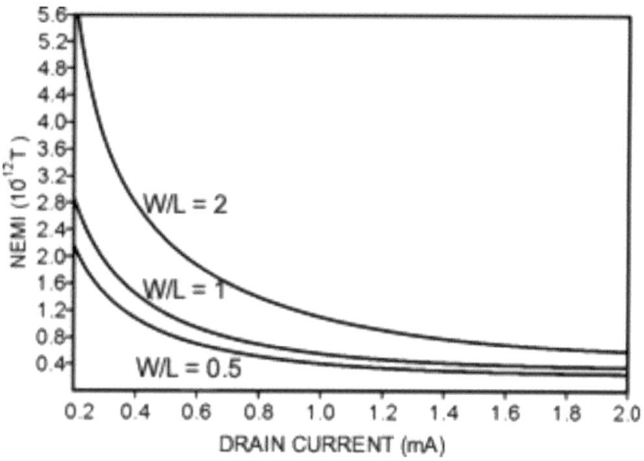

Fig. 2.12. The NEMI depending on the drain current for three devices of different geometry

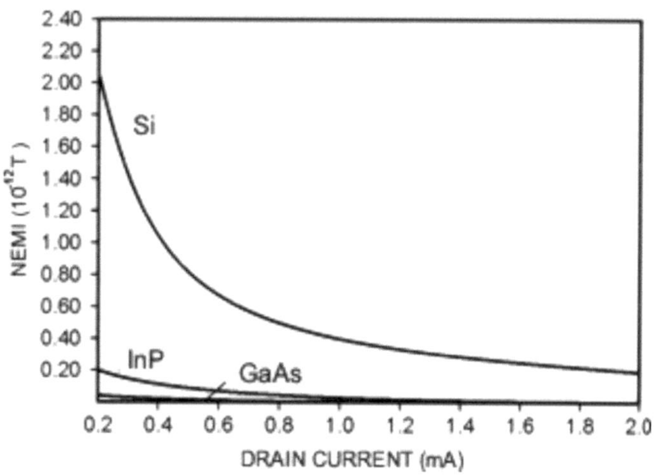

Fig. 2.13. The NEMI depending on drain current, for threedevices of different materials.

$$\text{MGT1: } Si,\ \mu_{HCh} = 0.07 m^2 V^{-1} s^{-1}$$

$$\text{MGT2: } InP,\ \mu_{HCh} = 0.23 m^2 V^{-1} s^{-1};$$

$$\text{MGT3:}GaAs,\quad \mu_{HCh}=0.42m^2V^{-1}s^{-1}$$

2.7 The $S_{NB}(f)$ for double- drain mosfet

From (2.11) it is obtained the noise-equivalent magnetic induction spectral density:

$$S_{NB}(f)=\frac{\partial<B_N^2>}{\partial f}=\frac{S_{NI}(f)}{S_A^2} \tag{2.14}$$

In case of shot noise, by analogy with (2.12) it results that:

$$S_{NB}(f)=2qI\cdot 4\left(\frac{W}{L}\right)^2\cdot\frac{1}{G^2\mu_{HCh}^2}\cdot\frac{1}{I_D^2}\leq 8q\left(\frac{W}{L}\right)^2\cdot\frac{1}{G^2}\cdot\frac{1}{\mu_{HCh}^2}\cdot\frac{1}{I_D} \tag{2.15}$$

Conclusions

Although magnetotransistors have a low magnetic sensitivity, very large signal-to-noise ratios are obtained, hence, a high magnetic induction resolution is resulting. A signal-to-noise ratio of about $8\cdot10^5$ at a magnetic induction of *200mT* has been obtained at double-drain magnetotransistors in case *GaAs*.

The analysis of the characteristics of magnetotreansistors structures shows that the $W/L=0.5$ ratio is theoretically favourable to high performance regarding the noise-equivalent magnetic induction.

The noise equivalent magnetic induction lowers with the increase of carriers mobility, this increase being significant for drain currents of relatively low values.

From double-drain MOSFET magnetotransistors, in case of shot noise, the $W/L=0.5$ structure provides superior *SNR* values, and smaller detection limit values. A detection limit of about $0{,}2\cdot10^{-6}T$ at a total drain-current of *0,5 mA* has been obtained at double-drain MOSFET magnetotransistor in case *GaAs*.

Also substituting the silicon technology by using other materials such as *GaAs* or *InSb* with high carriers mobility enables to manufacture higher characteristics devices.

2.8 The measurement of the torque at the naval engine shaft

Efficient operation of maritime ships and prevention of some considerable damages require supervision, measurement and adjusting of the main engine parameters together with other equipment and installations on board ship. Of a great importance is the permanent knowledge of the torque developed at the naval main engine shaft. The measurement of the mechanic torque M can be made based on the twisting angle Ψ that appears between two transversal sections of the shaft when this transmits mechanical power.

Following this purpose two disks S1, S2 are placed within those two sections which contain along their circumference, magnetic recording of two sinusoidal signals or rectangular of equal frequency.

Two transducers made with Hall magnetic microsensors positioned in the immediate vicinity of those two disks, allow during the rotation of the shaft to furnish information regarding the phase difference between those two signals, the rotation of the shaft to furnish information regarding the phase difference between those two signals, owing to its torque. The result of the measurement is exposed in numerical form.

2.8.1 Transducer based on the double-drain

Figure 2.15 shows the electrical diagram of a transducer based on double-drain magnetotransistors.
If the double-drain MOSFET works in saturation the differential output voltage is the following :

$$\Delta V_D = \mu_{HCh} \frac{L}{W} G V_R B_{\perp} \tag{2.16}$$

This voltage is applied to a comparator with hysteresis, which acts as a commutator. The existence of the two travel thresholds ensure the immunity at noise to the circuit. The monostable made with *MMC 4093* ensures the same duration for the transducers generated pulses.

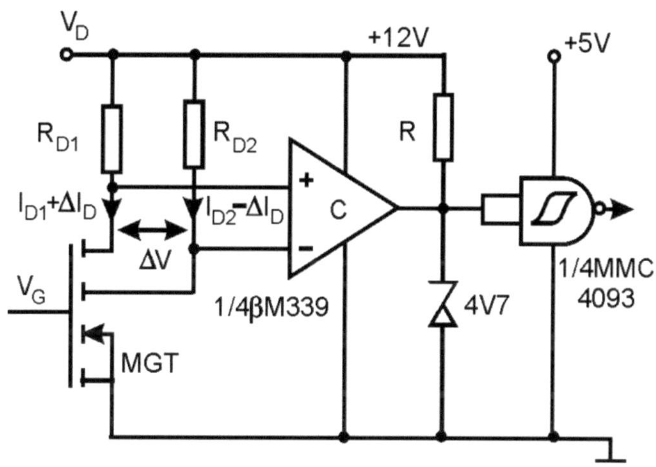

Fig. 2.15. The electrical diagram of transducer

2.8.2 Block diagram of the instalation and description of function

The disks with magnetical registration are distributed in such a way that the free rotation of the shaft, over the time when it is not transmitted the mechanical power, the signals produced by those two transducers are rigorously on phase.
At the power coupling, owing to the shaft torsion between those two sections S1 and S2 (figure 2.16) a twisting angle Ψ appears to which a phase difference between those two signals corresponds..
The work of installation may be supervised by means of the block diagram (figure 2.17) and by the forms of wave shown in figure 2.18.
The signals from the output of those two monostable CBM1, and CBM2 are applied to the differentiating circuits CD1 and CD2 which activate the bistable circuit CBB.
The positive impulses of the signal (b) put the flip-flop in the state 1 (high) and the positive impulses of the signal (b') bring it back to the state 0 (low).

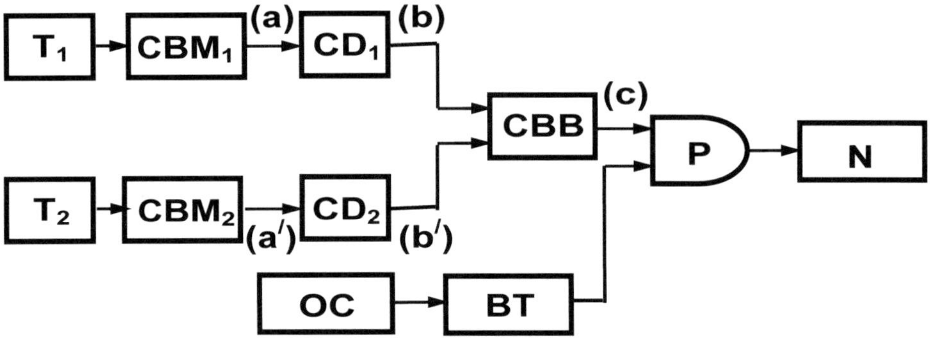

Fig. 2.17. Bloc diagram of the circuit for the measurement of mechanical torque

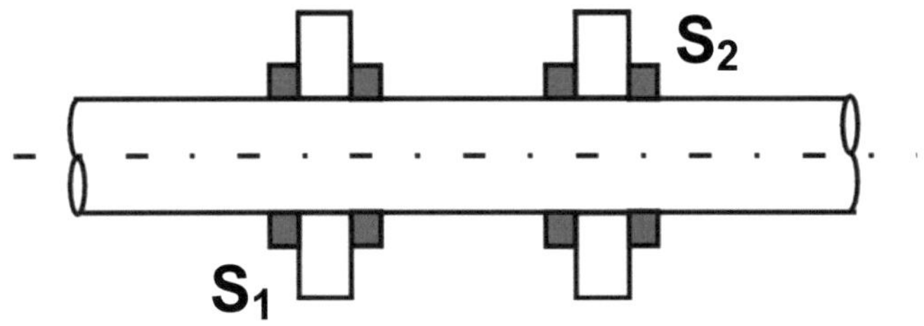

Fig. 2.16. Disc distribution on ship's engine shaft

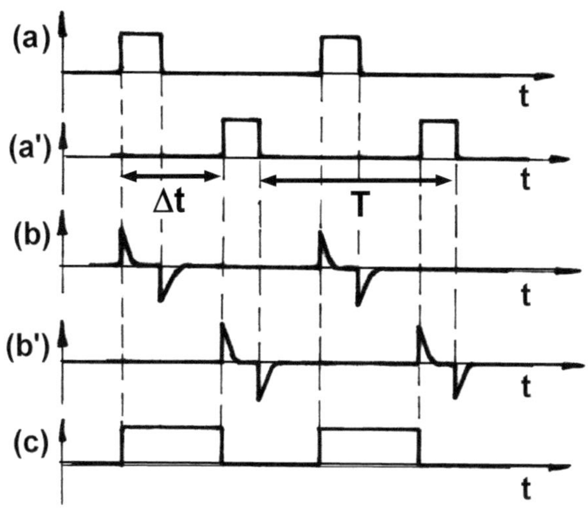

Fig. 2.18. Wave forms for the circuit measuring the torque

In this way at the output of flip-flop a right-angular signal (C) having the period T of magnetically registration and duration $\Delta t \sim \Psi$ is noticed description of the circuit gate P. The time interval Δt is measured by counting the signal periods of a quartz-oscillator, periods comprised within this interval.

Signal is applied to numerator N as long as gate P is open.
Numerator indications are modified at each impulse, so that the time interval between two states will be equal with the period of the given signal and it will represent the unit in which is expressed the result of measurement.

Remarks

In the case of very small value of the torsion twisting angle Ψ, the resolution capacity of the bi-state situation can be exceeded in which there are placed initially those two disks so as the transmitted signals by the traductors at free rotation of the shaft to be in opposition of phase.
Those two monostable circuits assure abrupt fronts to operational impulses of bi-stable resulting in the reduction of the level error of starting it.
The measurement precision depends on the relative error of numerator error of level starting of the bi-stable and the relative error of quartz oscillator.

3. The lateral bipolar magnetotransistor

3.1 General caracterisation of the lateral bipolar magnetotransistor

Figure 3.1 illustrates the cross section of a lateral bipolar magnetotransistor structure, operating on the current deflection principle, realized in MOS integrated circuits technology [8].
The n^+ regions of emitter E and primary collector C, are laterally separated on an L distance from base type p region. The two p^+ base contacts, allow for the application of the drift-aided field $\overline{E}_a$. On its action the most part of the minority carriers injected into the base drift to primary collector, producing collector current I_C . However some of electrons diffuse downwards to the n type substrate (the secondary collector) and thus produce the substrate parasitic current I_S .
In the presence a magnetic induction $B_{\perp}$, perpendicular to the plane of the section the ratio between I_C and I_S, change because of the current deflection.

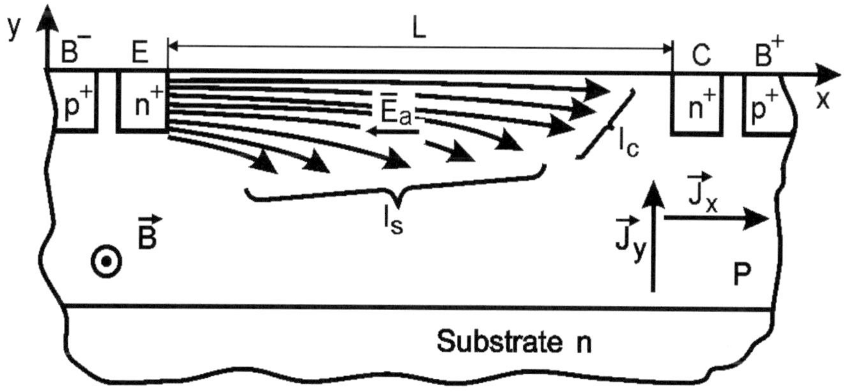

Fig. 3.1. Cross section of lateral magnetotransistor

In order to describe the qualitative operation of the device, let us assume that it is adequately biased for the forward active operation. If the very small magnetic field $\overline{E}_a$,is

oriented as shown in figure 3.1, the electrons are deviated to substrate junction. Only a few electrons will contribute to collector current.
The area from base region, between the emitter contact and collector contact, operates as a short Hall plate, and an induction field $\overline{B}$ causes the deflection of current lines. The transverse will be:

$$I_Y = J_Y(LW) \tag{3.1}$$

where W is the dimension along the axis Z.
In the absence of a induction $\overline{B}$, the current density along the axis X has the following expression:

$$J_X = \frac{I_C}{WY} \tag{3.2}$$

The Y parameter takes values in (y_{jn}, y_{jp}). Here y_{jn} and y_{jp} denote the junction depths of the collector region and the p-well respectively.
If it is considered the Hall angle expression $tg\theta_{Hn} = \mu_{Hn} \cdot B$ [6] then it is obtained:

$$J_Y = J_X \cdot tg\theta_{Hn} = \frac{(\mu_{Hn}B)I_C}{WY} \tag{3.3}$$

By substituting (3.3) into (3.l) it results:

$$I_H = I_Y = (L/Y)I_C\mu_{Hn}B_{\perp} = \Delta I_C \quad (3.4)$$

where μ_{Hn} is the Hall mobility of electrons in the base region.

3.2 The sensor response and the sensitivity

A magnetotransistor may be regarded as a modulation transducer that converts the magnetic induction signal into an electric current signal.
This current signal or output signal is the variation of collector current, caused by induction $\overline{B}_{\perp}$.
The sensor response is expressed by:

$$h(B) = \frac{\Delta I_C}{I_C} = \frac{L}{Y}\mu_{Hn} \cdot B \tag{3.5}$$

and it is linear for induction values which satisfy the condition: $\mu_H^2 \cdot B_{\perp}^2 << 1$. In figure 3.1 the geometry influence on $\overline{h(B)}$ values for three magnetotransistor structures can be seen. They are realized on silicon ($\mu_{Hn} = 0.15m^2V^{-1}s^{-1}$) and have different ratios L/Y $(L = 50\mu m)$.

$$\text{MGT}_1\text{:} \frac{L}{Y} = 0{,}5\,;$$

$$\text{MGT}_2\text{: } \frac{L}{Y} = 1\ ;$$

$$\text{MGT}_3\text{: } \frac{L}{Y} = 2\ ;$$

For the same geometry ($L/Y = 0.5$) the sensor response depends on material features. In figure 3.2 $\overline{h(B)}$ values for two sensor structures realized on Si ($\mu_{Hn} = 0.15 m^2V^{-1}s^{-1}$) and GaAs ($\mu_{Hn} = 0.80 m^2V^{-1}s^{-1}$)are shown .

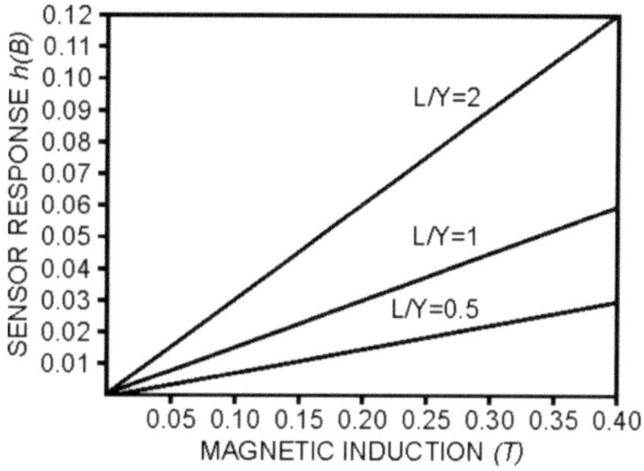

Fig. 3.2. The *h(B)* depending on *B* for three devices of different geometry

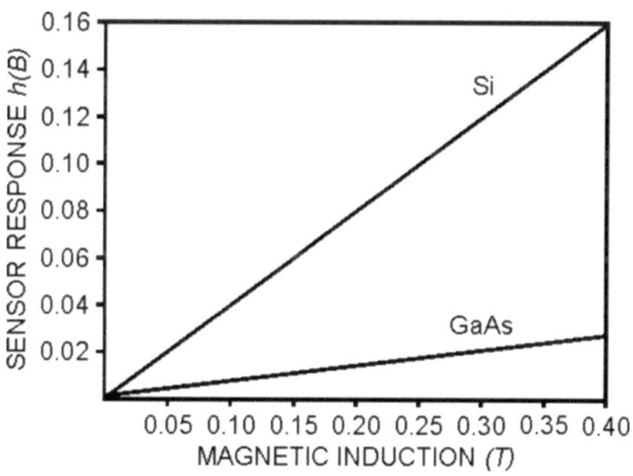

Fig. 3.3. The *h(B)* on *B* on the two sensors of different materials

We can see that the sensors made of high mobility materials have superior response. For the same magnetic induction $B = 0.2T$ at the GaAs device, $h(\overline{B})$ increases 5.6 times compared to that value for the silicon.

The magnetic sensitivity related to the devices current is defined as follows:

$$S_I = \frac{1}{I_C}\left|\frac{\Delta \mathrm{I}_C}{\mathrm{B}}\right| = \frac{L}{Y}\left|\mu_{Hn}\right| \tag{3.6}$$

For a given induction $(B = 0.4T)$ and at given collector current $I_C = 1mA$, the sensitivity depends of the device geometry and the material properties. In table 3.1 are presented the obtained values for five magnetotransistors structures.

MGT	L/Y	μ_{Hn} $(m^2V^{-1}s^{-1})$	$S_1(T^{-1})$
MGT_1 (Si)	3	0.15	0.45
MGT_2 (Si)	1	0.15	0.15
MGT_3 (Si)	0.5	0.15	0.075
MGT_4 (GaAs)	3	0.80	2.40
MGT_5 (GaSb)	3	0.50	1.50

Table 3.1 The numerical values of the supply-current related sensitivity

3.3 The offset equivalent magnetic induction

For bipolar lateral magnetotransistor presented in figure 4 the offset current consists in the flow of minority carriers which, injected into the base region in absence of magnetic field diffuse downwards and are collected by the secondary collector S.

The main causes of the offset are due to the misalignment of contacts to non-uniformity of the thickness and of the epitaxial layer doping. Also a mechanical stress combined with the piezo-effect, may produce offset.

To describe the error due to the offset to describe the error due to the offset the magnetic induction is determined, which is determined the magnetic induction, which produces the imbalance $\Delta I_C = \Delta I_{Coff}$. The offset equivalent magnetic induction is expressed by considering the relation (3.6):

$$B_{off} = \frac{\Delta I_{Coff}}{S_I I_C} = \frac{1}{\mu_{Hn}} \cdot \frac{\Delta I_{Coff}}{I_C} \quad (3.7)$$

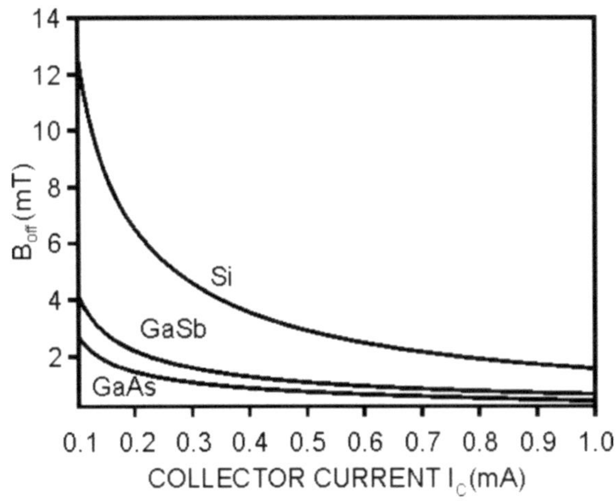

Fig. 3.4. The B_{off} depending on the I_C for three devices of different materials

Considering $\Delta I_{Coff} = 0.10\mu A$ and assuming that the low magnetic field condition is achieved, in figure 3.4 the dependence of B_{off} on I_c for three magnetotransistors is presented with the same geometry $L / Y = 0.5$ is presented, realized from different materials:

MGT1: Si with $\mu_{Hn} = 0.15 m^2 V^{-1} s^{-1}$;

MGT2: GaSb with $\mu_{Hn} = 0.50 m^2 V^{-1} s^{-1}$;

MGT3: GaAs with $\mu_{Hn} = 0.85 m^2 V^{-1} s^{-1}$;

The offset-equivalent magnetic induction lowers with the increase of carriers' mobility.
So for the same collector current $Ic=0.1mA$ the B_{off} value of the GaAs device decreases by 70% as compared to that of the silicon device.

3.4 Signal-to-noise ratio

The noise affecting the collector current of a magnetotransistors is shot noise and $1/f$ noise.
In case of *1/f* noise, and analogue with 1.11 it results:

$$SNR(f) \cong \frac{(nLYd)^{1/2}}{\alpha^{1/2}} \cdot \mu_{Hn} \left(\frac{f}{\Delta f} \right)^{1/2} \cdot \frac{L}{Y} B_{\perp} \qquad (3.8)$$

To illustrate the $SNR(f)$ dependence on device geometry three lateral magnetotransistor structures realised on silicon were simulated (figure 3.5).

$MGT_1 : L/Y = 0.5$;

$MGT_2 : L/Y = 1$;

$MGT_3 : L/Y = 4$.

It is considered that $f = 1.5Hz$, $\Delta f = 1Hz$, $\alpha = 10^{-7}$, $n = 4.5 \cdot 10^{21} m^{-3}$, $d = 10^{-5} m$ $q = 1.6 \cdot 10^{-19} C$ the device being biased in the linear region and the magnetic field having a low level.
For the same magnetic induction $B = 0,2T$ $SNR(f)$ is maximum in case $L = 4Y$. The increasing of the geometrical parameter Y causes the decreasing of *SNT(f)* with 50% for a square structure $Y = L$ and with 63.3% for $Y = 2L$.

In figure 3.6 it can be seen the material influence on *SNR(f)* values for three sensors MGT_1, MGT_2, MGT_3 realised on Si ($\mu_{Hn} = 0.15 m^2 V^{-1} s^{-1}$, $f = 1.2Hz$), GaSb ($\mu_{Hn} = 0.50 m^2 V^{-1} s^{-1}$, $f = 5Hz$) and GaAs ($\mu_{Hn} = 0.80 m^2 V^{-1} s^{-1}$, $f = 7.8Hz$); $L = 3Y$, $Y = 20\mu m$.
In case of shot noise (see equation1.8) is obtained:

$$SNR(f) = \frac{1}{\sqrt{2}} \mu_{Hn} \frac{L}{Y} \cdot \frac{I_C}{(q \cdot I \cdot \Delta f)^{\frac{1}{2}}} \cdot B_{\perp} \leq 0.707 \mu_{Hn} \frac{L}{Y} \left(\frac{I_C}{q \Delta f} \right)^{1/2} \cdot B_{\perp} \qquad (3.9)$$

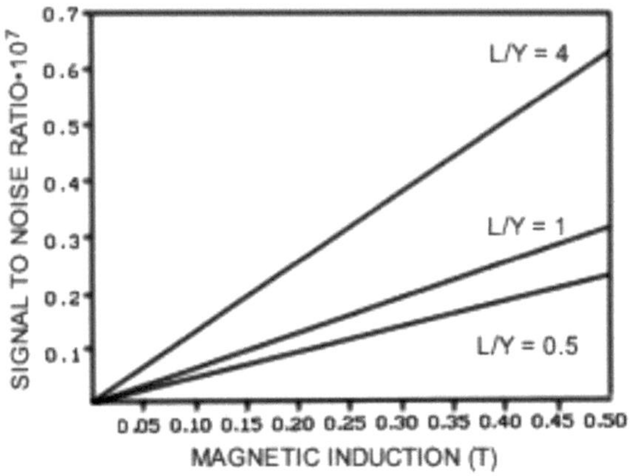

Fig. 3.5. *SNR(f)* depending on magnetic induction for three devices of different geometry

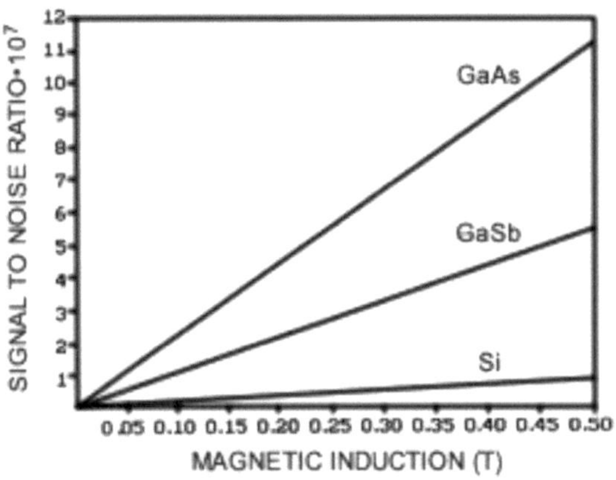

Fig. 3.6. SNR(f) depending on collector current for three devices of different materials

In figure 3.7 is shown the $SNR(f)$ dependence in collector current of three magnetotransistor structures of different materials $(L/Y = 5, \Delta f = 1Hz, B = 0.2T)$

$$MGT_1\text{: Si with } \mu_{Hn} = 0.15 m^2 V^{-1} s^{-1}$$

$$MGT_2\text{: Ga Sb with } \mu_{Hn} = 0.50 m^2 V^{-1} s^{-1}$$

$$MGT_3\text{: Ga As with } \mu_{Hn} = 0.80 m^2 V^{-1} s^{-1}$$

A high value of carrier mobility causes the increasing of $SNR(f)$. So for $I_c = 0.2mA$, $SNR(f)$ increases with 60% for Ga As comparative with GaSb.
To emphasise the dependence of $SNF(f)$ on device geometry there (figure 3.8) three magnetotransistor structures realised on silicon ($\mu_{Hn} = 0.15 m^2 V^{-1} s^{-1}$) were simulated having different ratios.

$$L/Y \ (L = 50\,\mu m; B = 0.2T;) \ \Delta f = 1Hz$$

$$MGT_1 : L/Y = 5$$

$$MGT_2 : L/Y = 3$$

$$MGT_3 : L/Y = 2$$

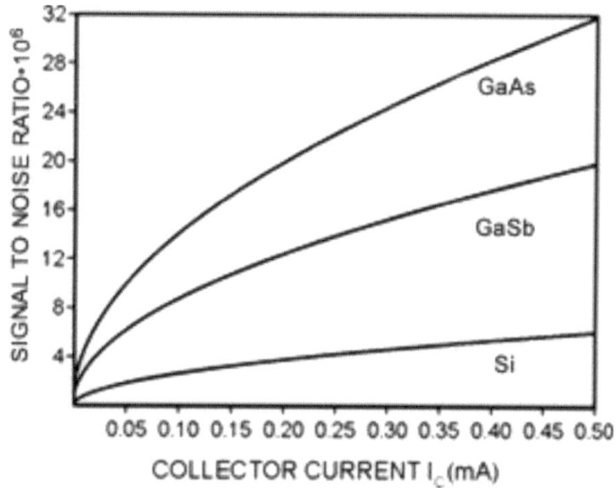

Fig. 3.7. SNR(f) depending on collector current for three devices of different materials

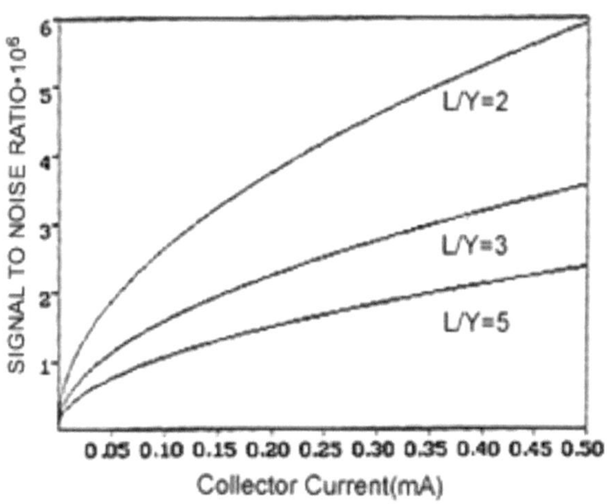

Fig. 3.8. SNR(f) depending on I_C for three devices of different geometry

3.5 The detection limit

A convenient way of describing the noise properties of a sensor is in terms of detection limit, defined as the value of the measured corresponding to a signal-to-noise ration of one. In case of shot noise, it is obtained from expression (3.9):

$$B_{DL} \geq \frac{(2q\Delta f)^{1/2}}{\mu_{Hn}} \cdot \frac{Y}{L} \cdot I_C^{-1/2} \tag{3.10}$$

In figure 3.9 are shown B_{DL} values obtained for three magnetotransistor structures made of different materials:

$$MGT_1: \text{ Si } (\mu_{Hn} = 0.15 m^2V^{-1}s^{-1}),$$

$$MGT_2: \text{GaSb } (\mu_{Hn} = 0.50 m^2V^{-1}s^{-1})$$

$$MGT_3: \text{GaAs } \mu_{Hn} = 0.80 m^2V^{-1}s^{-1}).$$

A high value carrier's mobility causes the decreasing of detection limit so B_{DL} decreases with 45% for *GaAs* comparative with *GaSb*.

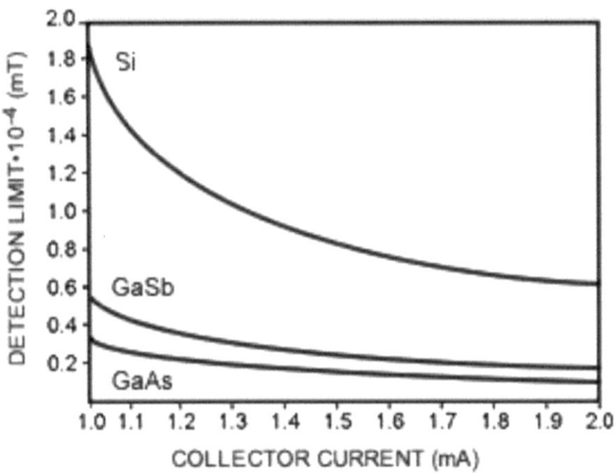

Fig. 3.9. B_{DL} depending on collector current for three devices of different materials

3.6 The noise equivalent magnetic induction

The noise current at the output of a magnetotransistor can be interpreted as a result of an equivalent magnetic induction. The mean square value of noise equivalent magnetic induction (NEMI) is defined by:

$$\langle B_N^2 \rangle = \frac{\int_{f1}^{f2} S_{NI}(f) \cdot df}{(S_I \cdot I_C)^2} \tag{3.11}$$

Here S_{NI} is the noise current spectral density in the collector current, and (f_1, f_2) is the frequency range.

In case of shot noise, the mean square value of noise equivalent magnetic induction (NEMI) is defined by similarity with relation (1.13):

$$\langle B_N^2 \rangle \le 2q \left(\frac{Y}{L} \right)^2 \frac{\Delta f}{\mu_{Hn}^2} \cdot \frac{1}{I_C} \tag{3.12}$$

In figure 3.10 NEMI values for three magnetotransistor structures made of different materials ($Y/L = 0.5; \Delta f = 1Hz$)are shown*MGT*$_1$: Si with $\mu_{Hn} = 0.15 m^2V^{-1}s^{-1}$

MGT_2: Ga Sb with $\mu_{Hn} = 0.50 m^2 V^{-1} s^{-1}$

MGT_3: Ga As with $\mu_{Hn} = 0.85 m^2 V^{-1} s^{-1}$

For the same collector current $I_C = 0,2mA$ the NEMI value of the Ga As device decreases by 25.6 times as compared to that of the silicon device.

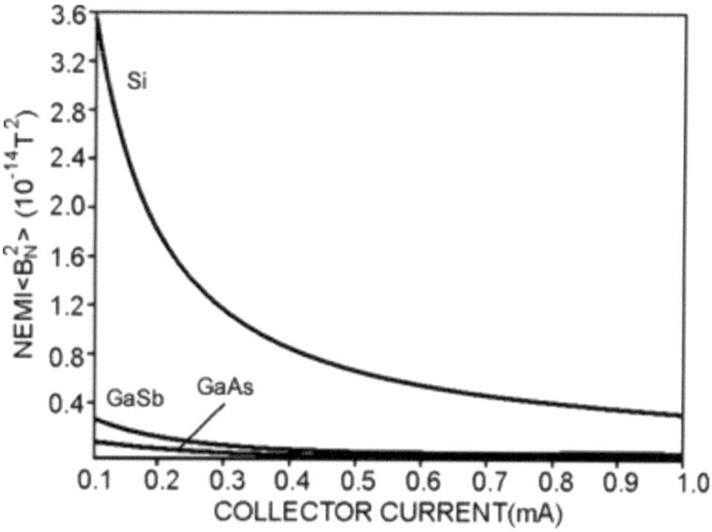

Fig. 3.10. NEMI depending on the collector current for three devices of different materials

To emphasise the dependence of NEMI device geometry, (figure 3.11) two magnetotransistor structures realised on silicon and having different ratios were simulated:

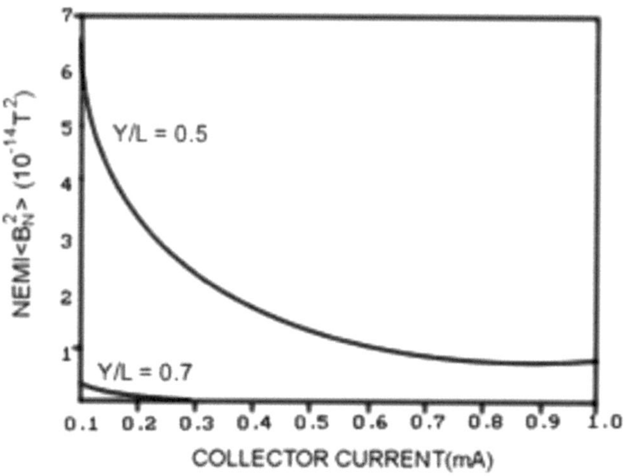

Fig. 3.11. NEMI depending on the collector current for two devices of different geometry

Y/L ($L = 50\,\mu m$,

$$MGT_1 : Y/L = 0.5\,;$$

$$MGT_2 : Y/L = 0.7\ ).$$

3.7 The noise-equivalent magnetic induction spectral density

From (3.11) the noise-equivalent magnetic induction spectral density is obtained:

$$S_{NB}(f) = \frac{\partial < B_N^2 >}{\partial f} = \frac{S_{NI}(f)}{S_A^2} \qquad (3.13)$$

In a narrow frequency band around the frequency *f*, it results [8]:

$$S_{NB}(f) \le 2q\left(\frac{Y}{L}\right)^2 \frac{1}{\mu_{Hn}^2 I_C} \qquad (3.14)$$

In figure 3.12 $S_{NB}(f)$ values for three magnetotransistor structures made of different materials ($Y/L = 0.5; \Delta f = 1Hz$) are shown

.MGT_1: Si,with $\mu_{Hn} = 0.15 m^2V^{-1}s^{-1}$

MGT_2: Ga Sb, with $\mu_{Hn} = 0.50 m^2V^{-1}s^{-1}$

MGT_3: Ga As, with $\mu_{Hn} = 0.80 m^2V^{-1}s^{-1}$

The noise-equivalent magnetic induction spectral density lowers with the increase of carriers mobility, this increase being significant for collector currents of relatively low values. So for the collector current $I_C = 0.1mA$, the offset equivalent magnetic induction value of the GaSb device decreases by 91.5% as compared to that of the silicon device

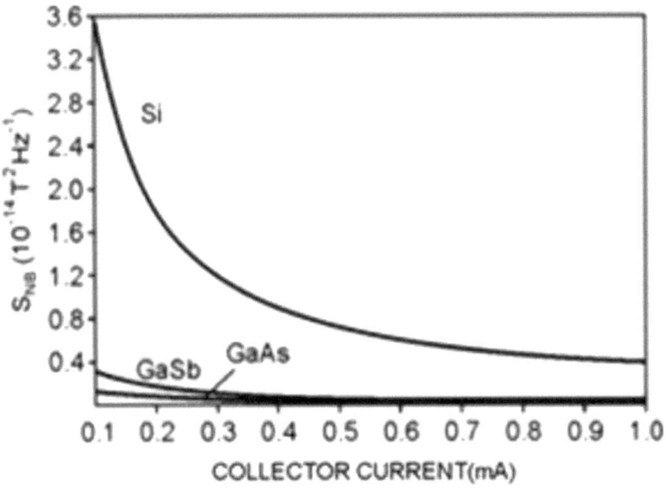

Fig. 3.12. The S_{NB} (f) depending on the I_C for three devices of different materials

3.8 A system to maintain the horizontal position of certain naval equipment

The present paper proposes an original solution to increase the efficiency of cardanic suspension which ensures the stabilization of horizontal position for gyrocompass and radar antenna.

A. Two platforms that can spin simultaneously, but independent of each other, driven by two direct current reversible motors are used.
B. The signals that determine the value of the engine supply voltage are given by two position transducers made up of with lateral bipolar magnetotransistors in differential connection.

On merchant ships, the establishment of both the horizontal position of the gyroscopic compass and the radar antenna is accomplished with the help of the suspension on 3 gimbals circles which eliminates the unwanted effect of rolling and pitching for the values included in the range -10° ÷ 10°.

An original solution wherewith the system of the gimbals suspension becomes capable for the pitching angles of the ship that oversteps the mentioned limits is the use of 2 superimposed platforms which are simultaneously rotating, but independently.

The driving shaft which constitutes at the same time the sustaining element of the first platform is horizontally disposed and parallel with the longitudinal axis of the ship. It is supported by bearings whose bolsters are mounted on a fixed element in the ship's structure. By rotating it this platform decreases the effect of the rolling.

The second platform which holds the suspension gimbals system is also sustained by her own shaft whose bearings have the holders jammed tight on the first platform. Being on the longitudinal axis of the ship, the leading shaft of this platform enables a rotating motion which decreases the effect of the pitch. Each platform is operated by a reversible direct current motor.

The signals which determine the bridge driving voltage polarity for the 2 engines are given by the position transducers made with magnetic transistors in differential connection.

3.9 The presentation of the Hall transducer

The magnetotransistor used in the construction of the displacement transducer (figure 3.13) has the structure of a MOS transistor with long channel but operates as a lateral bipolar transistor with a drift-aided field in the base region.

In the presence of a magnetic field adequately oriented the collector current is very small. If the magnetic induction decreases the device current increase which brings about the collector potential variation ΔV_C :

$$\Delta V_C = R_{C1} \cdot \Delta I_C = R_{C1} \frac{L}{Y} \mu_{Hn} I_C B \tag{3.15}$$

The outlet of the magnetic transistor is connected to the inlet of a logical gate of „trigger Schmitt" type (ex.CDB413 or MMC4093) so that it supplies logic level „0" signal, when the magnetic induction increases and logic level „1" when the magnetic induction decreases.

The description of the position transducer

Because by rotating one of the platforms eliminates the effect of rolling, and the other one the effect of pitching we use a transducer for every platform. The two sensors of the transducer (figure 3.14) are magnetic transistors with MOS structure that function as lateral bipolar transistors with supplementary drift field in the base region. For this kind of polarized device, the theoretical analysis shows that in case of a constant polarization, for a certain material and a geometry given to the device $\Delta I_C \approx B$. The two magnetic transistors are within the field of an magnetic pendulum. In the absence of the rolling or pitching, the pendulum is in a median position and the magnetic fields for the two magnetic transistors are equal. Therefore, $I_{C1} = I_{C2}$ and at the outlet of the transducer the voltage is $\Delta V_C = 0$.

At the inclination of the ship because of the rolling, to the port (or starboard) the induction value for a sensor is increasing, for exempt. MGT1, and decreasing for the other one. The balance of the two collector currents disappears even if $I_{C1} > I_{C2}$ the voltage at the outlet of the transducer is:

$$\Delta V_C = R_C(I_{C1} - I_{C2}) > 0 \tag{3.16}$$

If the ship is listed in the other way, then $I_{C1} < I_{C2}$ and it results:

$$\Delta V_C = R_C(I_{C1} - I_{C2}) < 0 \tag{3.17}$$

For the platform that can spin around an axis parallel to the longitudinal axis, the transducer pendulum moves in a plane perpendicular on the longitudinal axis of the ship, and for the other platform the magnetic pendulum of the transducer can move in a vertical plane in parallel with the longitudinal axis of the ship.

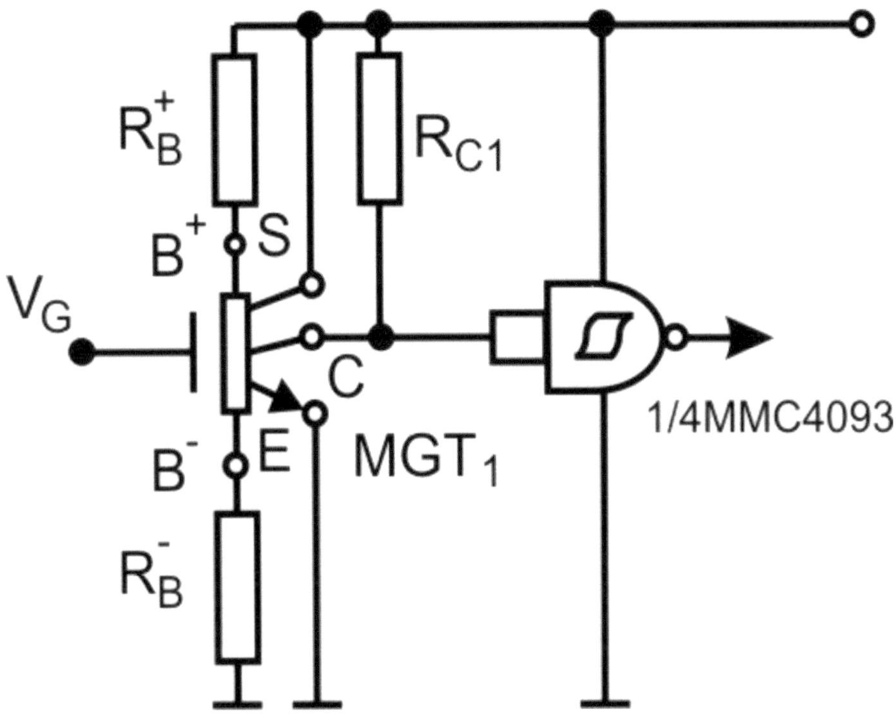

Fig. 3.13. The electric diagram of Hall transducer

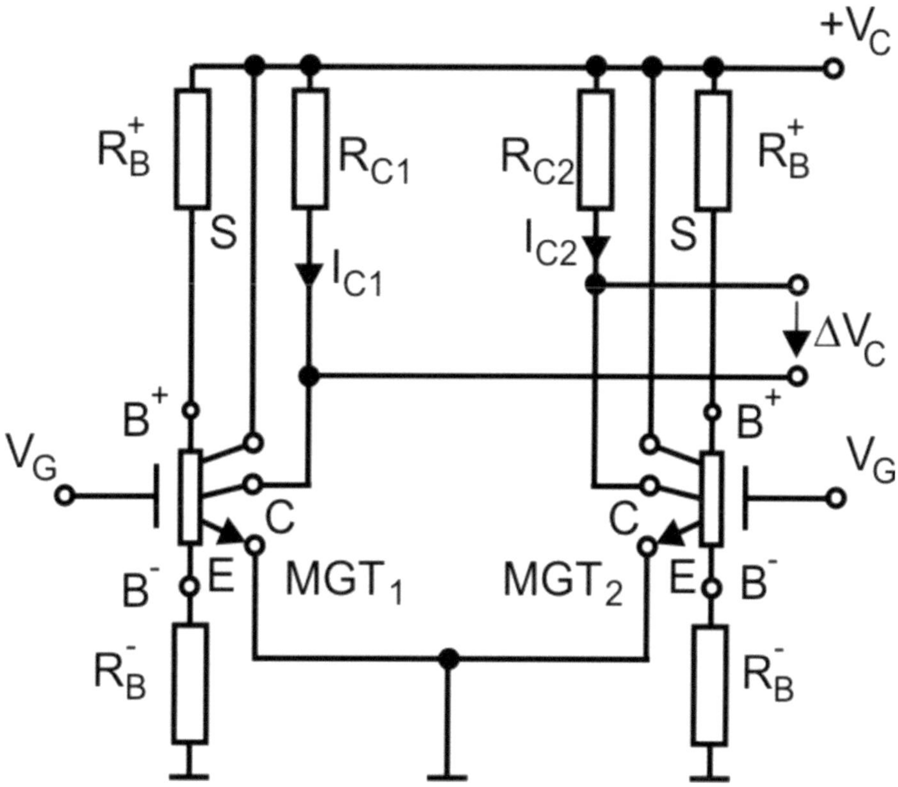

Fig. 3.14. The position transducer with magnetotransistor

Block diagram and the role of the component circuits

The block diagram (figure 3.15) contains: transducer T, integrator I, amplifier A, comparator C and the control assembly and energy supply of the motor BCA.

The level of signal to the outlet of the transducer is proportional to the ship's inclination angle, and its polarity shows the orientation of the ship's inclination.

The integrator eliminates the high frequency oscillations of the pendulum (the chip's oscillation frequency is reduced by fractions of *Hz*.

At the same time the integrator produces a small delay in the operating voltage variation of the engine, useful for the platform to reestablish the initial horizontal position. An essential contribution to this is the mechanic inertia of the system.

After amplification, the signal emitted by the transducer is compared with an additional reference transmission.

The comparator's threshold can be adjusted according to the delay produced by the integrator and the actuator mechanism inertia.

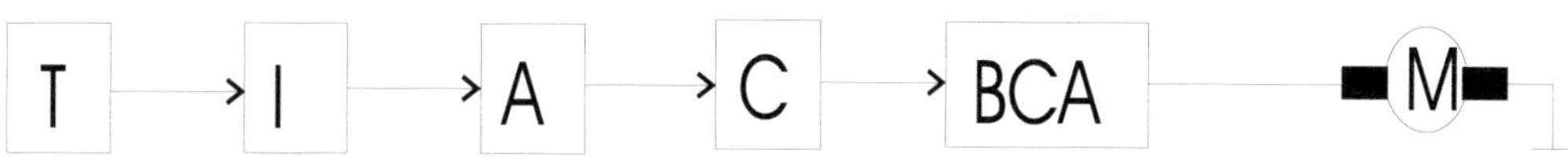

Fig. 3.15. The block diagram of the installation

The principle diagram and operating conditions

In figure 3.16 is presented the principle diagram of the transducer and amplifier. If the inclination of the ship to starboard brings the unbalance of the collector currents $I_{C1} > I_{C2}$ the transducer produces the signal:

$$\Delta V_C = R_C(I_{C1} - I_{C2}) = V_2 - V_1 > 0 \tag{3.18}$$

The operational is in a differential amplifier configuration and the outlet voltage is:

$$V_0 = -(R_2 / R_1)\Delta V_C < 0 \tag{3.19}$$

In figure 3.17 is presented the principle diagram of the motors' supply and control block.
When $V_0 < 0$ the voltage at the inputs of the two comparators have separate polarity. When $V_0 < 0$, $V_{i1} = V_0 - V_{rl} < 0$,and the output of the comparator C, is in DOWN state, therefore: $V_{01} = V_{0L} < 0$ which insures the blockage of the transistors T_1 and T_3.
In the same state , $V_0 < 0$ at the input of the comparator C_2 the voltage, is $V_{i2} = V_{r2} - V_0 > 0$, it passes in the UP state, $V_{02} = V_{0H} > 0$, which determines the conduction of the transistors T_2 and T_4, the polarity voltage at the jacks of the motor is the one indicated in the figure 3.17.
The direction of rotation is thereby given so that by moving the platform for the rolling compensation the balance of the two collector currents is re-established.
Clearly at the inclination of the ship to the port the outlet of the comparator C_1, passes in the UP state, conducts the transistors T_1 , T_3 the direction of the bridge driving voltage changes and the motor will rotate in the opposite direction re-establishing the horizontal position of the platform.

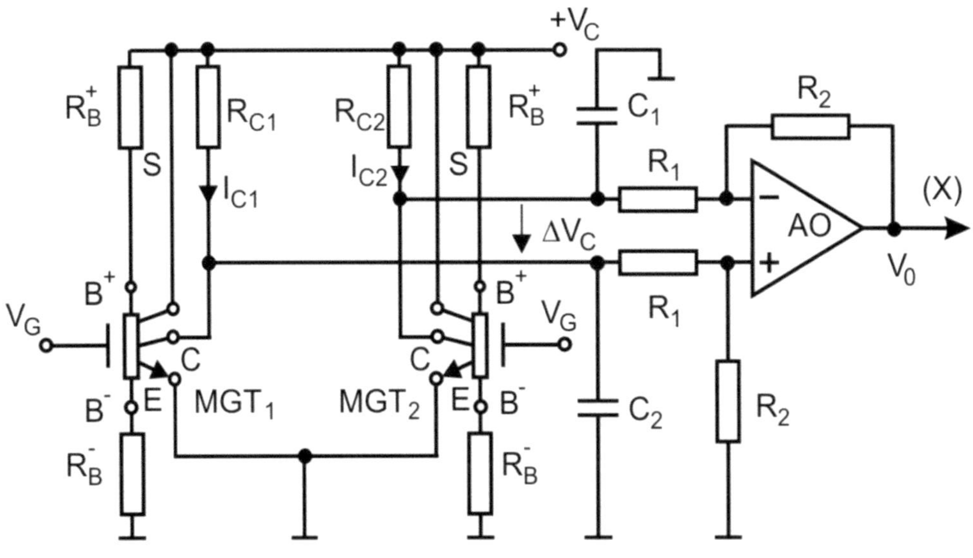

Fig. 3.16. The transducer and the differential amplifier

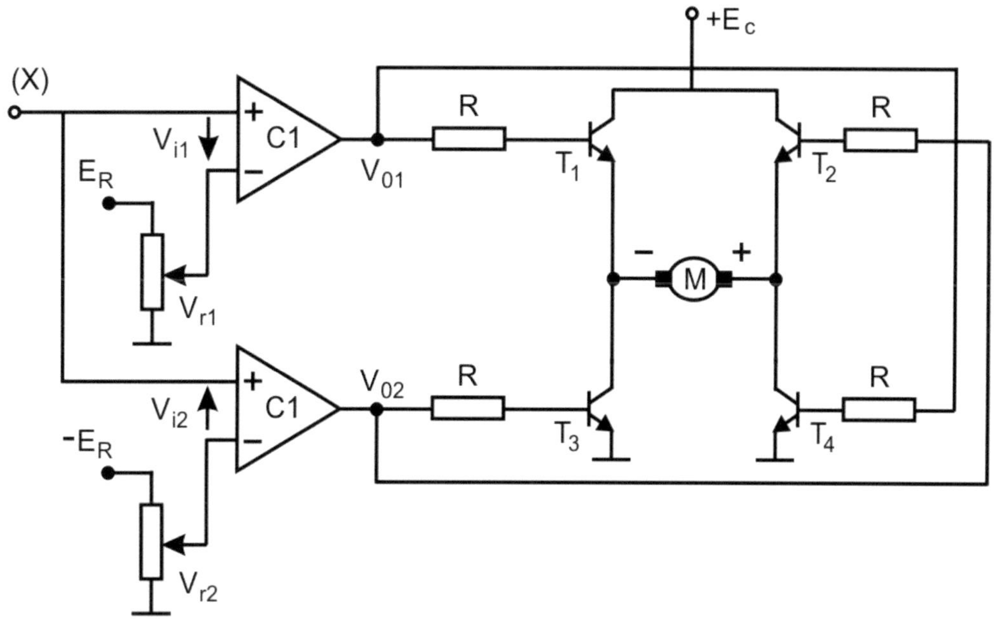

Fig. 3.17. The control and supply diagram of the motor

4. Conclusions

The presented system together with the cardanic suspension system eliminate the oscillation exceed of the ship in case of amplitudes that exceed $\pm 10^{o}$. The use of magnetotransistors as magnetic sensors allows for the achieving of same current-voltage conversion circuits, more efficient than the conventional circuits with Hall plates. Although the magnetotransistors have a low magnetic sensitivity, very large signal-to-noise ratios are obtained, hence, a high magnetic induction resolution is resulting. The possibility of having the sensor and the amplifier circuit on the same chip has lead to the achievement of transducers with high conversion efficiency as well as to increase in their range of practical work.

5. References

[1] Bodea M., "Traductoare integrate", Microelectronica, vol. 15, pp. 73-86, Editura Academiei R.S.Romania, Bucureşti, 1987

[2] Căruntu G., "Măsurări galvanomagnetice în dispozitive semiconductoare", Buletinul Sesiunii de Comunicări Ştiinţifice SECOMAR '99, Academia Navală "Mircea Cel Bătrân", Constanţa, vol. II, pp. 69-74, 1999

[3] Gray R.P., Meyer G.R., "Circuite integrate analogice. Analiză şi proiectare", Editura Tehnică, Bucureşti, 1973

[4] F.N. Hooge "1/f noise is no surface effect", Phys., 1969 Lett. 29A 139-40

[5] Middelhoek S., Audet S.A., "Physiscs of Silicon Sensors", Academic Press, London, 1989

[6] P.S. Kireev, "Fizica semiconductorilor", Editura Ştiinţifică şi Enciclopedică, Bucureşti 1977
[7] A. Nathan, H.P. Baltes, "Two dimensional numerical modelling of magnetic field sensors in CMOC technology", IEEE Trans Electron Devices ED-32 1212-19, 1985.
[8] R.S. Popovic , "Hall Effect Devices, Magnetic Sensors and Characterization of Semiconductors", Adam Hilger, Bristol , England , 1991

2

Photoelectronic Magnetic Microsensor with a Digit Readout

Hsing-Cheng Chang
Feng Chia University, Taichung, 40724
Taiwan

1. Introduction

The magnetic microsensor is a small detective device for sensing magnetic effects and transferring to measurable signals. Magnetic microsensors are important in various application areas that are biomagnetism, geomagnetism, nondestructive testing, automobile, field measurement, identification and communication. Eleven technologies have been described for magnetic field measurements that are search-coil, flux-gate, optically pumped, nuclear precession, SQUID, Hall-effect, magnetoresistive, magnetodiode, magnetotransistor, fiber optical, and magneto-optic (Lenz, 1990). Four-type classification of magnetic microsensors by principle are also summarised that are galvanic, conductimetric, voltaic, and acoustics (Gardner et al., 2001). The trend of sensor development is toward lower cost, small dimension, lower power consumption, and higher performance. A new physics phenomenon using new fabrication technology and improved materials for new applications should be the development trends in magnetic microsensors. Recent progress in applications of FBG sensors has reported (Lee, 2003, Rao, 1999). Different researches on optical magnetic sensors have been addressed by fiber-optic interferometric (Oh et al., 1997, Wang et al., 2008), cantilever bending (Keplinger et al., 2003), Lorentzian force (Okamura, 1990), or magnetic materials (Meng et al., 2001).
Operational principles based on electromagnetic systems, magnetic material properties, stress-induced magnetic interrelationship, fiber gratings, and superconductivity have been widely studied for magnetic detection or measurement (Ciudad et al., 2004; Dimitropoulos et al., 2003; Mapps, 1997; Sedlar et al., 2000; Seo et al., 2001). It must be noticed in magnetic sensors development to immunity the factors of temperature and humidity and optical fiber sensors can be designed to fit the requirement. To develop magnetic microsensor using optical fiber sensing method becomes popular due to the advantages of electromagnetic immunity, electronic isolation, low cost, light weight, small size, and anti-corrosive. Fiber grating sensors has been reviewed based on different grating sensing methods, including Bragg gratings, chirped gratings, long period-based gratings, and intragrating concepts (Kersey et al, 1997). The permalloy-on-membrane type of magnetic actuators with flexural cantilevers and torsion beams are built based on microelectromechanical system (MEMS) technology to satisfy large force and displacement requirements (Khoo & Liu, 2001, Liu & Yi, 1990). To develop fiber Bragg grating (FBG) sensors unaffected by temperature perturbations is important for practical applications. A precision optical fiber-based magnetic sensor requires temperature compensation design because of deformation of

inherent temperature dependence in fiber material. Several temperature compensation techniques in pressure or strain measurements have been established such as bimetal cantilevers, non-uniform or dual head FBGs, double shell cylinders and biomaterial effect (Hsu et al., 2006; Iadicicco et al., 2006; Khoo & Liu, 2001; Liu & Yi, 1990; Tian et al., 2005). These techniques guarantee stable measurements independent of temperature perturbation without any additional temperature-isolation or referencing process.

To develop a magnetic microsensor as a microsystem should contain environment sensing mechanism, data processing and storage modules, and automatic calibration and compensation functions. Because of very small magnetostriction rate below the order of 10^{-5}, the sensing range and reliability of magnetic field strength are limited by coating soft magnetic film on optical fibers directly. Accordingly, microelectromechanical system compatible FBG magnetic sensor can be designed to supply a wide measurement range with solid reliability. A photoelectronic magnetic microsensor with a temperature compensation function and a digital readout has been developed and fabricated as a smart sensing system. The batch of microfabrication technology used to deposit Ni/Cr permalloy flaps that can be driven to push the sensing FBGs by excitation magnetic force to supply capacitive and optical outputs. The finite element method (FEM) for equivalent model simulation was utilized to understand the coupling effect of magnetic and mechanical behaviors. The neodymium-iron-boron (Nd-Fe-B) magnets with residual surface magnetic flux density up to 1.26 Tesla (T) were used as excited power to investigate the influence of external magnetic fields on the density variation of the transmitting light signal. Measurement system and display in real-time mode were setup by connecting the designed microsensors with signal processing circuits and a PC display module.

2. Operation principle

2.1 Sensing theory

The operational principle of the FBG-based sensors is to monitor the central wavelength shift between input signal and back-reflected signal from the Bragg gratings. The first-order Bragg condition is given by the expression (Morey et al., 1989)

$$\lambda_B = 2n_{eff}\Lambda \tag{1}$$

where n_{eff} is the effective index of the core and Λ is the period of gratings. Bragg wavelength, λ_B, is the center wavelength of the back-reflected signal from the Bragg gratings.

Most of FBG sensing works have focused on the device design and fabrication for providing quasi-distributed sensing of temperature and strain which can be described as the following equation (Liu et al, 2007, Xu et al., 1993).

$$\frac{\Delta\lambda}{\lambda_B} = (1 - P_e)\Delta\varepsilon + (\alpha_f + \xi_f)\Delta T = K_\varepsilon \varepsilon + K_T T \tag{2}$$

where $\Delta\varepsilon$ is the change of strain; P_e is the effective photoelastic coefficient of the fiber glass; α_f is the thermal expansion coefficient of fiber; ξ_f is the thermal optic coefficient of the fiber; ΔT is the change of temperature; K_ε is the strain sensitivity; K_T is the temperature sensitivity. The strain response arises due to both the sensor elongation induced grating pitch variation and the photoelastic effect induced fiber index change. For measuring magnetic field strength accurately, magnetic force induced strain should be measured effectively with processing of temperature compensation. The relative Bragg wavelength shifts in response to axial strain change ($\Delta\lambda_\varepsilon/\lambda_B$) in a packaged FBG sensor can be described as the following equation (Liu et al., 2000).

$$\frac{\Delta\lambda_\varepsilon}{\lambda_B} = -(1-P_e)\frac{\Delta P}{E} = K_P\Delta P \tag{3}$$

where P_e is the effective photoelastic coefficient; K_p is the pressure sensitivity; E is the Young's modulus of the fiber; ΔP is the change of pressure.
When applying strain and temperature in the fiber, the effective index of the core and the uniform distribution of gratings will be affected to induce the shift in Bragg wavelength. It simply can be expressed using

$$\frac{\Delta\lambda_B}{\lambda_B} = C_1\Delta\varepsilon + C_2\Delta T \tag{4}$$

where C_1 and C_2 present sensing parameters found to be 0.78×10^{-12} and 6.67×10^{-6} respectively (Kersey et al., 1997). The factors are complicated from strain-optic effect that affected by the parameters of Poisson ratio, core effective index, Pockel's coefficient components, grating length variation, and total fiber grating length. It becomes comprehensible that the variation in wavelength is the sum of the strain and temperature terms. The sensitivities of normalized central wavelength shift to strain and temperature have been studied about 1.17×10^{-3} nm/με and 1×10^{-2} nm/°C respectively. Therefore, to design a photoelectronic magnetic microsensor with minimizing noise perturbation, the magnetic-actuated strain variation must be detected effectively, but the influence of temperature must be eliminated.

2.2 Major processes

Based on planar microfabrication technologies, silicon-based optical-electrical integration structures can be fabricated compatibly. The sensing platform contains a sensor-located U-shaped trench, an etched through sensing window, and an interdigitated magnetic reaction mechanism. Thermal oxidation layers as an etching mask for silicon bulk etching and electrical isolation are grown. Wet chemical etch and sacrificial technologies of bulk and surface micromachining are applied to form location U-shaped trenches and sensing windows and interdigitated actuated mechanism. To achieve magnetic measurement with high permeability, saturation flux, resistivity and low coercivity, interdigitated thin films of NiFe (μ_r = 2000) were deposited as magnetic manipulated mechanism by electroplating technology (Flynn, 2007). Each period of sensing grating in FBG devices has 10 mm long with the separated space of 4mm. Two different periods of fiber gratings, spectral peaks on 1550.25 nm and 1553.25 nm, are fabricated in a hydrogen-loaded single-mode fiber (SMF-28) by microlithographic writing technology using a phase-shifted mask and a 248 nm KrF excimer laser.

2.3 System operation

A flowchart to setup the magnetic flux measurement system with temperature compensation function is shown as Fig. 1. Main program contains three checking processes:

initial compensation value, effective magnetic force and measurement results. Information of effective magnetic flux measurement will display on a LCD screen or a universal asynchronous receiver transmitter (UART). Temperature compensation configuration on FBG magnetic sensing is shown in Fig. 2. Amplified spontaneous emission (ASE) is used as a lighting source and two FBG optical fibers are used as a sensor and a reference. The reflective spectra of both FBG signals are superposition on the spectrum of a long period grating (LPG) structure. Two photodiodes (PD_1 and PD_2) are used to detect FBG sensing signals. The PD_1 measures compound signal from magnetic and temperature effects and PD_2 measures temperature effect only. Initial temperature values of PD_1 and PD_2 and their difference value can be obtained for calculating temperature compensation.

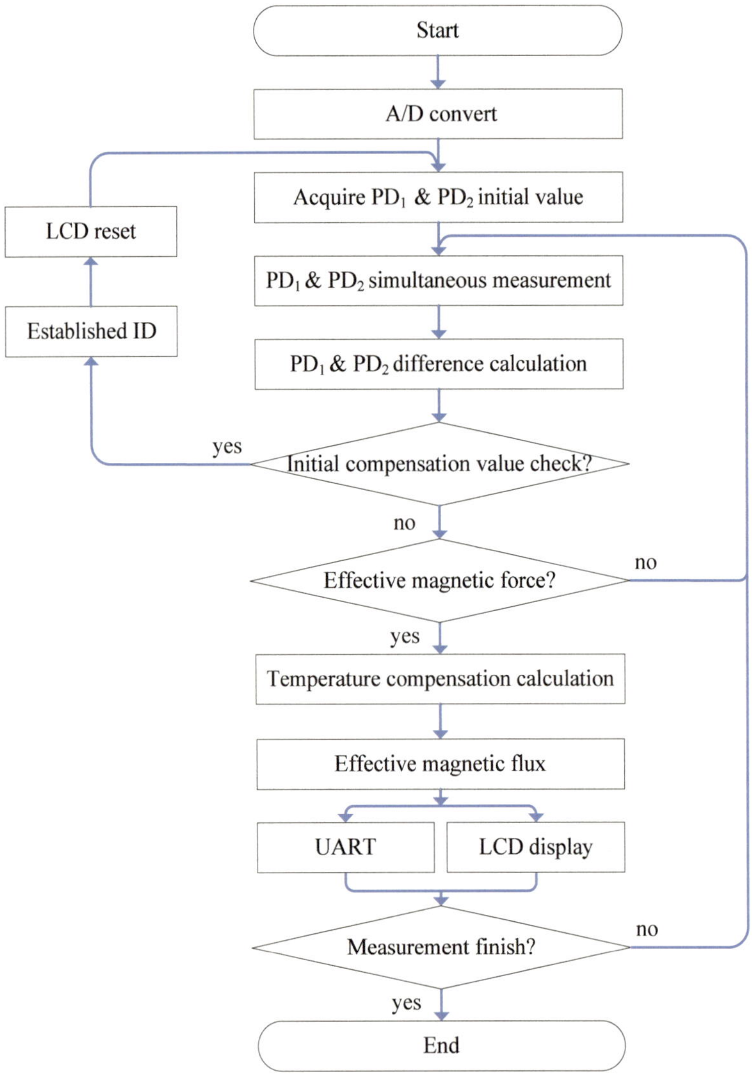

Fig. 1. A flowchart of magnetic flux measurements with temperature compensation

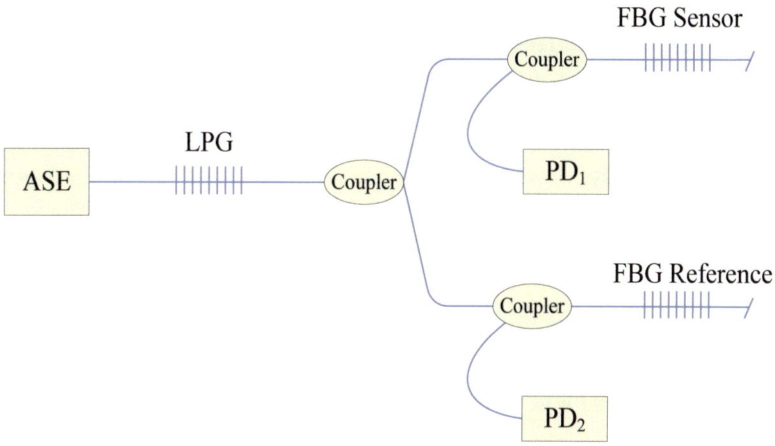

Fig. 2. A typical configuration for FBG temperature compensation

3. Design and fabrication

A schematic diagram of the developed photoelectronic magnetic microsensor with the temperature compensation mechanism is shown in Fig. 3. This magnetic microsensor has an optical fiber with two FBGs located on a bulk-etched silicon chip in which one with interdigitated cantilevers is for magnetic flux measurement and the other is for temperature compensation. These FBGs with the grating length of 10 mm and the separation of 4mm are fabricated in a hydrogen-loaded single-mode fiber (SMF-28). When the microsensor is applied to measures magnetic field strength, the cantilevers are attracted and deflected to push fiber a stretch which changes the grating period to induce peak shift of the Bragg wavelength.

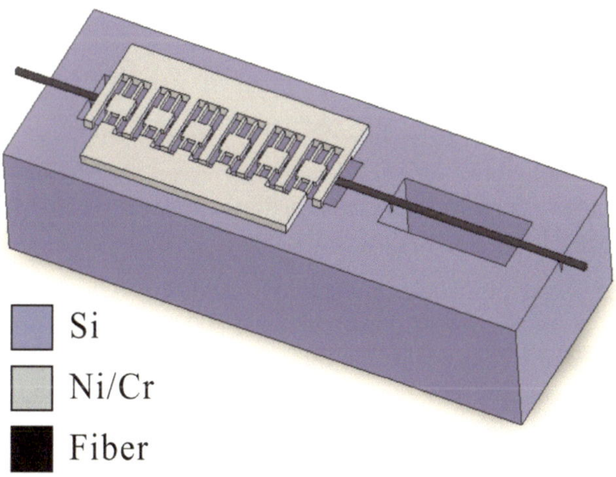

Fig. 3. Schematic diagram of a photoelectronic magnetic microsensor with temperature compensation mechanism

The major fabrication processes of the developed photoelectronic magnetic microsensor are shown in Fig. 4. (a) Double-sided polished silicon wafers grown a 1 μm thermal oxidation layer were used as supporting substrates. (b) Silicon wafer is patterned and etched in both sides to form sensing windows. Anisotropic silicon bulk etching was done in KOH at 70 °C to etching a 315 μm depth in bottom side and a 113 μm depth in top side. (c) Remove the top side SiO_2 layer. An adhesion Cr layer and the Ni seed layer were evaporated for increasing adhesion. The Ma-P1225 photoresist and a sacrificial photoresist AZ-4620 was spun coated. A two-electrode electroplating system was operated at room temperature. The Ni is electroplated in aqua solution. A low tensile stress 10 μm Ni layer was approached as magnetic actuated cantilevers with maximum permeability and minimum anisotropy field. During the process, wafers were removed from the solution for a short time every minute to desorption of H_2 bubbles to increase current. Then AZ-4620 was spun coated and Ni/Cr layer was evaporated. (d) The electroplating area is patterned for etching the Ni/Cr layer. (e) Sacrificial layer is released and the silicon base is etched out to release cantilevers.

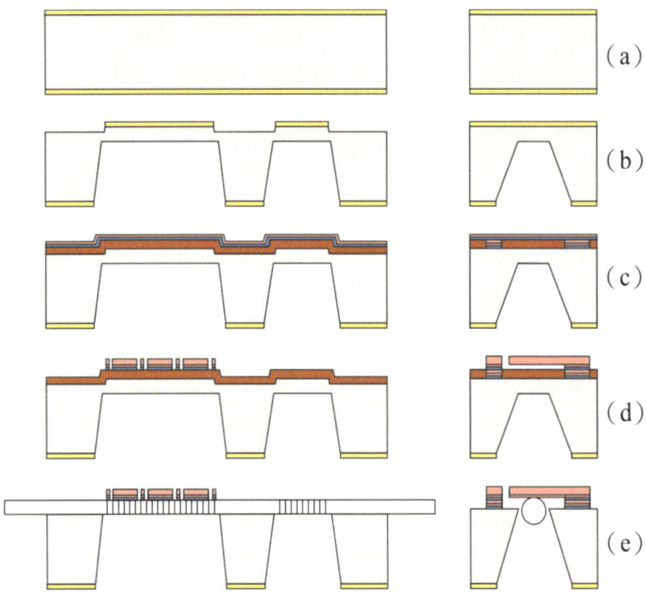

Fig. 4. Major processes of a photoelectronic magnetic microsensor with temperature compensation mechanism, longitudinal cross-sectional structures show in the left and transverse cross-sectional structures show in the right

Another developed photoelectronic magnetic microsensor with EM wave shielded packaging is shown in Fig. 5 (Chang et al., 2007). Fabrication processes of the photoelectronic magnetic microsensor are shown in Fig. 6. (a) A 1 μm oxidation layer is grown on silicon substrate. (b) SiO_2 layer is UV-lithographic patterned and etched to open silicon etching windows. (c) Silicon bulk is etched anisotropically in KOH at 60 °C to define a 30 μm thick sensing diaphragm. (d) Polyimide is spun onto the wafer and cured. A Cr layer (0.02 μm) and a Cu seed layer (0.1 μm) are deposited for increasing adhesion. A two-electrode electroplating system is operated at room temperature with a DC current density of 15 mA/cm^2. The Ni/Fe is electroplated in aqua solution of $NiSO_4$/$NiCl_2$/$FeSO_4$/ H_3BO_3/additives. A low tensile stress 5 μm $Ni_{0.8}Fe_{0.2}$ layer was approached as magnetic actuated flaps with maximum permeability and minimum anisotropy field. During the process, the wafers were removed from the solution for a short time every minute to desorption of H_2 bubbles to increase current efficiency. Thickness uniformity of permalloy layer is about 1.5 times in the corners because of the edge effect. AFM scanned rms roughness within 20 μm x 20 μm area was 90 nm for 5 μm $Ni_{0.8}Fe_{0.2}$ layer. The supporting silicon layer is etched out to release the membranes. From (e) to (f), second silicon wafer is patterned and etched anisotropically to define the measurement tunnel and a U-shaped trench for locating sensing fibers. The NiFe layer is electroplated for reducing interference. (g) The wafers are bonded and packaged with an Al-deposited cover as the fixed electrode of the capacitor. A soft silicone rubber layer is spun coated around measurement windows for increasing stability and safety. The interdigitated cantilevers as the actuated mechanism in developed photoelectronic magnetic microsensor are simulated and analyzed by the finite element modeling using the ANSYS® software. A typical FEM simulation with interdigitated cantilever size, 1000 × 1000 × 10 μm^3, deflected by external magnetic force is shown in Fig. 7. Different experimental diameters of the fibers have been achieved for optimal sensitivity using a selective wet etching technique with the solutions of trichloroethylene, xylene and hydrofluoric acid driven by an automatic instrumentation. Based on original 125 μm diameter of the single-mode fiber, related diameters of 65, 80, 95 and 110 μm in the parts of fiber gratings obtained. The simulation results that show the magnetic flux induced bending displacements related to different fiber diameters are shown in Fig. 8. Results show that the thinner the fiber diameter is, the higher the sensitivity obtains. A comparison of refraction spectra before and after side-polished FBG with iron coated films was studied (Tien et al., 2006).

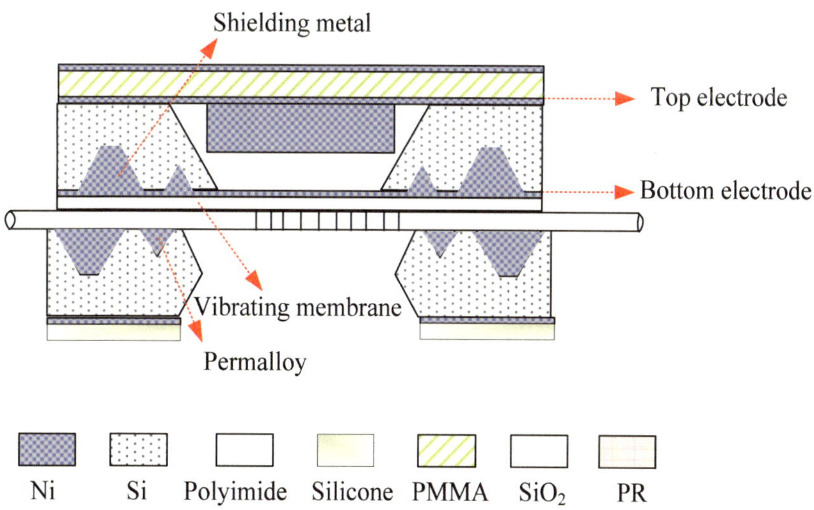

Fig. 5. Schematic diagram of a photoelectronic magnetic microsensor with EM wave-shielded packaging

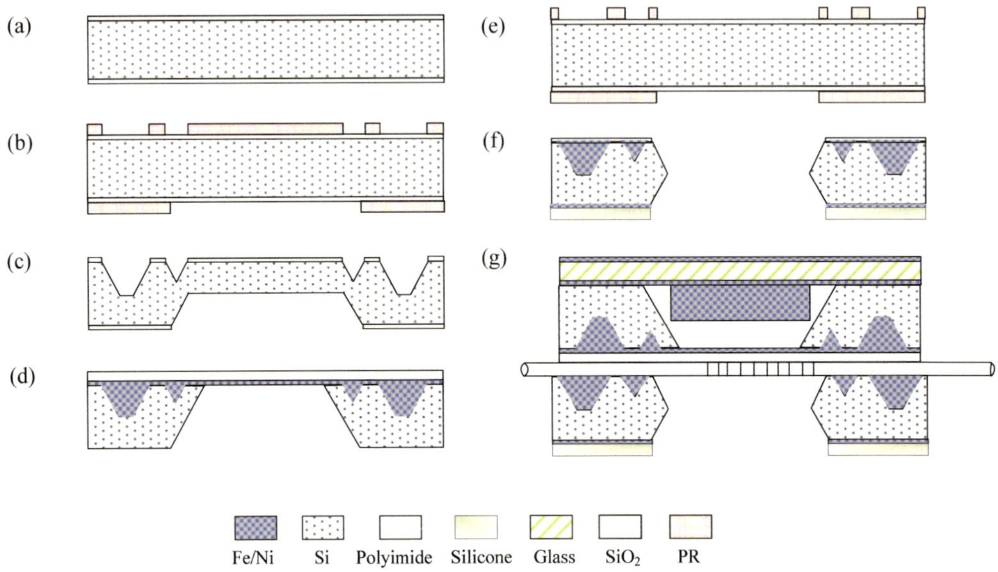

Fig. 6. Fabrication processes of the photoelectronic magnetic microsensor with EM wave-shielded packaging

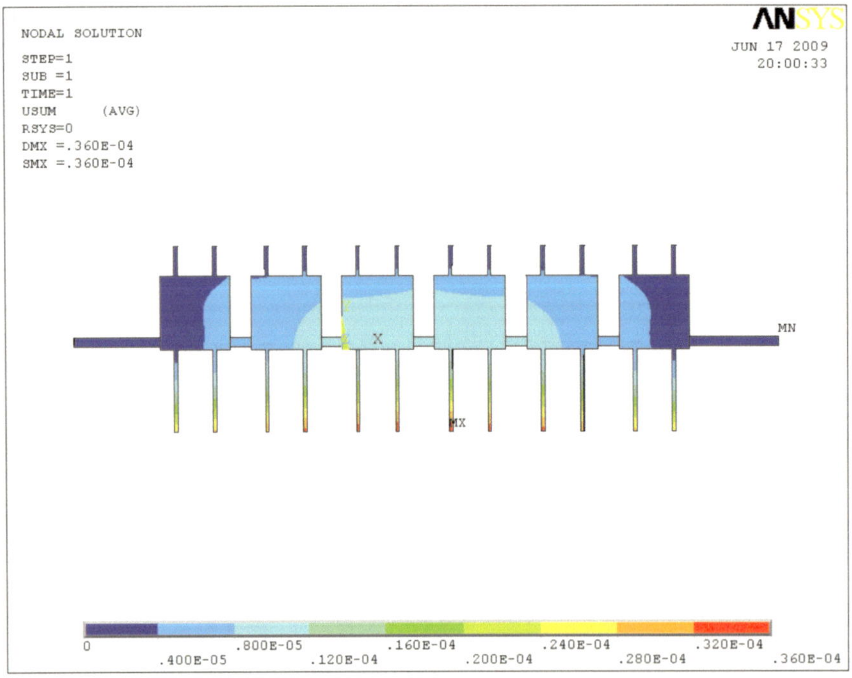

Fig. 7. FEM simulation using ANSYS to estimate cantilever bended conditions by magnetic force

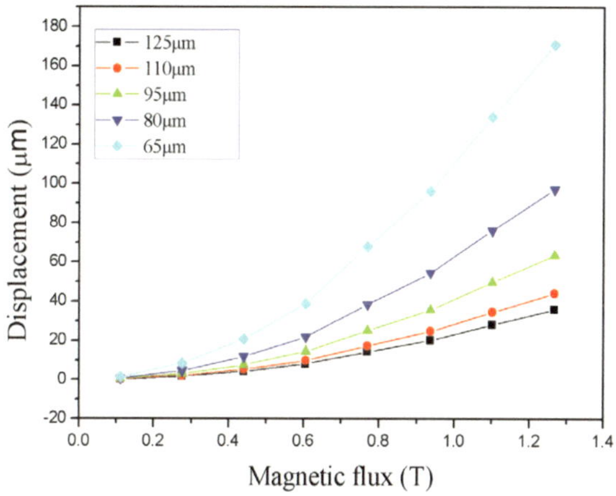

Fig. 8. Magnetic flux induced displacement related to different diameters of fibers

4. Measurement and analysis

The schematic cross-sectional view of the microsensor with interference reducing patterns connecting with measurement blocks is shown in Fig. 9. When actuated by the magnetic field, the membrane deflects to push fiber grating stretch that induce peak shift of the Bragg wavelength. The stretched fiber induces Ni-Fe film deformation actuated by external magnetic force that is the cause of fiber stretch to change the effective refraction index. The peak shift amount of Bragg wavelength is proportional to radial magnetic force measured using an optical spectrum analyzer with 0.01 nm resolution. Developed sensing structure of the microsensor includes a 300 nm Fe-film coated on FBG fiber and a parallel-plate capacitor both are deformed by a permalloy embedded polyimide membrane. Another measurement block diagram in the schematic cross-sectional view of the developed photoelectronic magnetic microsensor with temperature compensation is shown in Fig. 10. The sensing mechanism is the FBGs actuated by the covered cantilevers that a ferromagnetic material is topped on the surface. The cantilevers are attracted and bended to deform the optical gratings by magnetic flux density from external Nd-Fe-B magnets. The peak shift of the Bragg wavelength is produced from the major variations of the grating length and the fiber core effective refraction index. A precision LCR meter (Agilent E4980A) is used to measure magnetic induced capacitance variation because of the parallel interdigitated cantilever bends. The measured electrical response is in the range of 1.22 to 38.25 pF that is applied as a calibrating reference for comparing optical response excited by external magnetic flux. The signal of position-dependent capacitances is very weak that can be processed by designing capacitance-to-frequency transferred circuits to amplify magnetic response signal. Instruments used in the optical-magnetic measurements are very expensive commercial products. Therefore, to develop a precision, low cost, and portable measurement system is desired. The angular-orientation interference measurement of the permalloy surrounded packaging is shown in Fig. 11 to analyze environmental magnetic noise effect. The peak shift amount of Bragg wavelength is proportional to vertical radial magnetic force measured using an optical spectrum analyzer with 0.01 nm resolution. Original wave patterns of the reflection spectrum were calibrated as shown in Fig. 12. Experimental results show that noise can be reduced effectively when the photoelectronic magnetic microsensor with EM shielded packaging. Comparing with the responded pattern of temperature compensation, the central peak of wavelength of the sensing grating driven by bending cantilevers has shifted into the right side of long wavelength direction. The static wave patterns of the magnetic microsensor with an 80 % peak reflectivity and a 0.150 nm bandwidth, which wavelength peaks are 1550.24 nm and 1553.25 nm at 20 °C.

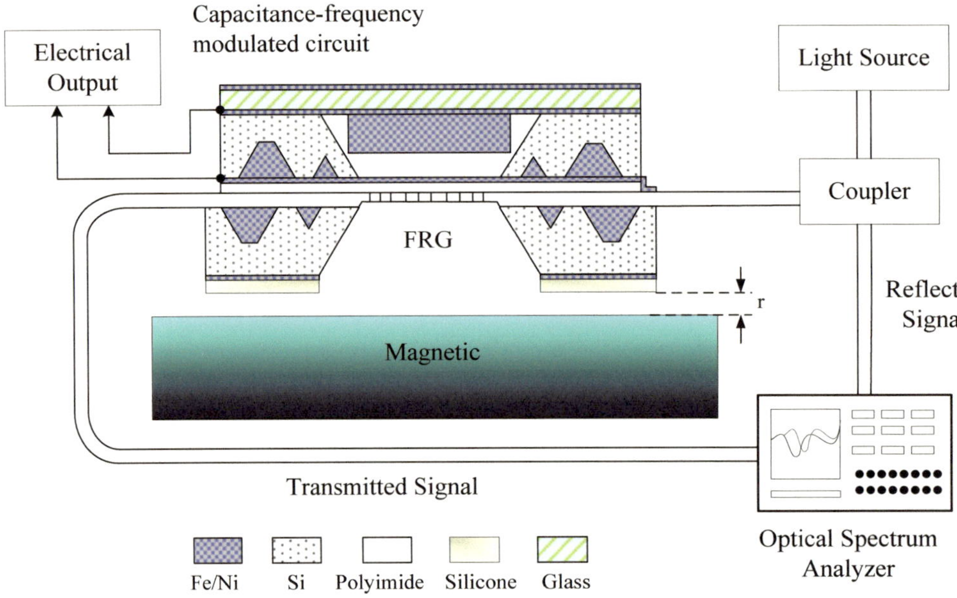

Fig. 9. Schematic diagram of the photoelectronic magnetic microsensor with measuring setup blocks

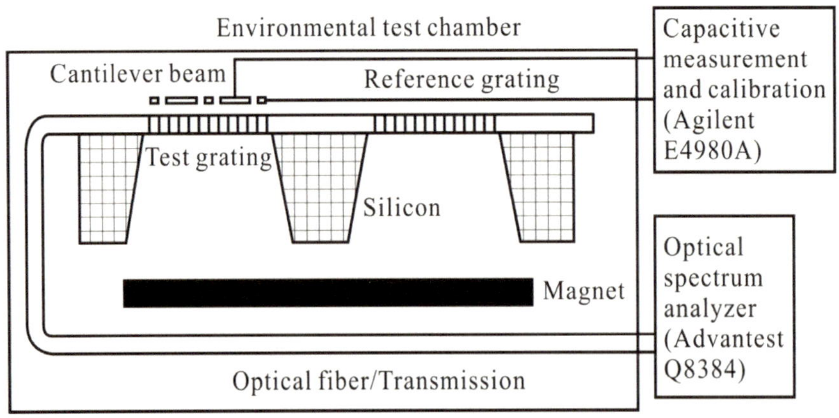

Fig. 10. Experimental setup of the photoelectronic magnetic microsensor with temperature compensation

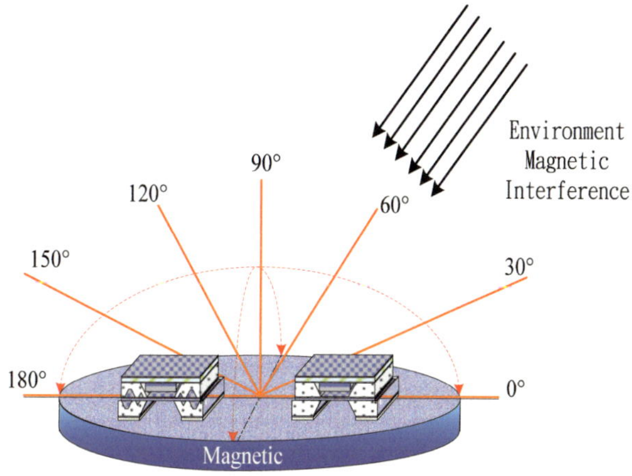

Fig. 11. Schematic of angular-orientation interference measurements for the packaged magnetic microsensors with and without permalloy surrounded

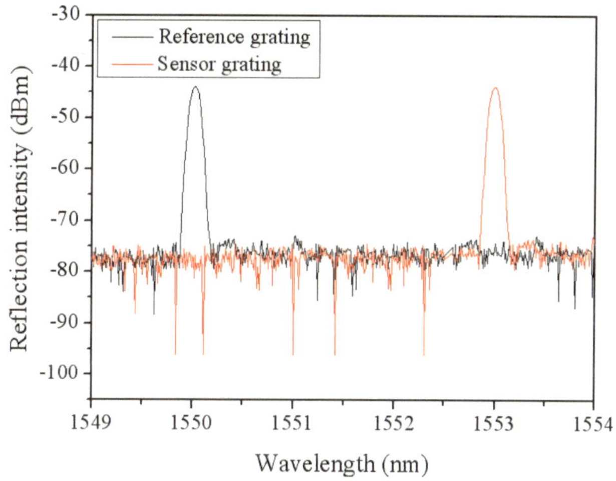

Fig. 12. Static reflection wave patterns at 20 °C, central peak wavelength are 1550.24 nm and 1553.25 nm for compensation gratings and sensor's gratings respectively

Temperature effects and related compensation results of the developed FBG microsensors without any magnetic loading reaction are analyzed and shown in Fig. 13. The temperature sensitivities of FBG sensors are 1.175×10^{-2} and 1.153×10^{-2} nm/deg for sensor and reference structures, respectively. There is temperature induced error less than 1 % that is much smaller than magnetic flux sensing responses induced variation, therefore a null deviation of temperature compensation can be obtained by calibrating. The residual magnetic strength of Nd-Fe-B magnets up to 1.26 T was used to achieve temperature-independent magnetic measurements by evaluating the Bragg wavelength shift with sensitivity about 2.145 T/nm. A linear measurement approach in the range of 0 to 450 mT with the average error less than 0.1% has been demonstrated that can be applied for precision magnetic measurement. The magnetic attractive force from a calibrated magnet is inverse proportional to the square of the distance between developed microsensor and the magnet.

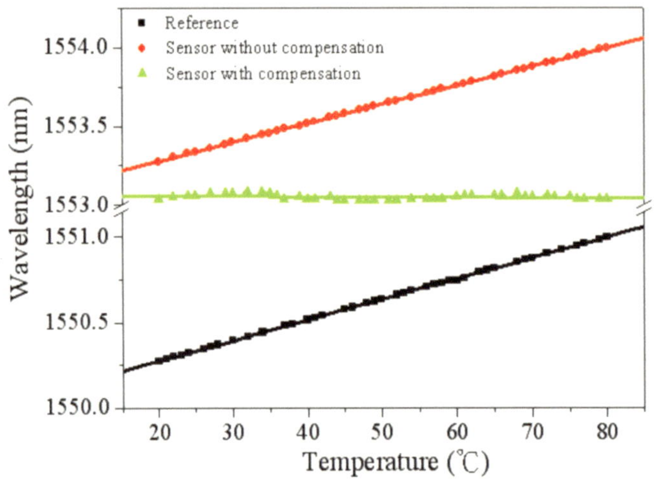

Fig. 13. Temperature induced responses and related compensation results of the developed FBG microsensor without any magnetic effect

The response curves in Fig. 14 present the measured center Bragg wavelength shifts and the net variations as the function of magnetic flux density. Responses without temperature compensation have different curves in separation to show magnetic response with temperature noise that have less sensitivity and high deviation at high temperature region. The curve slopes as magnetic sensitivity are smaller than the theoretical value. This is due to uncertainties of the FBG fiber properties in Young's modulus, Poisson's ration, and stress-optic coefficient with a high resistance in the optical-mechanical coupling reaction. The temperature independent sensitivity of the microsensors by evaluating the Bragg wavelength shift is about 2.238 T/nm, and all curves of the magnetic responses with temperature compensation are closely overlap with very small deviation measured in variable temperature environment.

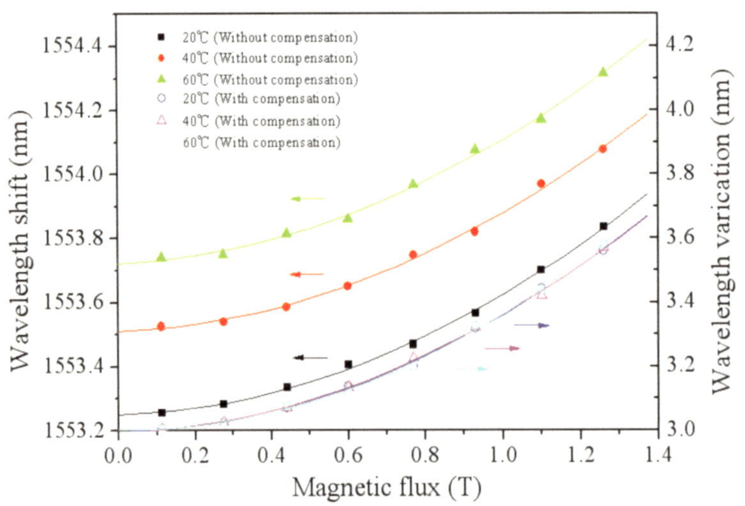

Fig. 14. Experimental results show reflected central wavelength shifts and the equivalent sensing variation induced by different magnetic fluxes of magnets

Based on simple FBG signal measurement technology using expensive instrumentation (Huang et al., 2007, Kersey et al, 1997), a reliable low cost and portable measurement system is designed in the research for measuring photoelectronic magnetic field or even for universal optical sensing applications shown as Fig. 15. The photodiodes are used as the receivers to detect FBG microsensor signals and a non-biased resistance is optimal matched for better sensitivity. The signal of photocurrent is converted to sinusoidal voltage via selected resistance as input signal for amplifier, filter, and peak detection circuits. Analog voltage signals converted into digital signals by an A/D converter that is sent to a microprocessor to perform the functions of data acquisition, calculation, storage and display control in real-time mode. Measured and analyzed information is shown on LCD display or a computer human-machine interface (HMI) through a RS232 serial transmission.

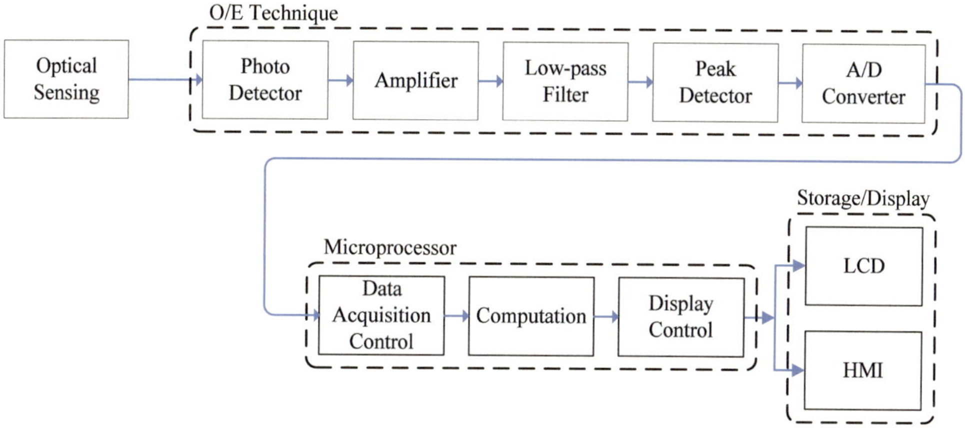

Fig. 15. The block diagram of a measurement system for developed photoelectronic sensors

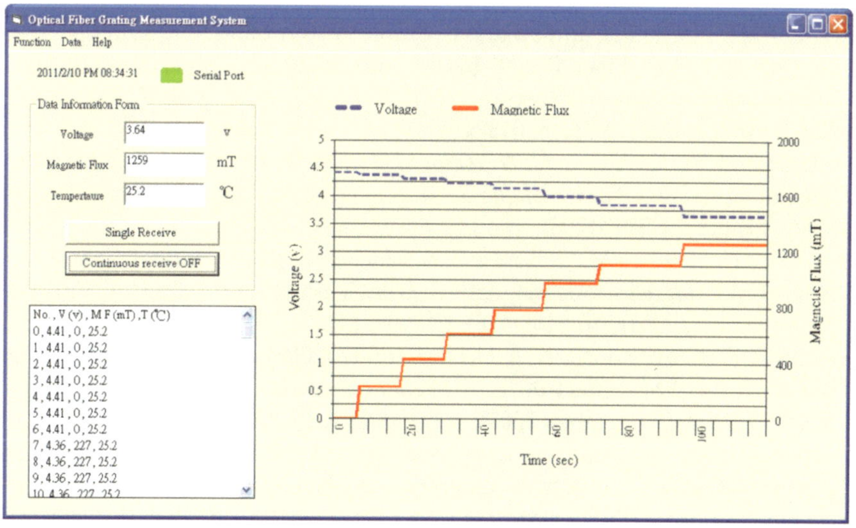

Fig. 16. A typical digital readout of human-machine interface for the photoelectronic magnetic microsensor

Fig. 16 shows a typical HMI used in a photoelectronic magnetic microsensor measurement. Dynamic measurement and analyzed information, data acquisition and storage do show on the screen in real-time mode.

5. Conclusion

The novel FBG-based magnetic microsensors with temperature compensation and EM shielded packaging have been fabricated and tested. Computer simulation has been successfully applied to optimize design parameters of microsensing structures. Electroplated permalloy cantilevers interact with magnets to provide bending force for expanding fiber gratings. Experiments have demonstrated the external magnetic flux on the order of 1.26 T can be provided 350 μm displacements of the cantilevers and 0.59 nm wavelength differences between the dual FBG. The test and reference gratings have same temperature sensitivity of 0.012 ± 0.001 nm/°C to cancel temperature-induced deviation. The magnetic sensitivity of 2.145 T/nm has been achieved using Nd-Fe-B magnets with residual magnetic strength up to 1.26 T. The signal processing circuits and the display HMI for digital readout in real-time mode are designed for static and dynamic magnetic flux measurements. Developed reliable low cost and portable measurement system can be applied to universal strain-induced photoelectrical sensing mechanism.

6. Acknowledgment

The National Science Council of Republic of China financially supporting is appreciated. The author would like to thank Dr. W. H Liu of Electrical Engineering Department for his help in FBG fabrication. Graduate students of the MEMS and Automation Lab. of Automatic Control Engineering Department provided invaluable technical assistance.

7. References

Ambrosino C. Capoluongo P. Campopiano S. Cutolo A. Giordano M. Devino D. Visone C. & Cusano A. (2007). Fiber Bragg grating and magnetic shape memory alloy: novel high-sensitivity magnetic sensor, *IEEE Sensors Journal*, Vol. 7, No. 2, (February, 2007), pp. (228-229), ISSN1530437x

Chang H. C. Hwang C. C. Liu C. C. & Tseng C. L. (2007), An opto-mechatronic magnetic field microsensor, *Physica Status Solidi (a)*, Vol. 204, No. 12, (December, 2007), pp. (4079-4082), ISSN39134208

Ciudad D. Aroca C. Sánchez M. C. Lopez E. & Sánchez P. (2004), Modeling and fabrication of a MEMS magnetostatic magnetic sensor, *Sensors and Actuators A*, Vol. 115, No. 2-3, (September, 2004), pp. (408-416), ISSN09244247

Dimitropoulos P. D. Avaritsiotis J. N. & Hristoforou E. (2003), A novel micro-fluxgate sensor based on the AMR effect of ferromagnetic film-resistors, *Sensors and Actuators A*, Vol. 107,No. 3, (November, 2003), pp. (238-247), ISSN09244247

Flynn D. Toon A. Allen L. Dhariwal R. & Desmulliez M. P. Y. (2007), Characterization of core materials for microscale magnetic components operating in the megahertz frequency range, *IEEE Transactions on Magnetics,* Vol. 43, No. 7, (July, 2007), pp. (3171-3180), ISSN00189464

Gardner J. W. Varadan V. K. & Awadelkarim O. O. (2001), *Microsensors, MEMS, and Smart Devices*, John Wiley & Sons, Inc. ISBN047186109x, New York, USA

Hsu Y. S. Wang L. Liu W. F. & Chiang Y. J. (2006), Temperature Compensation of Optical Fiber Bragg Grating Pressure Sensor, *IEEE Photonics Technology Letters*, Vol. 18, No. 7, (April, 2006), pp. (874-876), ISSN10411135

Huang X. F. Sheng D. R. Cen K. F. & Zhou H. (2007) Low-cost relative humidity sensor based on thermoplastic polyimide-coated fiber Bragg grating, *Sensors and Actuators B*, Vol. 127, No. 2, (November, 2007), pp. (518-524), ISSN09254005

Iadicicco A. Campopiano S. Cutolo A. Giordano M. & Cusano A. (2006), Self temperature referenced refractive index sensor by non-uniform thinned fiber Bragg gratings, *Sensors and Actuators B*, Vol. 120, No. 1, (December, 2006), pp. (231-237), ISSN 09254005

Keplinger F. Kvasnica S. Hauser H. & Grossinger R. (2003), Optical readouts of cantilever bending designed for high magnetic field application, *IEEE Transactions on Magnetics*, Vol. 39, No. 56, (September, 2003), pp. (3304-3306), ISSN00189464

Khoo M. & Liu C. (2001), Micro Magnetic Silicone Elastomer Membrane Actuator, *Sensors and Actuators A*, Vol. 89, No. 3, (April, 2001), pp. (259-266), ISSN09244247

Kersey A. D. Davis M. A. Patrick H. J. LeBlanc M. Koo K. P. Askins C. G. Putnam M. A. & Friebele E. J. (1997), Fiber Grating Sensors, *Journal of Lightwave Technology*, Vol. 15, No. 8, (August, 1997), pp. (1442-1463), ISSN07338724

Lee B. (2003), Review of the present status of optical fiber sensors, *Optical Fiber Technology*, Vol. 9, No. 2, (April, 2003), pp. (57-59), ISSN10685200

Lenz J. E. (1990). A Review of Magnetic Sensors, Proceedings *IEEE of Magnetics*, Vol. 78, No. 6, (June 1990), PP. (973-989), ISSN00189219

Liu C. & Yi Y. W. (1990), Micromachined Magnetic Actuators Using Electroplated Permalloy, *IEEE Transactions on Magnetics*, Vol. 35, No.3, (1990), pp. (1976-1985), ISSN 00189464

Liu Y. Guo Z. Zhang Y. Chiang K. S. & Dong X. (2000), Simultaneous pressure and temperature measurement with polymer-coated fiber Bragg gratings, *Electronic Letter*, Vol. 36, (March, 2000), pp. (564-566), ISSN00135194

Mapps D. J. (1997), Magnetoresistive sensors, *Sensors and Actuators A*, Vol. 59, No. 1-3, (April, 1997), pp. (9-19), ISSN09244247

Meng Z. Hu Y. Chen Z. & Long X. (2001), Miniature optical fiber sensor for alternating magnetic fields based on In-BiCa vanadium iron garnet crystal, *IEEE Transactions on Magnetics*, Vol. 37, No.5, (September, 2001), pp. (3807-3809), ISSN00189464

Morey W. W. Meltz G. & Glenn W. H. (1989), Fiber Bragg grating sensors, *Proc. SPIE Fiber Optic & Laser Sensors VII*, Vol. 1169, (January, 1989), pp. (98-107), ISBN 9780819402059

Oh K. D. Ranade J. Arya V. Wang A. & Claus R. O. (1997), Optical fiber Fabry-Perot interferometric sensor for magnetic field measurement, *IEEE Photonics Technology Letters*, Vol. 9, No. 6, (June, 1997), pp. (797-799), ISSN10411135

Rao Y. J. (1999), Recent progress in applications of in-fiber Bragg grating sensors, *Optical Lasers Engineering*, Vol. 31, No. 4, (April, 1999), pp. (297-324), ISSN01438166

Sedlar M. Matejec V. & Paulicka I. (2000), Optical fibre magnetic field sensors using ceramic magnetostrictive jackets, *Sensors and Actuators A*, Vol. 84, No. 3, (September, 2000), pp. (297-302), ISSN09244247

Seo Y. H. Han K. H. & Cho Y. H. (2001), New plasma Hall effect magnetic Sensors: macrosensors versus microsensors, *Sensors and Actuators A*, Vol. 92, No. 1-3, (August, 2001), pp. (123-131), ISSN09244247

Tian K. Liu Y. & Wang O. (2005), Temperature-independent fiber Bragg grating strain sensor using bimetal cantilever, *Optical Fiber Technology*, Vol. 11, No. 4, (October, 2005), pp. (370-377), ISSN10685200

Tien C. L. Hwang C. C. Chen H. W. Liu W. F. & Lin S. W. (2006), Magnetic sensor based on side-polished fiber Bragg grating coated with iron film, *IEEE Transactions on Magnetics*, Vol. 42, No. 10, (October, 2006), pp. (3285-3287), ISSN00189464

Wang X. Chen S. Du Z. Shi C. & Chen J. (2008), Experimental study of some key issues on fiber-optic interferometric sensors detecting weak magnetic field, *IEEE Sensors Journal*, Vol. 8, No. 7, (July, 2008), pp. (1173-1179), ISSN1530437x

Xu M. G. Reekie L. Chow Y. T. & Darkin J. P. (1993), Optical in-fiber grating high pressure sensor, *Electronic Letter*, Vol. 29, (February, 1993), pp. (398-399), ISSN00135194

3

Development of Resonant Magnetic Field Microsensors: Challenges and Future Applications

Agustín L. Herrera-May[1,2], Luz A. Aguilera-Cortés[2], Pedro J. García-Ramírez[1], Nelly B. Mota-Carrillo[1], Wendy Y. Padrón-Hernández[1] and Eduard Figueras[3]
[1]*Centro de Investigación en Micro y Nanotecnología, Universidad Veracruzana*
[2]*Departamento Ingeniería Mecánica, Universidad de Guanajuato*
[3]*Instituto de Microelectrónica de Barcelona/CSIC*
[1,2]*Mexico*
[3]*Spain*

1. Introduction

Microelectromechanical Systems (MEMS) integrate electrical and mechanical components with feature sizes in the micrometer-scale, which can be fabricated using integrated circuit batch-processing technologies (Gad-el-Hak, 2001). The development of devices using MEMS has important advantages such as small size, light weight, low-power consumption, high sensitivity and high resolution (Herrera-May et al., 2009a). MEMS have allowed the development of several microdevices such as accelerometers (L. Li et al., 2011), gyroscopes (Che et al., 2010), micromirrors (Y. Li et al., 2011), and pressure sensors (Mian & Law, 2010). Recently, some researchers (Mohammad et al., 2010, 2011a, 2011b; Wang et al., 2011) have integrated acceleration, pressure or temperature sensors using MEMS. A potential market for MEMS will include magnetic field microsensors for applications such as automotive industry, telecommunications, medical and military instruments, and consumer electronics products (Lenz & Edelstein, 2006).

The most sensitive magnetic field sensor is the Superconducting Quantum Interference Device (SQUID), which has a resolution on the order of several femptoteslas (Josephs-Franks et al., 2003). It operates at low temperature based on two effects: flux quantization and Josephson effects. This sensor needs a sophisticated infrastructure that increases its size and cost, which limits its commercial applications.

Hall effect sensors have a low cost, small size, and a power consumption from 100 to 200 mW. They are fabricated on standard Complementary Metal-Oxide Semiconductor (CMOS) technology and can measure either constant or varying magnetic field between temperature ranges from -100 to + 100 °C (Ripka & Tipek, 2007). Nevertheless, Hall effect sensors have a low resolution from 1 to 100 mT and require temperature compensation circuits (Popovic, 2004). Fluxgate sensors can measure static or low frequency magnetic field with a resolution of 100 pT (Ripka & Tipek, 2007). They have a size of several millimeters and a power consumption close to 100 mW. These sensors require a complex fabrication of the magnetic core and coils (Baschirotto, 2006); in addition, the reduction of their mass and power decreases both their sensitivity and stability (Diaz-Michelena, 2009). In order to miniaturize fluxgate sensors is needed to solve two problems: the miniaturization of the coils and the integration of the magnetic core (Perez et al., 2004).

Anisotropic Magnetoresistive (AMR) sensors have a resolution of 10 nT, a size on order of millimetres, and a power consumption of a few milliwatts (Ripka, 2000). AMR sensors employ the anisotropic magnetoresistive effect of ferromagnetic transition metals, whose electrical resistance depends on the angle between the electrical current and the direction of magnetization. However, these sensors require a complex resetting procedure and are saturated at low magnetic fields (close to several milliteslas).

Giant Magnetoresistive (GMR) sensors can detect magnetic fields form 10 to 10^8 nT (Lenz & Edelstein, 2006) and have a size close to 1 mm. They suffer a large change in the electrical resistance when a magnetic field is applied on their thin layers of ferromagnetic and non-magnetic materials. GMR sensors can be damaged under magnetic fields close to 1 T and have higher both temperature dependence and offset than AMR sensors (Ripka, 2008).

Fiber optic sensors can detect magnetic fields from 10^{-2} to 10^6 nT and are immune to Electromagnetic Interference (EMI) (Bahreyni, 2006). These sensors exploit the magnetostrictive effect, which changes the dimensions of the magnetostrictive material under an external magnetic field. The magnetostrictive material is bonded over a piece of optical fiber that operates as a Mach-Zender interferometer. It measures the strain of the optical fiber under an external magnetic field. Unfortunately, fiber optic sensors are affected by both temperature and pressure shifts.

Resonant magnetic field microsensors based on MEMS are a new alternative in order to detect magnetic fields with a resolution close to nanoteslas (Herrera-May et al., 2009a). These microsensors employ resonant structures for monitoring magnetic fields through Lorentz force and use capacitive, piezoresistive or optical sensing techniques. They offer advantages such as a small size (on order of micrometers), low power consumption (around of milliwatts), high functionality, wide dynamic range, and low cost through batch fabrication. They can be more compact and lighter than the SQUIDs, fiber optic sensors, and search coil sensors. Magnetic field sensors based on MEMS can be placed closer to low-magnetic field sources in order to increase their output signals. These sensors need to be subjected to more studies of reliability in order to guarantee a safe operation in future commercial market. Several of they have problems such as nonlinear response, necessity of vacuum packaging, post-processing steps for releasing their resonant structures, and complex signal conditioning circuits.

This work presents the development of resonant magnetic field microsensors based on MEMS that exploit the Lorentz force principle. It describes the general performance, advantages, drawbacks, challenges and future applications of the resonant magnetic field microsensors.

2. Development of resonant magnetic field microsensors

This section presents the development of several resonant magnetic field microsensors based on MEMS. It describes the general performance and sensing techniques of these microsensors.

2.1 Operation principle and materials

This section describes the general performance of resonant magnetic field microsensors. A resonant magnetic field microsensor uses a resonant structure in order to detect external magnetic fields. Due to the resonant structure, the sensitivity of this sensor can be increased by a parameter equal to its quality factor. The structure of the magnetic field microsensor is excited at its fundamental resonant frequency through Lorentz forces or electrostatic forces. Generally, this structure is integrated by beams and/or plates, which are fabricated using surface or bulk micromachining. An external magnetic field alters the deflections of the resonant structure that can be measured by means of capacitive, piezoresistive or optical sensing techniques.

Figure 1 shows the SEM image of a resonant magnetic field microsensor formed by thin silicon beams and a Wheatstone bridge with four p-type piezoresistors. It has been developed by the MEMS group from the Research Center for Micro and Nanotechnology (MICRONA) of the Veracruzana University into collaboration with the Microelectronics Institute of Barcelona (IMB-CNM, CSIC). This microsensor exploits the Lorentz force and uses a piezoresistive sensing technique. It has a silicon structure that oscillates at its first resonant frequency under an external magnetic field, as shown in Fig. 2(a). A Lorentz force is generated by the interaction between an external magnetic field B_x (parallel a length of the resonant structure) and a sinusoidal excitation current (I_e) flowing through an aluminium loop. The Lorentz force is given by

$$F_L = I_e L_y B_x \tag{1}$$

$$I_e = \sqrt{2} I_{RMS} \sin(2\pi f t) \tag{2}$$

where I_{RMS} is the root-mean square (RMS) of I_{Al}, L_y is the overall wide of the aluminium loop perpendicular to B_x, t is the time and f is the frequency.

The Lorentz force causes a bending moment on the resonant structure, which strains four bending beams as well as two piezoresistors placed on them. The maximum bending moment (M_{max}) and longitudinal strain (ε_L) over these beams are calculated as

$$M_{\max} = 2F_L L_x \tag{3}$$

$$\varepsilon_l = \frac{3M_{\max}}{2Ebh^2} Q \tag{4}$$

where L_x is the distance between a free end of the aluminum loop to the connection point of a bending beam with the supporting beam, Q is the quality factor of the resonant structure, E is the elastic modulus of the silicon, b and h are the width and thickness of the bending beam, respectively.

The longitudinal strain (ε_l) modifies the initial resistances (R_i) of two piezoresistors, which changes the output voltage (V_{out}) of the Wheatstone bridge (see Fig. 2(b)). The V_{out} can be determined as

$$V_{out} = \frac{\Delta R_i}{2R_i} V_{bias} \tag{5}$$

where ΔR_i is the resistance variation of the piezoresistors and V_{bias} is bias voltage of the Wheatstone bridge.
Thus, the magnetic input signal is converted to an electrical output signal through the Wheatstone bridge.

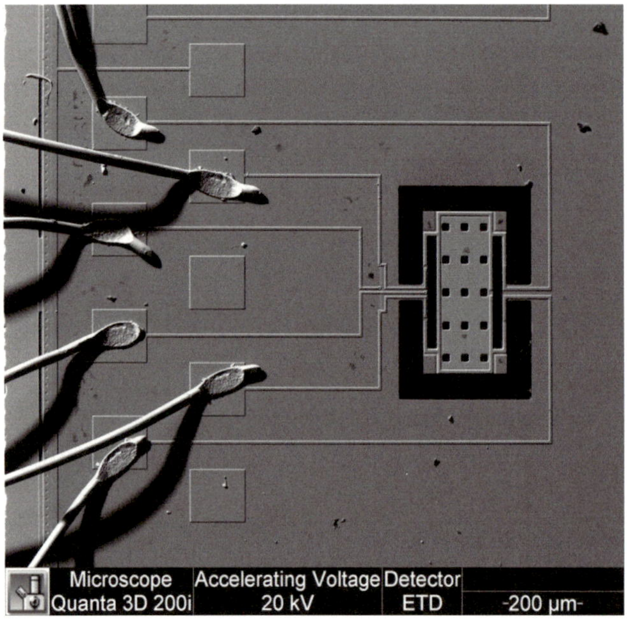

Fig. 1. SEM image of a resonant magnetic field microsensor based on MEMS.

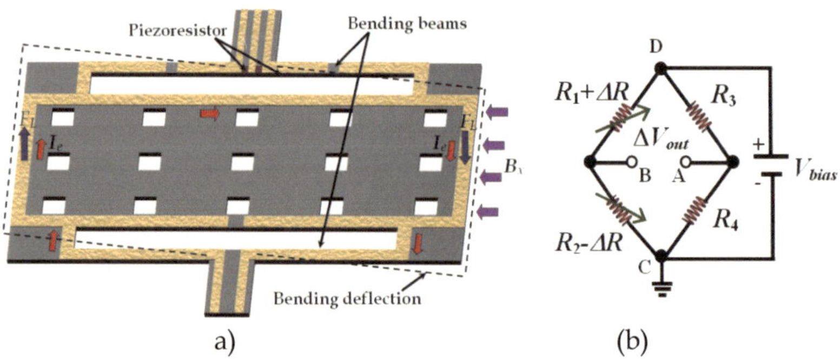

Fig. 2. A schematic view of the (a) operation principle of a resonant magnetic field microsensor with piezoresistive detection and (b) its associated Wheatstone bridge.

For the two piezoresistors under a longitudinal strain (ε_l), their ΔR_i is determined by (Clark and Wise, 1979)

$$\frac{\Delta R_i}{2R_i} = \pi_l E \varepsilon_l \tag{6}$$

where π_l is the piezoresistive coefficient parallel to the piezoresistor length.

The sensitivity (S) of the resonant magnetic field microsensor can be determined as the ratio of the output voltage (V_{out}) to the range of the external magnetic field (ΔB_x).
The sensitivity of the microsensor is limited by the quality factor of its resonant structure. The quality factor is defined as the ratio of the total energy stored in the resonant structure (E_T) to the energy lost per cycle (E_C) due to the damping effect:

$$Q = 2\pi \frac{E_T}{E_C} \tag{7}$$

The quality factor of the resonant structure depends of three damping sources (Beeby et al., 2004): a) energy lost to surrounding fluid; b) energy dissipated internally within the material of the resonant structure; and c) energy coupled through the structure's support to a surrounding solid.
Quality factor (Q_f) related to energy dissipated to surrounding fluid depends on the type of the fluid, vibration mode of the structure, fluid pressure, size and shape of the resonant structure, and its distance (gap) respect to adjacent surfaces. Q_f increases significantly when the fluid pressure decreases to magnitudes close to vacuum (Li & Fang, 2009). At atmospheric pressure, Q_f of the resonant magnetic field microsensor is due to the interactions of its resonant structure with the air.
Thermoelastic damping is the principal source of internal energy loss inside the material of the magnetic field microsensor. The oscillating strain gradient due to the vibration of microsensor's structure generates an oscillating temperature gradient. This temperature gradient causes a thermal energy loss known as thermoelastic damping (Vangallatore, 2005; Tunvir et al., 2010). The quality factor (Q_i) related to this damping is dominant when the resonant structure operated close to vacuum.
Quality factor (Q_l) due to support damping of the resonant structure represents the vibration energy dissipated by transmission through its support. The bending vibration of the resonant structure generates both vibrating moment and shear force on its clamped ends that excite elastic waves propagating into the supports (Hao et al., 2003). The supports of the resonant structure absorb part of the vibration energy.
Total quality factor (Q_T) of the resonant magnetic field microsensor can be obtained as

$$\frac{1}{Q_T} = \frac{1}{Q_f} + \frac{1}{Q_i} + \frac{1}{Q_l} \tag{8}$$

Most of the resonant magnetic field microsensors have been fabricated using silicon and polysilicon materials. The mechanical properties of these materials are affected by variations in their temperature. For instance, the silicon elastic modulus E(T) as function of the temperature from 250 K to 1500 K can be expressed as (Hull, 1999)

$$E(T) = E_i\left(1 - 9.4 \times 10^{-5} T\right) \tag{9}$$

where E_i is the elastic modulus of silicon at room temperature and T is the temperature in Kelvin.
The thermal expansion coefficient of silicon (α_{Si}) for temperature range from 293 K to 1025 K can be determined as (Watanable, 2004)

$$\alpha_{Si}(T) = -3.0451\times10^{-6} + 0.035705\times10^{-6}T - 7.981\times10^{-11}T^2 + 9.5783\times10^{-14}T^3 - 5.8919\times10^{-17}T^4 + 1.4614\times10^{-20}T^5 \quad (10)$$

The thermal conductivity of silicon (k_{Si}) for temperature range from 300 K to 400 K can be obtained as (Hull, 1999)

$$k_{Si}(T) = 309 - 0.51T \quad (11$$

Temperature fluctuations affect the elastic modulus, thermal expansion coefficient, and thermal conductivity of silicon. Thus, the fundamental resonant frequency of a magnetic field microsensor changes when its elastic modulus is modified. In addition, temperature shifts cause internal stresses on the resonant structure, which alter its resonant frequency. High temperatures may generate wear-out, corrosion, and performance degradation of a resonant structure (Darrin, 2006). Therefore, the temperature control is important issue in order to obtain a stability in the fundamental frequency of the magnetic field microsensor. It can be got through compensation electronic circuits.

2.2 Capacitive sensing

This section presents several resonant magnetic field microsensors based on MEMS that use capacitive sensing techniques.

Kádár et al. (1998) developed a resonant magnetic field microsensor with capacitive sensing. It was fabricated using a combination of bipolar processing, micromachining, glass processing and glass-to-silicon anodic bonding. This microsensor consists of a resonant silicon plate (2800× 1400 μm) with electrodes and a rectangular coil deposited on its surface. The interaction of an external magnetic field with a sinusoidal current generates a Lorentz force that causes a seesaw motion of the silicon plate. The seesaw motion modifies the distance (gap) between the silicon plate and the electrodes located into its packaging, which changes the capacitance value. Thus, the magnetic field signal is related with the capacitance shift detected by the electrodes. This microsensor can reach a resolution of 1 nT, but requires a vacuum packaging and a complex electronic circuit for the signal processing. For a single loop, it has a sensitivity of 500 $\mu V \mu \cdot T^{-1}$, a resonant frequency close to 2.4 kHz, a quality factor of 700 at 5 Pa, and a power consumption of a few milliwatts.

Emmerich and Schöfthaler (2000) fabricated a resonant magnetic field microsensor using a Bosch's standard surface micromachining process. It constains a suspended polysilicon beam with movable and fixed fingers electrodes. The interaction of a current and external magnetic field displaces the suspended beam, which changes the capacitance between the movable and fixed electrodes. This capacitance shift depends of the magnitud and direction of the external magnetic field. The microsensor operates in its resonant frequency and into a vacuum ambient in order to increase its sensitivity. It presents a sensitivity of 820 $\mu V \cdot \mu T^{-1}$ for an excitation current of 930 μA, a quality factor close to 30 at 101 Pa, a resonant frequency around 1.3 kHz, and a resolution of 200 nT considering a frequency diference of 10 Hz. This microsensor needs a vacuum packaging and a complex fabrication process. In addition, it has an offset close to 60 μT caused by the magnetic background field, voltage offsets and unbalanced parasitic capacitances.

Tucker et al. (2002) designed a xylophone microbar to detect external magnetic fields. It is the resonant structure of a magnetic field microsensor that was fabricated using a commercial 0.5

micron CMOS process. Post-CMOS processing was needed to release the xylophone microbar. Fig. 3 illustrates the operation principle of this microsensor. The microbar is supported by four arms at the nodes of its fundamental vibration mode in order to decrease the support damping. A Lorentz force is generated when a sinusoidal current flows through the microbar under an external magnetic field. This force causes a microbar vibration, which is measured capacitively. The microsensor has a die area close to 0.5 mm^2, a resonant frequency around 100 kHz, a quality factor about 1000, a power consumption of 7.5 mW, and a noise of 0.5 $nT{\cdot}Hz^{-1/2}$. It must be vacuum packaging in order to increase the vibration amplitudes of the microbar.

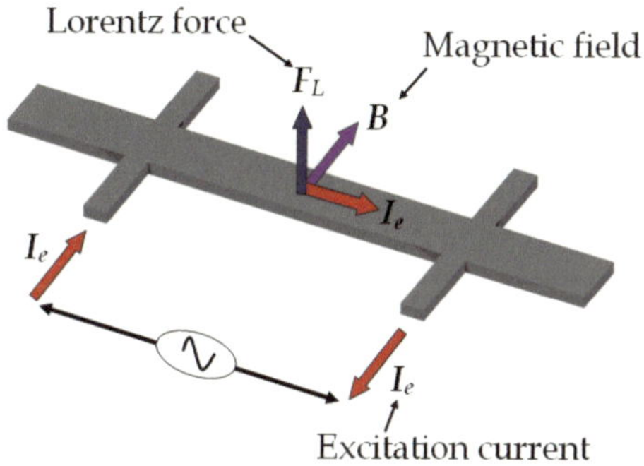

Fig. 3. Operation principle of a resonant magnetic field microsensor based on xylophone microbar proposed by Tucker et al. (2002).

Bahreyni & Shafai (2007) designed a magnetic field microsensor based on an electrostatic resonator (shuttle) fabricated in MicraGEM technology available from Micralyne Inc. The shuttle is connected to two crossbars (520 × 9 × 10 μm) by four microbeam springs (200 × 3 × 10 μm), as shown in Fig. 4. The crossbars are designed to be much stiffer in the y direction than the x direction. The shuttle is driven and kept into resonance by means of electrostatic actuation and sensing. During the microsensor operation, a DC current (I_{XB}) flowing through the crossbars. The interaction of I_{XB} with a magnetic field (B_z) normal to the microsensor's surface generates a Lorentz force (F_L) in the x direction of the crossbars. This force is axially transferred to the four microbeam springs, modifying the total stiffness of the four springs. The stiffness variation is proportional to B_z and I_{XB}, which alters the resonant frequency of the microsensor's structure. Then, the resonant frequency shift is monitored using signal processing electronics. This microsensor has a sensitivity of 69.6 $Hz{\cdot}T^{-1}$, a resolution of 217 nT for I_{XB} equal to 10 mA, a resonant frequency close to 27 kHz, a quality factor of 15,000 at 2 Pa. Alterations in the environment temperature and the heat generated by I_{XB} affect the microsensor's performance.

Ren et al. (2009) reported a magnetic field microsensor based on a silicon resonator, which was fabricated using conventional MEMS technology and a silicon-to-glass anodic bonding process. The resonator is formed by a low-resistivity silicon plate (3000 × 2000 × 60 μm) suspended over a glass substrate by two torsional beams (500 × 20 × 60 μm), as shown in Fig. 6. This silicon plate acts as electrode of sensing capacitances, which simplifies the fabrication process of the microsensor. Au capacitance plates are fabricated on the glass substrate and a multi-turn coil (Cr and Au layers) is deposited on silicon-plate surface. A Lorentz force is generated when a sinusoidal current flows into the excitation coil under an external magnetic field. This force causes an oscillating motion of silicon plate around the

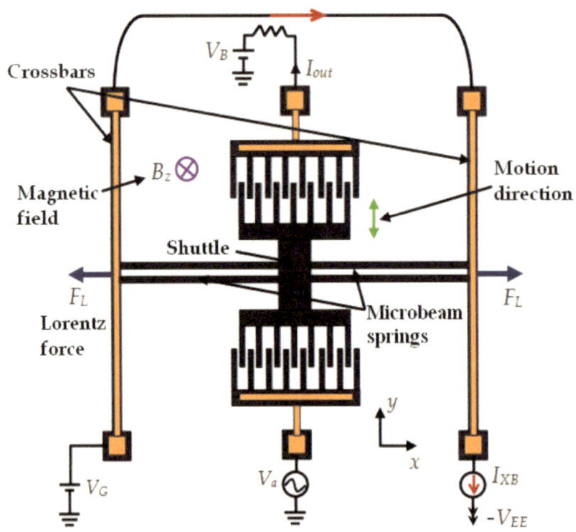

Fig. 5. Operation principle of the resonant magnetic field microsensor designed by Bahreyni & Shafai (2007).

torsional beams. This motion produces a capacitance shift between the Au electrodes and the silicon plate. A capacitance detection circuit measured the capacitance change that depends of the magnitude and direction of the external magnetic field. The microsensor needs a vacuum packaging to increase its performance. For a pressure of 10 Pa and 150 mV driving voltage amplitude, the microsensor has a resolution of 30 nT, a sensitivity of 0.40 $V \cdot T^{-1}$, a resonant frequency close to 1380 Hz, and a quality factor around 2500. Nevertheless, it presented a non-linear response from 0 to 3 μT.

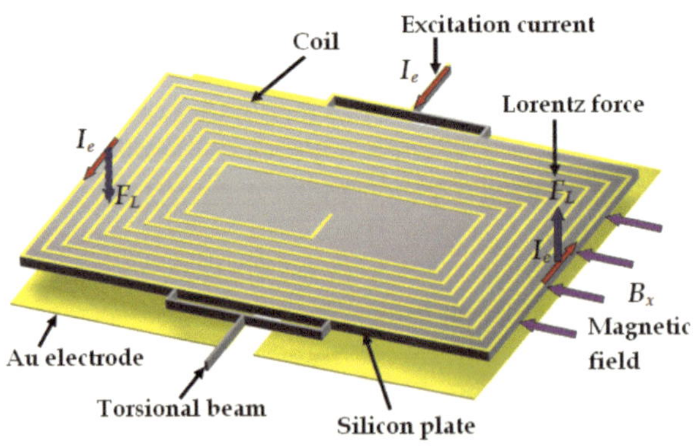

Fig. 6. Operation principle of the magnetic field microsensor designed by Ren et al. (2009).

Brugger & Paul (2009) developed a novel magnetic field microsensor (dimensions of 7.5 × 3.2 mm) integrated by an electrostatically driven silicon resonator, an amorphous magnetic concentrator, and a pair of planar coils. Interdigitated combs (200 × 14 × 25 μm) are employed for the electrostatic excitation of the resonator and the capacitive detection of its resonance (see Fig. 7). The silicon resonator is suspended by four straight flexural springs (1000 × 14 × 25 μm). The magnetic concentrator (5000 × 250 × 18 μm) is constructed using a

Metglas 2714A ribbon and is cut into three segments separated by two narrow gaps. The inner segment is attached to the resonator surface and both outer segments are fixed. The magnetic concentrator saturates to magnetic field above of 713 μT. An auxiliary magnetic field parallel to the magnetic concentrator is generated with the planar coils. Each coil comprises 12 windings with a linewidth of 55 μm and a line-to-line spacing of 15 μm. The resonator is excited at its resonant frequency applying a dc voltage (V_{dc}) to it and ac voltage (V_{ac}) to the half of the interdigitated combs. Its oscillating motion is monitoring using the other half of the interdigitated combs. A magnetic force acts on the resonator when an external magnetic field parallel to the magnetic-concentrator axis is applied. This force counteracts the restoring force exerted on the resonator by its four flexural springs, which decrease both the total spring constant and resonant frequency of the resonator. Then, a quadratic relationship between the resonant frequency and a magnetic field is obtained. In order to achieve a linear output signal of the microsensor, the external magnetic field is combined with an auxiliary magnetic field generated by the planar coils. For a coil current of 80 mA, V_{dc} = 20V, V_{ac} = 404 mV and a pressure of 10^{-5} mbar, this microsensor presents a sensitivity of 1.0 MHz·T^{-1}, a resolution of 400 nT, and a quality factor around 2400. It does not need a complex feedback and modulation electronics; however, it requires a vacuum packaging and a complex fabrication process. This process combines the following: 1) MEMS technology based on a silicon-on-insulator (SOI) substrate for the MEM structure; 2) the epoxy-resin-based attachment of a thin amorphous magnetic ribbon subsequently structured using wet chemical etching; and 3) micropatterning of the magnetic concentrator by UV-laser.

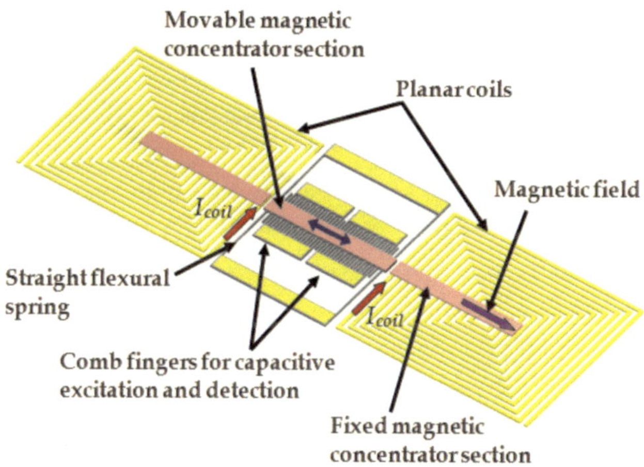

Fig. 7. Operation principle of a magnetic field microsensor based on a magnetic concentrator developed by Brugger & Paul (2009).

Generally, a resonant magnetic field microsensor with capacitive detection needs a vacuum packaging in order to increase its sensitivity and resolution. It suffers from parasitic capacitances, which can be reduced through the monolithic integration of the micromachined resonators with electronic circuits.

2.3 Optical sensing

This section includes the resonant magnetic field microsensors based on MEMS that use optical sensing.

Zanetti et al. (1998) fabricated a magnetic field microsensor formed by a xylophone microbar, which is supported (at the nodes of its fundamental vibration mode) by four microbeams. The microsensor uses optical detection for measuring the deflections of the xylophone microbar under an external magnetic field parallel to its length, as shown in Fig. 8. The interaction of this field with ac excitation current (*I*) flowing into the microbar produces a Lorentz force that deflects the microbar. If the ac excitation current is supplied with one frequency equal to the bending resonant frequency of the microbar, then, its deflection is increased. This deflection depends of the magnitude of the external magnetic field and is measured using a miniature laser and a position sensitive detector. Thus, the magnetic signal is proportional to the deflections of the xylophone microbar. The microsensor has dimensions of 5000 × 500 × 250 μm, a quality factor close to 700, a resolution around 1 nT and a power consumption of few milliwatts.

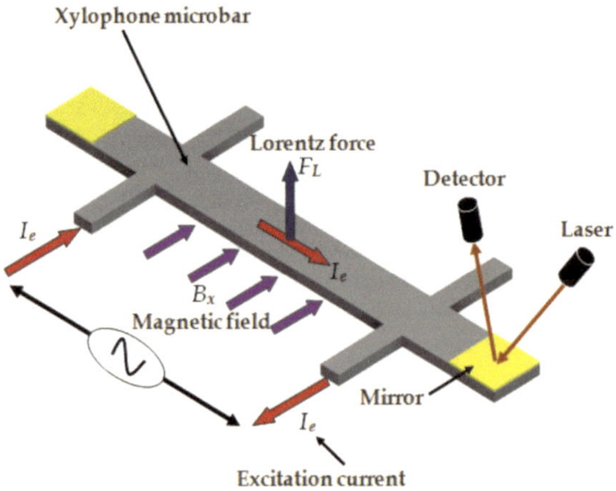

Fig. 8. Operation principle of a magnetic field microsensor with optical detection based on a xylophone microbar designed by Zanetti et al. (1998).

Keplinger et al. (2003, 2004) proposed a resonant magnetic field microsensor based on a micromachined U-shaped silicon cantilever and an optical sensing system. The cantilever (1100 × 100 × 10 μm) contains on its surface a 0.5-μm-thick Au layer, which flows an ac excitation current. A Lorentz force is generated due to the interaction of the current with a magnetic field parallel to the length of the U-shaped cantilever. This force causes a bending motion of the cantilever that is measured through optical fibers. Thus, the magnetic input signal is converted into a movement of the cantilever, which is increased when the cantilever operates at its resonant frequency. The optical sensing uses two-fiber arrangements for avoiding the problem of the interfering reflected light. In the first design the emitted light beam is reflected only once at the cantilever front side, as shown in Fig. 9. Nevertheless, it requires a large lateral space for the chip and microsensor. The second design considers a cracked cantilever to allow a parallel alignment of the fibers (see Fig. 10), which requires a perfect vertical front side of the cantilevers. This optical detection system allows the efficient transmission of the microsensor signal in an electromagnetically noisy environment. This microsensor detects magnetic fields from 10 mT to 50 T and has a resonant frequency close to 5 kHz, a quality factor of 200 at atmospheric pressure, a resolution of 10 mT, and a power consumption of a few milliwatts.

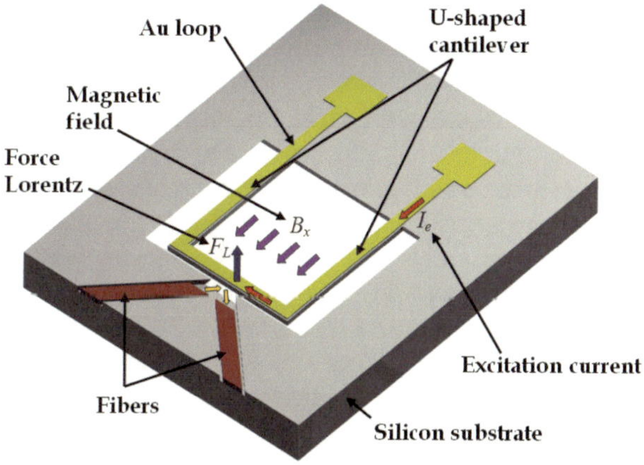

Fig. 9. Operation principle of a resonant magnetic field microsensor integrated by a U-shaped silicon cantilever and two fibers placed in curved channels, which was developed by Keplinger et al. (2003, 2004).

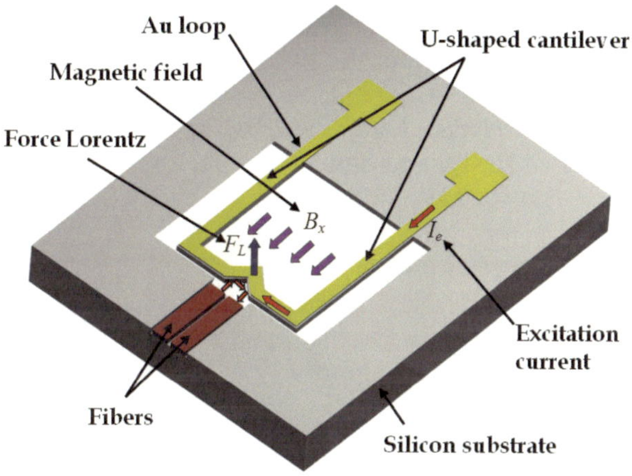

Fig. 10. Operation principle of a resonant magnetic field microsensor integrated by a U-shaped cracked cantilever and two parallel fibers, which was developed by Keplinger et al. (2003, 2004).

Wickenden et al. (2003) reported a resonant magnetic field microsensor formed by a polysilicon xylophone microbar (500 × 50 × 2 μm), which has a similar performance to microsensor designed by Zanetti et al. (1998). A magnetic input signal is converted into an oscillating motion of the xylophone microbar. This motion is detected by an optical readout system based on a laser diode beam and a position sensitive detector. This microsensor has a resonant frequency of 78.15 kHz, a quality factor around 7000 at 4.7 Pa, ac current of 22 μA, a thermal noise of 100 $pT \cdot A \cdot Hz^{-1/2}$, and a resolution on the order of nanoteslas. However, it presents a linear response up to 150 μT and its performance is affected by variations of pressure and temperature. The optical readout system used in resonant magnetic field microsensors allows a reduction in their electronic circuitry and weight. Furthermore, the microsensors with optical sensing have immunity to EMI.

2.4 Piezoresistive sensing

This section includes the description of several resonant magnetic field microsensors with piezoresistive sensing.

Beroulle et al. (2003) reported a magnetic field microsensor integrated by a resonant U-shaped cantilever (520 μm length and 80 μm thick), a planar aluminium coil of 8 turns, and a Wheatstone bridge of polysilicon strain gauges. This microsensor exploits the Lorentz force principle in order to measure an external magnetic field, converting it into an electrical output signal through Wheatstone bridge. The microsensor was fabricated using an industrial CMOS processes with a post-processing etch step to release its resonant structure. It has a sensitivity of 0.53 $V \cdot T^{-1}$, a theoretical resolution of 2 μT, a thermal noise of 5.3 $nV \cdot Hz^{-1/2}$, a resonant frequency of 8.97 kHz, a quality factor of 59, and a mass of 750 ng.

Sunier et al. (2006) developed a magnetic field microsensor using a resonant silicon structure, a planar aluminium coil, two heating resistors, and P-channel Metal Oxide Semiconductor (PMOS) transistors (see Fig. 11). This microsensor operates based on Lorentz force principle and provides a frequency output signal. The silicon structure vibrates at its resonant frequency because of the thermal actuation of two heating resistors. The oscillating motion of the silicon structure is detected using PMOS transistors connected in a Wheatstone bridge. A Lorentz force (generated by the interaction between an electrical current flowing through coil and an external magnetic field) changes the total spring constant of the resonant structure. This modifies the resonant frequency value of the silicon structure. Thus, the resonant frequency shift is related to the magnitude of the external magnetic field. This microsensor has an efficient continuous offset cancelation, high robustness, and low cross sensitivity. In addition, it presents a sensitivity of 60 $Hz \cdot T^{-1}$, a resolution of 1 μT, a resonant frequency of 175 kHz, a quality factor of 600, and a power consumption close to 5 mW.

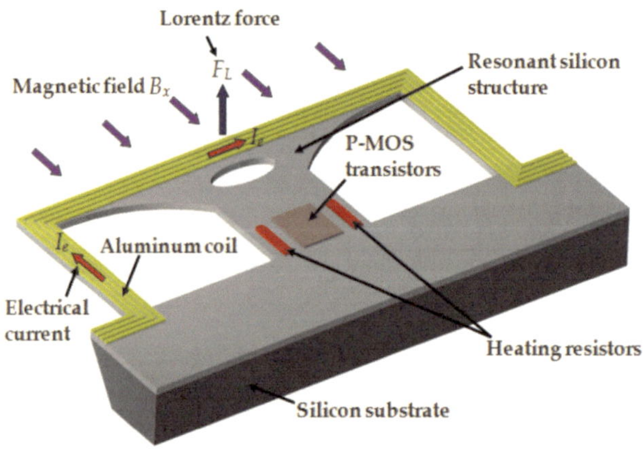

Fig. 11. Operation principle of a resonant magnetic field microsensor with frequency output designed by Sunier et al. (2006).

Herrera-May et al. (2009b) developed a resonant magnetic field microsensor with piezoresistive sensing, which contains a resonant silicon microplate (400 × 150 × 15 μm) supported by four bending microbeams, an aluminium loop, and a Wheatstone bridge of four p-type piezoresistors (see Fig. 12). It exploits the Lorentz force for monitoring external magnetic fields. This force causes an oscillating motion of the microplate and microbeams that strains two piezoresistors. This changes the piezoresistors resistance, modifying the output voltage of the Wheatstone bridge. Then, the magnetic input signal is converted into an electrical signal. This microsensor was designed for Tenaris TAMSA Corporation for measuring residual magnetic fields in welded steel tubes. It was fabricated using the bulk micromachining technology of the Microelectronics Institute of Barcelona (IMB-CNM, CSIC). Its characteristics are: sensitivity of 0.403 $V \cdot T^{-1}$, resolution of 143 nT, resonant frequency of 136.52 kHz, quality factor of 842, and a power consumption below 10 mW. However, this microsensor presents a offset and a nonlinear response for low magnetic fields.

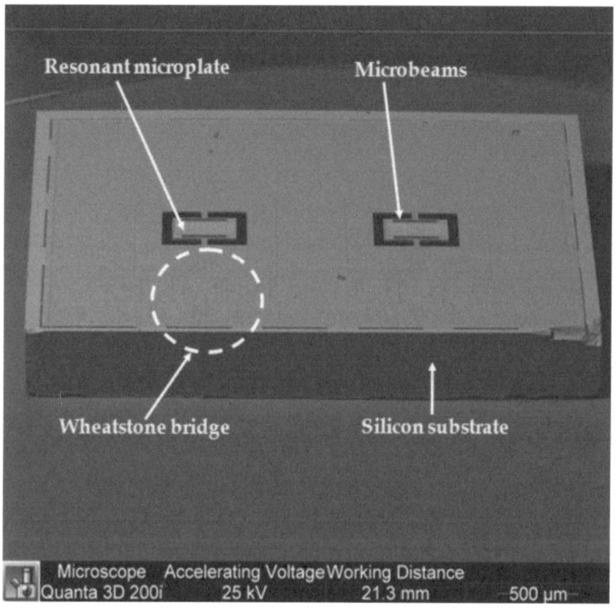

Fig. 12. SEM image of a die with two resonant magnetic field microsensors with piezoresistive sensing designed by Herrera-May et al. (2009b).

The MEMS group (Herrera-May et al., 2011) from MICRONA-UV into collaboration with IMB-CNM developed a resonant magnetic field microsensor integrated by a rectangular loop (700 × 450 × 5 µm) of silicon beams suspended with four bending beams (130 × 12 × 5 µm), an aluminium loop (1 µm thick), and four p-type piezoresistors connected into a Wheatstone bridge (see Fig. 13). This microsensor operates based on Lorentz force and has a linear response under weak magnetic fields from 40 to 2000 µT. In addition, it has sensitivity 4.8 times larger than the microsensor reported by Herrera-May et al. (2009b). This microsensor presents a resolution close to 43 nT, a resonant frequency of 22.99 kHz, a quality factor of 96.6 at atmospheric pressure, and a low power consumption of 16 mW. Later, researchers of MICRONA-UV designed other resonant magnetic field microsensor with piezoresistive sensing (see Fig. 14). It was used for monitoring magnetic fields generate by an electronic neuron developed by researchers from Institute of Physiology-BUAP (Tapia et al., 2011). This microsensor exploits the Lorentz force and is formed by an arrangement of silicon microbeams (700 × 450 × 5 µm) suspended by four bending microbeams, an aluminium loop (1 µm thick), and a Wheatstone bridge with four p-type piezoresistors. It has a compact structure, a response linear to low magnetic fields, a sensitivity of 1.2 $V \cdot T^{-1}$, a resolution of 80 nT, a resonant frequency of 13.87 kHz, a quality factor of 93 at atmospheric pressure, a power consumption of 2.05 mW, and a simple signal processing.

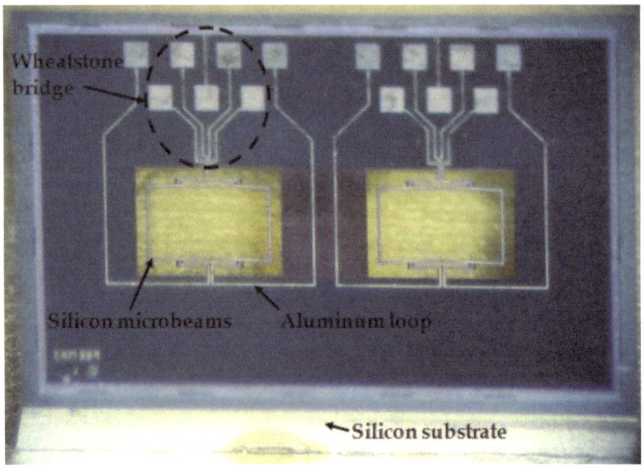

Fig. 13. Microphotograph of a die with two resonant magnetic field microsensors designed by the MEMS group of MICRONA-UV.

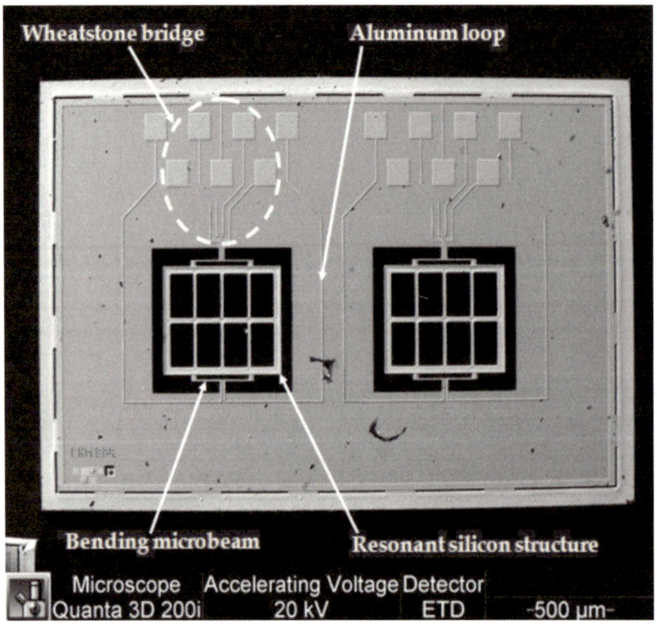

Fig. 14. SEM image of a die with two resonant magnetic field microsensors with piezoresistive detection used for neural applications (Tapia et al., 2011).

A piezoresistive sensing system used into magnetic field microsensors can allow them a simple signal processing, high sensitivity, and low-cost fabrication. Nevertheless, it is affected by temperature variations, which can be reduced through compensation circuits.

2.5 Comparison of resonant magnetic field microsensors

Most the resonant magnetic field microsensors have a compact structure, small size, and high sensitivity, wide dynamic range, and easy signal processing. Several of them can measure low magnetic fields around nanoteslas with low consumption power. Their dynamic range can be adjusted changing the magnitude of the excitation current, allowing them to measure lower or higher magnetic fields. A comparison of the main characteristics of these microsensors are indicates in Table 1. They could compete with conventional magnetic sensors for monitoring magnetic fields with resolution on the order of nanoteslas.

Microsensor	Sensing technique	Resolution (nT)	Resonant frequency (kHz)	Quality factor	Size of resonant structure (μm × μm)
Kádár et al. (1998)	Capacitive	~1	2.40	700 @ 5 Pa	2,800 × 1,400
Emmerich & Schöfthaler (2000)	Capacitive	~200	1.30	30 @ 101 Pa	1,300 × 500
Tucker et al. (2002)	Capacitive	$<10^3$	100	~1000 @ 1 P_{atm}	<1,000 × 1,000
Bahreyni et al. (2007)	Capacitive	217	~27	15000 @ 2 Pa	~520 × 400
Ren et al. (2009)	Capacitive	30	1.38	2500 @ 10 Pa	3000 × 2000
Brugger & Paul (2009)	Capacitive	400	2.5	2400 @ 10^{-5} Pa	7500 × 3200
Zanetti et al. (1998)	Optical	1	---*	~7000 @ P_{atm}	5,000 × 500
Keplinger et al. (2003, 2004)	Optical	10^7	5	200 @ P_{atm}	1,100 × 1,000
Wickenden et al. (2003)	Optical	$<10^3$	78.15	7000 @ 4.7 Pa	500 × 50
Beroulle et al. (2003)	Piezoresistive	2×10^3	8.97	59 @ P_{atm}	520 × 520
Sunier et al. (2006)	Piezoresistive	10^3	175	600 @ P_{atm}	~400 × 185
Herrera-May et al. (2009b)	Piezoresistive	143	136.52	842 @ P_{atm}	400 × 150
Herrera-May et al. (2011)	Piezoresistive	43	22.99	96.6 @ P_{atm}	700 × 450

* Data not available in the literature

Table 1. Main characteristics of resonant magnetic field microsensors.

3. Challenges and future applications

This section considers some challenges and potential applications of the resonant magnetic field microsensors.

Future commercial markets will require the development of multifunctional sensors on a single chip for monitoring several input signals such as magnetic field, pressure, acceleration, and temperature. These sensors will be integrated by several microstructures, transducers, and electronic circuits on a same substrate using monolithic fabrication. They will have important advantages such as small size, compact structure, low power consumption, and high functionality. Therefore, the integration on a single chip of the resonant magnetic field microsensors with other type of sensors is an important challenge.

Another challenge is the suppression of any background noise in order to increase the magnetic resolution. The performance of the future magnetic field microsensors will be limited by the variations in Earth's magnetic field because of geological effects. In addition, the reduction of the output response offset and the temperature dependence of future microsensors are needed. Potential applications into the automotive industry, telecommunications, and consumer electronics products demand magnetic field microsensors of low cost and high reliability. Also, a challenge is the optimization of their performance and the decrease of the design-phase time. Investigations on new materials with better electrical and mechanical properties than silicon could be used into the future microsensors.

Automotive industry could be a future market of the magnetic field microsensors in order to detect the speed and size of vehicles (see Fig. 15). A traffic's detection system may be formed by two microsensors (separated by 1 meter distance from each other) placed in parallel beside the road. The microsensors will measure the change of the Earth's magnetic field due to the vehicle motion, which will be proceeded to A/D converter and digital data processing system. The magnetic variation will depend of the vehicle's size and speed, and it will be detected by the microsensors in different times (t_1 and t_2). Then, the vehicle's speed will be determined through the ratio of the separation distance between the two microsensors to the time difference t_1-t_2. This system could be applied with an intelligent signal control to decrease traffic congestion on roads.

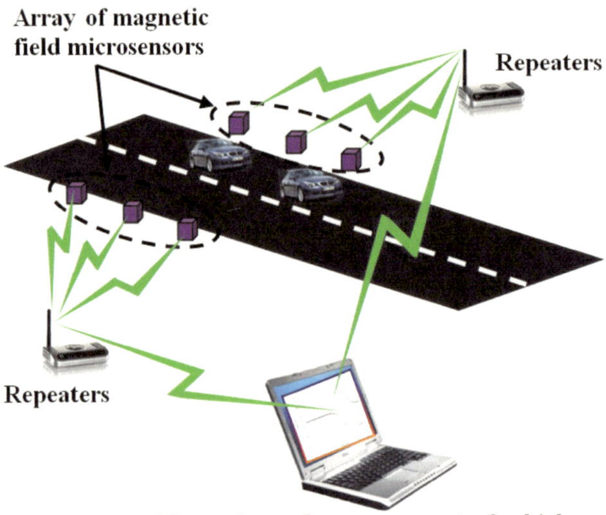

Fig. 15. Schematic diagram of a traffic's detection system based on magnetic field microsensors.

Also, magnetic field microsensors could be employed at the electronic stability program (EPS), which keeps the vehicle dynamically stable in critical situations such as hard braking and slippery surfaces. ESP systems needs data about steering-wheel angle, lateral accelerations, yaw rate, and wheel speed. These parameters could be measured through accelerations, gyroscopes, pressure sensors, and magnetic field microsensors.

Another potential application of the magnetic field microsensors is the monitoring of the corrosion and geometrical defects in ferromagnetic pipeline. Fig. 16 depicts an inspection platform for oil pipeline walls reported by Nestleroth & Davis (2007). It is integrated by a rotating permanent magnetic exciter, which may induce uniform eddy currents in the pipe wall. The eddy currents are deflected pipeline defects such as corrosion and axially sligned cracks. The variation of the current densities (that causes a magnetic flux leakage in the pipe wall) could be measured by magnetic field microsensors. Therefore, the defects location could be reached with these microsensors.

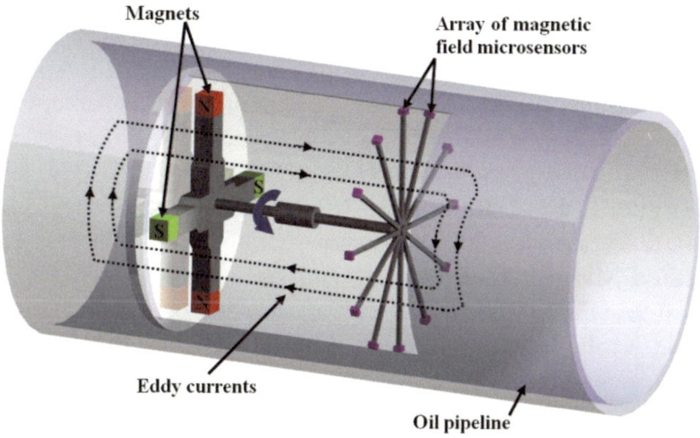

Fig. 16. Schematic view of an inspection platform of oil pipeline walls that consists of a rotating permanent magnetic exciter and an array of magnetic field microsensors.

Magnetic field microsensors could detect cracks, geometrical defects or stress concentration zones in ferromagnetic structures using passive magnetic techniques such as Metal Magnetic Memory (MMM). This technique relies on the self magnetization of ferromagnetic structures by ambient magnetic fields such as the Earth's field (Wilson et al., 2007). It measures changes in the self magnetic leakage field of the ferromagnetic structures due to geometrical discontinuities and high density dislocations.

New cell phones could use resonant magnetic field microsensors, accelerometers, and gyroscopes integrated on a single chip for their global positioning system (GPS). This could reduce the size, cost, and power consumption of the cell phones.

The important advantages of resonant magnetic field microsensors will allow their incorporation in future commercial markets, principally into the automotive sector, telecommunications, and consumer electronics products.

4. Conclusion

The development of resonant magnetic field microsensors based on MEMS has been presented. These microsensors exploit the Lorentz force for measuring magnetic fields and can use different sensing types such as: capacitive, optical, or piezoresistive. Their main advantages are small size, compact structure, light weight, low power consumption, high sensitivity, and high resolution. Most microsensors with piezoresistive detection have had an easy signal processing and a straightforward fabrication process. However, temperature fluctuations have affected their performance. Optical readout systems have allowed microsensors with a reduction in the electronic circuitry and immunity to electromagnetic interference. Microsensors with capacitive sensing have presented little dependence on the temperature, but have needed vacuum packaging and complex electronic circuitry. Future commercial markets will need multifunctional sensors on a single chip for measuring several parameters such as magnetic field, pressure, acceleration, and temperature. several

5. Acknowledgment

This work was supported by CONACYT through grant 84605. The authors would like to thank B. S. Fernando Bravo-Barrera of LAPEM for his assistance with the SEM images.

6. References

Bahreyni, B. (2006). Design, Modeling, Simulating, and Testing of Resonant Micromachined Magnetic Field Sensors. Ph. D. Thesis, University of Manitoba, Winnipeg, Canada.

Bahreyni, B. & Shafai, C. (2007). A Resonant Micromachined Magnetic Field Sensor. IEEE Sensors Journal, Vol. 7, No. 9, (September 2007), pp. 1326-1334, ISSN 1530-437X.

Baschirotto, A.; Borghetti, F.; Dallago, E.; Malcovati, P.; Marchesi, M.; Melissano, E.; Siciliano, P. & Venchi, G. (2006). Fluxgate Magnetic Sensor and Front-End Circuitry in a Integrated Microsystem. Sensors and Actuators A, Vol. 132, No. 1, (November 2006), pp. 90-97, ISSN 0924-4247.

Beeby, S.; Ensell, G.; Kraft, M. & White, N. (2004). MEMS Mechanical Sensors, Artech House, ISBN 978-1-58053-536-6, Norwood, USA.

Beroulle, V.; Bertrand, Y.; Latorre, L. & Nouet, P. Monolithic Piezoresistive CMOS Magnetic Field Sensors. Sensors and Actuators A, Vol. 103, No. 1-2, (January 2003), pp. 23-32, ISSN 0924-4247.

Brugger, S. & Paul, O. (2009). Field-Concentrator-Based Resonant Magnetic Sensor With Integrated Planar Coils. Journal of Microelectromechanical Systems, Vol. 18, No. 6, (December 2009), pp. 1432-1443, ISSN 1057-7157.

Che, L.; Xiong, B.; Li, Y. & Wang, Y. (2010). A Novel Electrostatic-Driven Tuning Fork Micromachined Gyroscope With a Bar Structure Operating at Atmospheric Pressure. Journal of Micromechanics and Microengineering, Vol. 20, No. 1, (January 2010), pp. 015025, ISSN 0960-1317.

Clark, S. K. & Wise, K. D. (1979). Pressure Sensitivity in Anisotropically Etched Thin-Diaphragm Pressure Sensors. IEEE Transactions Electron Devices, Vol. 26, No. 12, (December 1979), pp. 1887-1896, ISSN 0018-9383.

Darrin, M. A. G. (2006). Impact of Space Environment Factors on Microtechnologies, In: MEMS and Microstructures in Aerospace Applications, R. Osiander, M. A. G. Darrin & J. L. Champion, (Eds.), pp. 67-82, Taylor & Francis Group, ISBN 978-0-8247-2637-9, Boca Raton, USA.

Diaz-Michelena, M. (2009). Small Magnetic Sensors for Space Applications. Sensors, Vol. 9, No. 4, pp. 2271-2288, ISSN 1424-8220.

Emmerich, H. & Schöfthaler, M. (2000). Magnetic Field Measurement With a Novel Surface Micromachined Magnetic-Field Sensor. IEEE Transactions on Electron Devices, Vol. 47, No. 5, (May 2000), pp. 972-977, ISSN 0018-9383.

Gad-el-Hak, M. (2001). Introduction, In : The MEMS Handbook, M. Gad-el-Hak, (Ed.), ch. 1, CRC Press, ISBN 9780849321061, Florida, USA.

Hao, Z.; Erbil, A. & Ayazi, F. (2003). An Analytical Model For Support Loss in Micromachined Beam Resonators With In-Plane Flexural Vibrations. Sensors and Actuators A, Vol. 109, No. 1-2, (December 2003), pp. 156-164, ISSN 0924-4247.

Herrera-May, A. L.; Aguilera-Cortés, L. A. ; García-Ramírez, P. J. & Manjarrez, E. (2009a). Resonant Magnetic Field Sensors Based on MEMS Technology. Sensors, Vol. 9, No. 10, (September 2009), pp. 7785-7813. ISSN 1424-8220.

Herrera-May, A. L.; García-Ramírez, P. J. ; Aguilera-Cortés, L. A.; Martínez-Castillo, J. Sauceda-Carvajal, A.; García-González, L. & Figueras-Costa, E. (2009b). A Resonant Magnetic Field Microsensor With High Quality Factor at Atmospheric Pressure. Journal of Micromechanics and Microengineering, Vol. 19, No. 1, (January 2009), pp. 015016, ISSN 1057-7157.

Herrera-May, A. L.; García-Ramírez, P. J.; Aguilera-Cortés, L. A.; Figueras, E.; Martínez-Castillo, J.; Manjarrez, E.; Sauceda, A.; García-González, L. & Juárez-Aguirre, R.

(2011). Mechanical Design and Characterization of a Resonant Magnetic Field Microsensor With Linear Response and High Resolution. Sensors and Actuators A, Vol. 165, No. 2, (February 2011), pp. 399-409, ISSN 0924-4247.

Hull, R. (1999). Properties of Crystalline Silicon, INSPEC, ISBN 0-85296-933-3, London, UK.

Josephs-Franks, P.; Hao, L.; Tzalenchuck, A.; Davies, J.; Kasokova, O.; Gallop, J. C.; Brown, L. & Macfarlane, J. C. (2003). Measurement of the Spatial Sensitivity of Miniature SQUIDs Using Magnetic-Tipped STM. Superconducting Science Technology, Vol. 16, No. 12, (December 2012), pp. 1570-1574, ISSN 0953-2048.

Kádár, Z.; Bossche, A.; Sarro, P. M. & Mollinger, J. R. (1998). Magnetic-Field Measurements Using an Integrated Resonant Magnetic-Field Sensor. Sensors and Actuators A, Vol. 70, No. 3, (October 1998), pp. 225-232, ISSN 0924-4247.

Keplinger, F.; Kvasnica, S.; Hauser, H. & Grössinger, R. (2003). Optical Readouts of Cantilever Bending Designed For High Magnetic Field Application. IEEE Transactions on Magnetics, Vol. 39, No. 5, (September 2003), pp. 3304-3306, ISSN 0018-9464.

Keplinger, F.; Kvasnica, S.; Jachimowicz, A.; Kohl, F.; Steurer, J. & Hauser, H. (2004). Lorentz Force Based Magnetic Field Sensor With Optical Readout. Sensors and Actuators A, Vol. 110, No. 1-3, (February 2004), pp. 112-118, ISSN 0924-4247.

Lenz, J. & Edelstein, A. S. (2006). Magnetic Sensors and Their Applications. IEEE Sensors Journal, Vol. 6, No. 3, (June 2006), pp. 631-649, ISSN 631-649.

Li, P. & Fang, Y. (2009). A New Free Molecular Model for Squeeze-Film Damping of Flexible Microbeam in Low Vacuum. Micro and Nanosystems, Vol. 1, No. 1, (March 2009), pp., ISSN 1876-4029.

Li, L.; Stankovic, V.; Stankovic, L.; Li, L.; Cheng, S. & Uttamchandani, D. (2011). Single Pixel Optical Imaging Using a Scanning MEMS Mirror. Journal of Micromechanics and Microengineering, Vol. 21, No. 2, (February 2011), pp. 025022, ISSN 1057-7157.

Li, Y.; Zheng, Q.; Hu, Y. & Xu, Y. (2011). Micromachined Piezoresistive Accelerometer Based on a Asymmetrically Gapped Cantilever. Journal of Microelectromechanical Systems, Vol. 20, No. 1, (February 2011), pp. 83-94, ISSN 1057-7157.

Mian, A. & Law, J. (2010). Geometric Optimization of a Van Der Pauw Structure Based MEMS Pressure Sensor. Microsystem Technologies, Vol. 16, No. 11, (November 2010), pp. 1921-1929, ISSN 0946-7076.

Mohammadi, A. R.; Graham, T. C. M.; Bennington, C. P. J. & Chiao M. (2010). Development of a Composated Capacitive Pressure and Temperature Sensor Using Adhesive Bonding and Chemical-Resistant Coating for Multiphase Chemical Reactors. Sensors and Actuators A, Vol. 163, No. 2, (October 2010), pp. 471-480, ISNN 0924-4247.

Mohammadi, A. R.; Bennington C. P. J. & Chiao M. (2011a). Development of a Combined Piezoresistive Pressure and Temperature Sensor Using a Chemical Protective Coating for Kraft Pulp Digester Process Monitoring. Journal of Micromechanics and Microengineering, Vol. 21, No. 1, (January 2011), pp. 015009, ISSN 1057-7157.

Mohammadi, A. R.; Bennington C. P. J. & Chiao M. (2011b). A Hybrid Capacitive Pressure and Temperature Sensor Fabricated by Adhesive Bonding Technique for Harsh Enviroment of Kraft Pulp Digesters. Microsystem Technologies, Vol. 17, No. 1, (January 2011), pp. 149-160, ISSN 0946-7076.

Nestleroth, J. B. & Davis, R. J. (2007). Application of Eddy Currents Induced by Permanent Magnets For Pipeline Inspection. NDT & International, Vol. 40, No. 1, (January 2007), pp. 77-84, ISSN 0963-8695.

Perez, L.; Aroca, C.; Sánchez, P.; López, E. & Sánchez, M. C. (2004). Planar Fluxgate Sensor With an Electrodeposited amorphous core. Sensors and Actuators A, Vol. 109, No. 3, (January 2004), pp. 208-211, ISSN 0924-4247.

Popovic, R. S. (2004). Hall Effect Devices, (2nd ed.), Institute of Physics Publishing, ISBN 0-7503-0855-9, Bristol, USA.

Ren, D.; Wu, L.; Yan, M.; Cui, M.; You, Z. & Hu, M. (2009). Design and Analysis of a MEMS Based Resonant Magnetometer. Sensors, Vol. 9, No. 9, (September 2009), pp. 6951-6966, ISSN 1424-8220.

Ripka, P. (2000). New Directions in Fluxgate Sensors. Journal of Magnetism and Magnetic Materials, Vol. 215-216, No. 2, (June 2000), pp. 735-739, ISSN 0304-8853.

Ripka, P. & Tipek, A. (2007). Modern Sensors Handbook, ISTE Ltd, ISBN 9781905209668, Wiltshire, UK.

Ripka, P. (2008). Improvening the accuracy of magnetic sensors, In: Sensors: Advancements in Modeling, Design Issues, Fabrication and Practical Applications, S. C. Mukhopadhyay & R. Y. M. Huang, (Eds.), 45-60, Springer-Verlag, ISBN 978-3-540-69030-6, Berlin, Germany.

Sunier, R.; Vancura, T.; Li, Y.; Kay-Uwe, K.; Baltes, H. & Brand, O. (2006). Resonant Magnetic Field Sensor With Frequency Output. Journal of Microelectromechanical Systems, Vol. 15, No. 5, (Octuber 2006), pp. 1098-1107, ISSN 1057-7157.

Tapia, J. A.; Herrera-May, A. L.; García-Ramírez, P. J.; Martínez-Castillo, J.; Figueras, E.; Flores, A. & Manjarrez, E. (2011). Sensing Magnetic Flux Density of Artifical Neurons With a MEMS Device. Biomedical Microdevices, Vol. 13, No. 2, (April 2011), pp. 303-313, ISSN 1387-2176

Tucker, J.; Wesoleck, D. & Wickenden, D. (2002). An Integrated CMOS MEMS Xylophone Magnetometer With Capacitive Sense Electronics, NanoTech 2002, pp. 5723, Houston, Texas, USA, September 9-12, 2002.

Tunvir, K.; Ru, C. Q. & Mioduchowski, A. (2010). Thermoelastic Dissipation of Hollow Micromechanical Resonators. Physica E: Low-dimensional Systems and Nanostructures, Vol. 42, No. 9, (July 2010), pp. 2341-2352, ISSN 1386-9477.

Vengallatore, S. (2005). Analysis of Thermoelastic Damping in Laminated Composite Micromechancial Beam Resonators. Journal of Micromechanics and Microengineering, Vol. 15, No. 12, (December 2005), pp. 2398-2404, ISSN 1057-7157.

Wang, Q.; Li, X.; Li, T.; Bao, M. & Zhou, W. (2011). On-Chip Integration of Acceleration, Pressure, and Temperature Composite Sensor With a Single-Sided Micromachining Technique. Journal of Microelectromechanical Systems, Vol. 20, No. 1, (February 2011), pp. 42-52, ISSN 1057-7157.

Watanabe, H.; Yamada, N. & Okaji, M. (2004). Linear Thermal Expansion Coefficient of Silicon From 293 to 1000 K. International Journal of Thermophysics, Vol. 25, No. 1, (January 2004), pp. 221-236, ISSN 0195-928X.

Wickenden, D. K.; Chamption, J. L.; Osiander, R.; Givens, R. B.; Lamb, J. L.; Miragliotta, J. A.; Oursler, D. A. & Kistenmacher, T. J. (2003). Micromachined Polysilicon Resonating Xylophone Bar Magnetometer. Acta Astronautica, Vol. 52, No. 2-6, (January-March 2003), pp. 421-425, ISSN 0094-5765.

Wilson, J. W., Tian, G. Y. & Barrans, S. (2007). Residual Magnetic Field Sensing For Stress Measurement. Sensors and Actuators A, Vol. 135, No. 2, (April 2007), pp. 381-387, ISSN 0924-4247.

Zanetti, L. J.; Potemra, T. A.; Oursler, Lohr, D. A.; Anderson, B. J.; Givens, R. B.; Wickenden, D. K.; Osiander, R.; Kistenmacher, T. J. & Jenkins, R. E. (1998). Miniature Magnetic Field Sensors Based on Xylophone Resonators, In: Science Closure and Enabling Technologies for Constellation Class Missions, V. Angelopoulos & P. V. Panetta, (Eds.), 149-151, University of California Press, ISBN 0-9670138-0-1, CA, USA.

Part 2

Chemical Microsensors

4

A Heat Flux Microsensor for Direct Measurements in Plasma Surface Interactions

Dussart Rémi, Thomann Anne-Lise and Semmar Nadjib
GREMI, University of Orleans/CNRS
France

1. Introduction

The energy transfer from a plasma to a surface always plays an important role in low pressure plasma material processing (deposition, etching, surface treatment...) [1, 2]. Three different types of plasma species interact with the surface: charge carriers, neutrals and photons [3]. The energy due to charged particles (mainly ions and electrons) represents a significant contribution, especially when the substrate is biased. The energy coming from neutrals can be divided into different contributions: gas conduction, metastable de-excitation, fast neutrals (sputtered atoms, charge transfer mechanisms,…) and reactions at the surface (e.g. chemical etching…) In argon, for example, the energy due to neutrals is shared between gas conduction and metastable de-excitation since no reaction occurs at the surface. In reactive plasmas, the energy contribution of chemical reactions between radicals and substrate materials can be very high and has to be considered as well.
From the results of conventional plasma diagnostics (knowledge of flux and energy carried by interacting species), it is possible to estimate the maximum energy that can be transferred to a surface through energy balances. But the true energy delivered during plasma/surface interaction is difficult to evaluate. Thus, it might be more accurate to perform direct measurements of the energy influx. Most of the techniques used until now only lead to indirect estimations (eg. time evolution of the substrate temperature) [3]. These methods only gives a posteriori values averaged over several minutes, although for most processes (especially time resolved ones) real time measurement of the energy flux would be of interest.
To make direct heat flux measurements in plasma processes, we proposed to use a commercially available heat flux microsensor (HFM) [4]. This HFM is composed of hundreds of integrated micro thermocouples, which form a thin thermopile having a very good time resolution (<10 ms). In the following section, we present in details the diagnostic and the experimental setup we used to make measurements. Then, we will explain the method we used to calibrate it. The third section will describe the different contributions in the total energy transfer from a plasma to a surface. In the fourth section, measurements of the energy transfer from an inductively coupled plasma of argon to the HFM will be presented. Special diagnostics such as Langmuir probe and diode laser absorption have been used to evaluate the contribution of the different species (eg. charged particles, neutrals, metastables,…) in the total measured energy flux. In section five, we show an example of the evaluation of the energy flux due to chemical reactions between fluorine

radicals produced by an SF_6 plasma and a substrate of silicon. In the last section, some measurements of the energy flux in a plasma sputtering deposition experiment are presented and show the good sensitivity of the diagnostic.

2. Detailed description of the diagnostic

2.1 Heat flux sensor

The Heat Flux Microsensor is produced by Vatell Corporation based in Virginia in the United States [5]. The sensor mounted on the rod is shown in figure 1(a). The active surface, which is shown in the inset of figure 1(a), has a 6 mm diameter. It is composed of two distinguished sensors. The first one is a thermopile made in Nichrome and Constantan [5] based on Seebeck effect. A simple drawing is shown in figure 1(c) to explain this effect. Thermocouples are mounted in series. The junctions are located on two different levels of the sensor (figure 1(b)). Each thermocouple produces a voltage which is proportional to the heat flux, which is transferred from the top surface to the bottom of the sensor. The HFM proposed by Vatell is composed of hundreds of thermocouples (1600 cm^{-2}) fabricated by thin film deposition processes. When submitted to an energy influx, a very low temperature gradient appears between both levels of thermocouples which results in a very low voltage for each thermocouple. But, since there is a quite high density of these thermocouples, the resulting voltage is high enough to be measured by a nanovoltmeter. The second sensor is a Pt100 temperature sensor surrounding the thermopile. The PT100 is used to control the sensor temperature. Note that this second sensor is not necessary in our experiment to measure the energy flux. Moreover, by making this temperature measurement, some heat is produced which can perturb the heat flux evaluation.

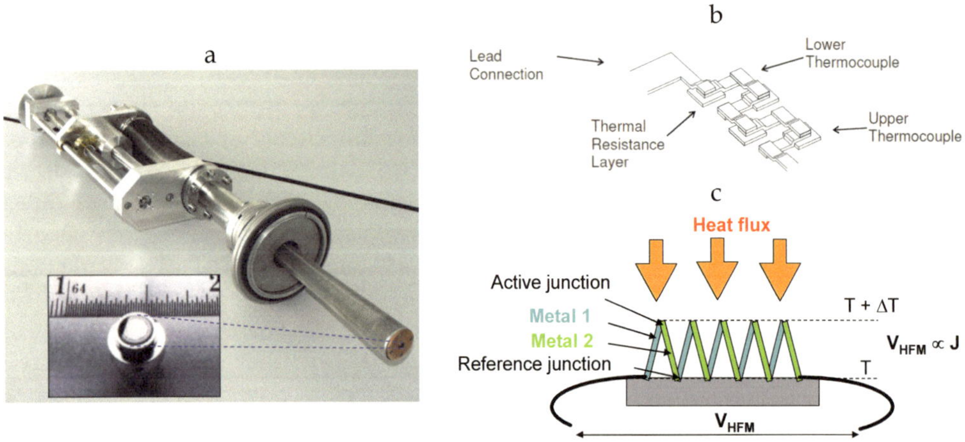

Fig. 1. (a) Heat flux microsensor mounted on the translating rod (inset) Picture of the active surface, (b) Schematic of the thermopile at the microscopic scale, (c) drawing showing the seebeck effect principle

Both sensors are inserted in a copper chamber cooled by water and controlled in temperature. For our experiments, we used the HFM-7 model, which can hold a temperature as high as 700°C. The intrinsic response of the thermopile sensor is 17 μs. However, the sensor is coated with a black paint in order to ensure radiation absorption. The presence of this coating increases the time response up to 300 μs.

Before placing the HFM inside a low pressure plasma, a calibration of the sensor has to be carried out, which is described in the next section.

2.2 Calibration procedure

For the calibration procedure, the whole sensor is placed in front of a cylinder black body (BB) (Figure 2). The temperature of the HFM (T_1) is fixed by the controller and the BB temperature (T_2) is varied from ambient to 250 °C. A thermo-radiative energy transfer takes place between the BB and the HFM, that can be calculated using the Stefan law [6]. The temperature of the back side heat flux sensor is maintained to 5°C and does not deviate during the calibration. This was done to avoid the increase in the sensor temperature, which would lead to radiation loss by IR emission.

Fig. 2. Picture of the black body (left) and top view of the calibration experiment (right)

According to the NIST Protocole we used, the calibration is first conducted in the equilibrium regime: the radiative transfer exchange is considered only when the BB averaged temperature and output signal of the HFM lead to stabilized values. Then, using the radiative heat exchange balance, the correlation between the radiative heat flux density and the output electrical signal provided by the HFM can be made. We have also conducted a calibration procedure under a dynamic regime, which gives similar results as in the equilibrium regime (Figure 3). When we investigated the calibration procedure, we showed that gas convection took place even at relatively low pressure, which prevent's from the knowledge of the transferred energy [4]. Thus, calibration of the sensor must be carried out in vacuum (10^{-5} Pa).

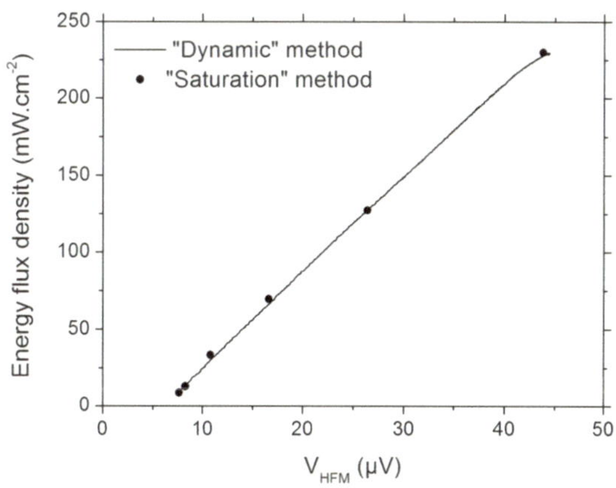

Fig. 3. Calibration curve under vacuum: comparison between dynamic measurements (line) and values taken at saturation (circles).

2.2.1 Calibration with thin samples

It is known that the energy transferred during interaction of particles with a surface widely depends on the surface characteristics (chemical composition, morphology etc.). It is thus of particular relevance to perform heat flux measurements on true samples.

The calibration curve plotted in figure 3 was recorded on the bare HFM surface. When a sample is placed in front of the HFM, assumption has to be made on its surface temperature in order to find the relationship between the HFM signal and the corresponding radiative heat flux at the sample surface. In a first approach to determine the real surface temperature, a 1D conduction model of a radiatively heated thin solid is computed using SCILAB®. We demonstrate that the temperature difference between front and back faces of 0.5mm thin solids does not exceed 5.10^{-3} K for both, Si and Cu samples. Secondly a 3D conduction simulation of the process is performed with COMSOL multiphysics to compute temperature and heat flux fields over the whole sample. The main studied parameter is the thermal contact resistance (R_{ctc}) appearing between the sample and the HFM [7]. Two important points are particularly detailed: the time for samples to be at thermal steady state and the heat flux values.

As explained above before the calibration procedure consists in heating the black body at different temperatures with different voltages and evaluating the radiative heat flux density using the Stefan law [6]

$$\varphi_{BB/HFM} = \sigma.(a_{HFM}.\varepsilon_{BB}.T_{BB}^4 - a_{BB}.\varepsilon_{HFM}.T_{HFM}^4) \quad (1)$$

T, a, ε are respectively the absolute temperature, absorptivity and emissivity of the black body (BB) and the HFM microsensor (HFM). σ is the stefan-boltzmann coefficient and ϕ the flux density of the radiative heat transfer.

The corresponding HFM voltage is displayed on the nanovoltmeter while the BB temperature increases. In case of bare HFM sensor, absorption and emission coefficients are assumed to be unity since the active surface is black painted [8]. However absorptivity and emissivity must be adjusted when a sample is fixed against the HFM probe. The view factor is set equal to 1 because of the short distance between the blackbody and the fluxmeter and because their active areas are similar. The net radiative heat flux density exchanged between a blackbody and a grey surface is deduced from the following expression [6]

$$\varphi_{BB/sample} = \sigma.\varepsilon_S.(T_{BB}^4 - T_S^4) \quad (2)$$

S label stands for sample surface

Samples are thin solids of copper and (001)-silicon of 15×15 mm^2 surface and 0.5 mm thickness (e). The former is frosted in order to remove the major part of the oxide layer whereas silicon has no particular treatment. The thermo-physical properties used in our calculations are summarized in table 1:

	Density ρ ($kg.m^{-3}$)	Surface emissivity ε_S	Thermal conductivity λ ($W.m^{-1}.K^{-1}$)	Heat capacity C_v ($J.Kg^{-1}.K^{-1}$)
Copper	8960	0.1	385	385
Silicon	2330	0.65	120	740

Table 1. Thermo-physical properties of Cu and Si at low temperatures

Thermal properties of samples are extracted from values provided in literature [7]. Taking into account their temperature dependence, these values were extrapolated for room temperature of sample surfaces.
HFM voltage is plotted versus time in figure 4. The blackbody temperatures given in the legends correspond to the thermal equilibrium with the HFM or the sample if it is present. On this figure, it is clearly seen that the HFM voltage at saturation, i.e. the heat flux density, is different depending on both the BB temperature and the sample nature. Furthermore, for a given BB temperature, the HFM voltage is related to the presence of a sample in front of its surface. This behaviour, that seems irrational at first glance, can be explained by the difference in collecting areas with or without sample. As a matter of fact, the microsensor active area is only 28.3 mm^2 whereas the collecting one in presence of a sample is 225 mm^2. The difference between Si and Cu curves is mainly due to the emissivity ratio and is in agreement with the expected values.
The offsets, different from zero, indicates that heat flux is collected even at ambient temperature. This is explained by the the radiative flux between the vessel at room temperature and the HFM at lower temperature (298K).

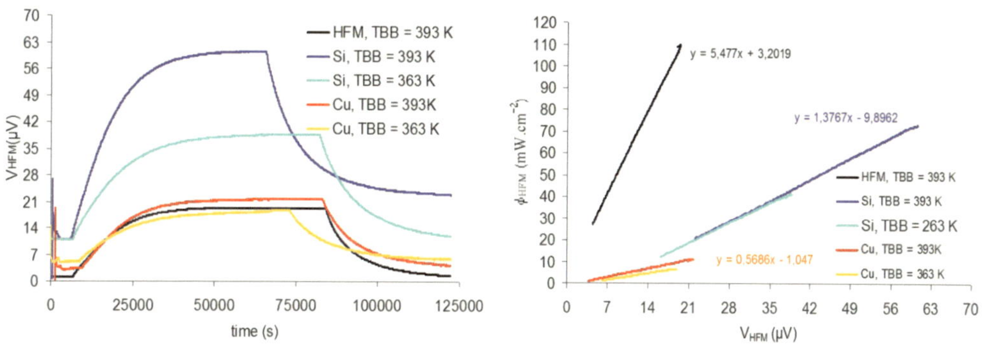

Fig. 4. Experimental measurements of heat flux density (V_{HFM}) on the left and dynamic calibration curves on the right side with respect to the black body temperatures (T_{BB}) at low pressures (~10^{-5} Pa)

As explained above and according to the NIST calibration protocol, the calibration curves are also plotted on the right graph of figure 4. Thanks to the calibration line equations deduced from this graph, one can measure the heat flux density arriving on the HFM surface or on thin solid samples, whatever the nature of the heat source (collisions, radiation…). However these results have to be used with care since rough assumptions on the sample emissivity have been made (theoretical values). Moreover the sample surface temperature has been supposed to be the same as the cooled HFM, which might be not true in the present configuration of the system. In order to get more accurate results, sample surface temperature is estimated from the resolution of a 1-D heat balance.

2.2.2 1D Modelling of thin sample on HFM

At stationary state, the heat flux crossing a thin solid sample could roughly be supposed in equilibrium with the heat exchanged at the surface of the solid. Furthermore, no heat loss is considered through the lateral faces of the sample, which is surrounded by an insulator.

$$\phi = \frac{\lambda}{e}.(T_S - T_b).S_m = \varepsilon.\sigma.(T_{BB}^4 - T_S^4).S_{BB} \tag{3}$$

where S_{BB} stands for the surface radiated by the black body, T_b the temperature at the backside of the sample and S_m is an averaged surface smaller than S_{BB} but greater than the active area of the heat probe as shown in figure 5.
We have then to resolve a 4-order polynomial

$$\begin{aligned} &P_4(T_S) = \varepsilon.\sigma.S_{BB}.T_S^4 + \frac{\lambda}{e}.S_m.T_S - (\varepsilon.\sigma.S_{BB}.T_{BB}^4 + \frac{\lambda}{e}.S_m.T_b) = 0 \\ &with\ \ T_{BB} = f(t)\ \ and\ \ T_b = T_{HFM} = 278.15K \end{aligned} \tag{4}$$

The curves reported hereafter are results obtained when solving relation (4) with SCILAB code. The BB and HFM temperatures are the time dependent input data derived from experiments whereas the surface temperature of the sample is the unknown parameter. Note that its value is determined at saturation. The results presented in figure 5a demonstrate that the temperature(T_S) difference between the front and the back (T_{HFM}) side of the thin solid is of the order of 10^{-3}K for silicon and around 10^{-4}K for Cu samples. These values are so insignificant that the approximation made in previous chapter, stating $T_S = T_{HFM}$, is justified.

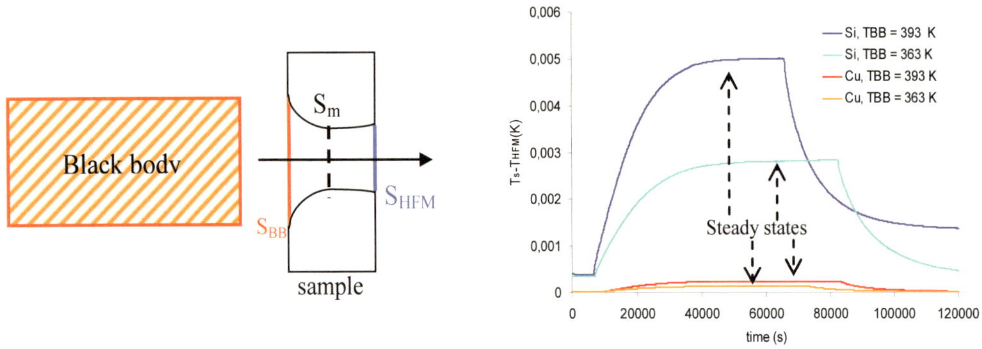

Fig. 5.a Scheme used for the 1D model (left) and temperature difference for Cu and Si obtained from 1-D model simulation (right).

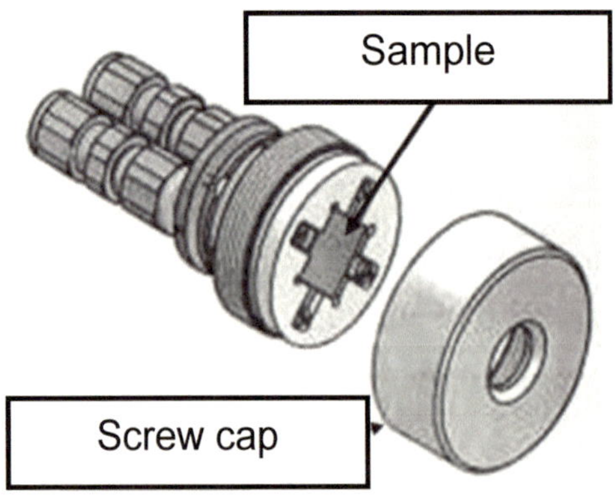

Fig. 5.b Schematic view of the sample fixation of the HFM using a screw element

However, it is important to notice that the model does not consider the existence of a thermal contact resistance between the HFM surface and the sample, since T_b has been taken equal to T_{HFM}. Unfortunately, this contact may not be perfect. Thus, a 3-dimensionnal study of the problem is necessary to have the best approach of our system.

2.2.3 Full 3D modelling of thin sample on HFM

Our thermal system is simulated with COMSOL multiphysics. In fact, only the sample is built, whereas the heat fluxes and temperatures are given as boundary conditions. Since the thermal contact resistance is not directly accessible in the code, it is simulated as a particular convective boundary condition at the HFM/substrate interface, where the convective heat coefficient and the thermal contact resistance are related by the following equation:

$$h = \frac{1}{R_{ctc}} \tag{5}$$

Boundaries	Heat flux density	Parameters
	Rarefied gas only $\varphi = h.((T+2)-T)$	$h = 3.\beta.\sqrt{\frac{k_B.P_{gas}}{2.\pi.m_{gas}.T}}\, W.m^{-2}.K^{-1}$ *
	Cooling bath $\varphi = \frac{1}{R_{ctc}}.(T_{HFM}-T)$	$T_{HFM} = 278.15K$; $R_{ctc} = 10^0, 10^{-1}...10^{-7}\, m^2.K.W^{-1}$
	Black body radiated area $\varphi = h.((T+3)-T)+\varepsilon.\sigma.(T_{BB}^4 - T^4)$	$T_{BB} = 373.15K$
	Screw cap presence $\varphi = \frac{\lambda_{cap}}{e_{cap}}.((T+2)-T)$	$\lambda_{cap} = 0.25 W.m^{-1}.K^{-1}$; $e_{cap} = 3.10^{-3} mm$
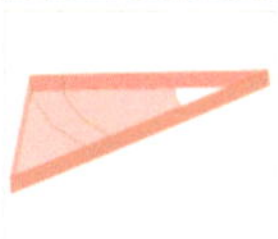	Insulated areas $\varphi = 0$	Either external boundaries or symmetry axis

* formula extracted from [9]

Table 2. Thermo-physical parameters used in this work

In the vicinity of the surface sample, the temperature of the rarefied gas is close to the solid one. To estimate the maximum heat loss from the surface,the gas temperature is here voluntarily increased by 2K or 3K, depending on how far the heat source (BB) is located. For the same reason, the screw cap temperature is also set at 2K over the solid surface one (fig. 5-b).

Owing to the plans of symmetry existing in the squared sample, the geometry of the problem has been reduced at one eighth for the sake of finer meshing and fast computer calculations. The whole boundaries conditions are summarized in table 2.
As shown in figure 6, the low temperature difference between the front and the back side of the thin sample (ΔT_{sample}), already obtained from 1D model is confirmed in a full 3D modelling, whatever the thermal contact resistance value. Moreover, the temperature difference in silicon sample is found approximately ten times higher than in copper , again in good agreement with the 1D model..
One can also see on figure 6 the effect of R_{ctc} on the delay to reach the steady state. As expected, the lower is the thermal contact resistance; the faster the equilibrium regime is reached. Note that the time expressed here could not be compared with the experiment one, which strongly depends on the blackbody inertia. In simulations, the blackbody temperature being immediately set at 373 K, the time evolution is only characteristic of the thermal response of the system (cooled HFM with sample)

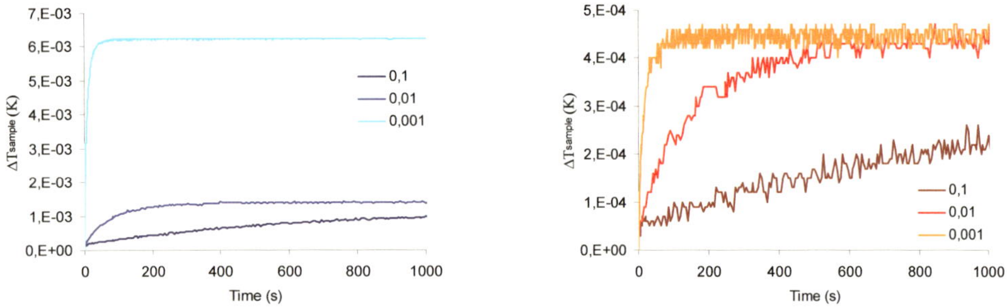

Fig. 6. Difference of front and back temperatures for Si (left) and Cu (right), at various thermal contact resistances (in $m^2.K.W^{-1}$) obtained from full 3D- model

The front (T_S) and the back (T_b) side temperatures of the sample presented in figure 7 are strongly dependent on the thermal contact resistance. It is seen that the temperature of the surface sample may be different from that of the cooling bath i.e. 278 K, even for weak thermal contact resistances.

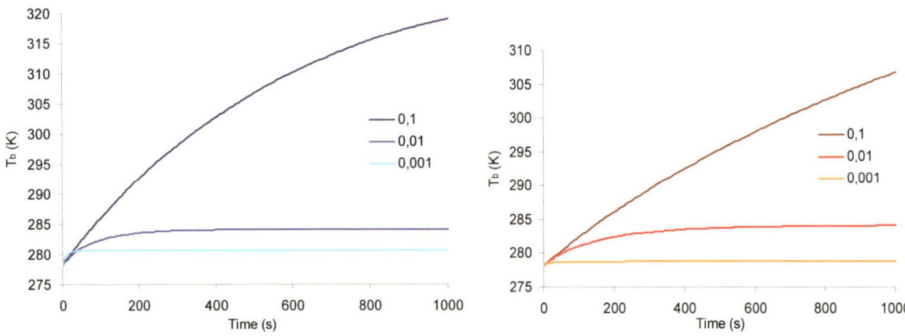

Fig. 7. Simulation of the temperature time evolution at the back face of the sample for Si (left) and Cu (right) for several thermal contact resistances, from full 3D- modelling

Simulated heat flux reaching the fluxmeter is plotted in figure 8. In the full 3D-computations, the heat fluxes are calculated for a black body radiating at 373 K. One could easily notice that measured heat fluxes are close to the calculated ones. This result indicates that thermal contact resistance values are in the range 10^{-3} to 10^{-1} $m^2.K.W^{-1}$, which is in good agreement with values given in literature for solid-solid thermal contact resistances [6].

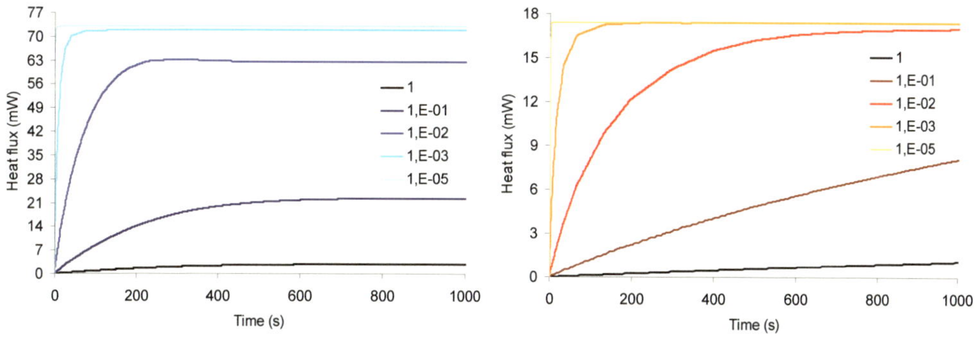

Fig. 8. Simulation of heat flux time evolution at the HFM surface for Si (left) and Cu (right) samples, at various thermal contact resistances (in $m^2.K.W^{-1}$), from full 3D- modelling

It is interesting to compare heat fluxes deduced from experimental curves with those calculated by 3D-simulations in similar conditions. This is summarized in table 3.

Samples	T_{BB} (°C)	Experiments	3D simulations for T_{BB} = 373K			
			R_{ctc} ($m^2.K.W^{-1}$) =	10^{-1}	10^{-2}	10^{-5}
Copper	363K	12.5 mW		13 mW	17 mW	17.5 mW
	393K	16 mW				
Silicon	363K	53 mW		21 mW	63 mW	74 mW
	393K	92 mW				

Table 3. Comparisons of measured and 3D simulated heat fluxes (values given for saturation states) for Cu and Si samples

2.3 Comparison with other heat flux probes

Since the 1960s many authors tested various techniques to measure the energy influx [10-12]. Results provided by the literature most often come from calculations based on temperature measurements [13, 14]. Among them, calorimetric probes, based on an original idea of Thornton [10], were successfully applied to plasma science [15-18]. Some sophisticated thermal probes have been developed [19-21], such as for example the one designed in the IEAP Kiel, which consists of a thermocouple brazed to a metal plate (substrate dummy). This probe has been used by Kersten et al to characterize many kinds of low pressure plasmas used for powder generation, space propulsion, PECVD, etc [17, 21].
Nevertheless, with this kind of probes, the total energy flux is always estimated a posteriori from thermograms recorded during the heating and cooling steps. Mathematical treatments are then employed to estimate the heat flux, which introduce systematic deviations. Moreover, with these kinds of probes, it is not possible to evidence transfer mechanisms of different kinetics such as transfer by collision (instantaneous) or transfer involving a heating step (IR emission). Detection of transient or small energetic contributions (several $mWcm^{-2}$) could not be reasonably achieved.

To illustrate results that can be obtained by calorimetric probes and by the HFM, typical signals recorded in an RF argon discharge are presented in Fig. 9. Even if HFM measurements last about 100s, it is seen on the graph that a stabilized voltage is reached within several seconds. The corresponding energy influx value is directly deduced from the calibration curve. In the case of the calorimetric probe, the thermogram has to be recorded at least during 120s in order to be further treated by a software to calculate the influx. The offset value (close to 2.5 μV) observed on the HFM graph between two signals is due to radiative transfer between the chamber and the sensor kept at 298K. To determine the energy influx only due to the RF plasma, the voltage difference between the offset and the plasma-on signal has been taken into consideration.
Due to the sensitivity of the thermopile (thin film design) the noise on the HFM voltage signal is very low, even at low energy flux density values. Consequently, the corresponding energy influxes are determined with minor errors. In comparison, the signals obtained by the calorimetric probe for RF power less than 60W (e.g. energy influxes less than $35 mWcm^{-2}$) are rather noisy. This fact induces an additional source of error. The increase of the background temperature in this case may also lead to errors on the determination of the influx.

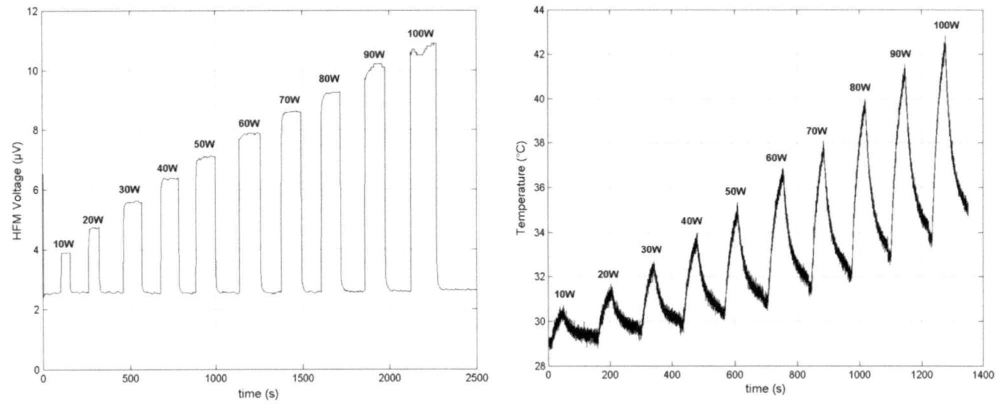

(a) (b)

Fig. 9. (a) Temporal evolution of the HFM voltage in a asymmetric RF discharge for different input powers, (b) temporal evolution of the temperature for the calorimetric probe in an asymmetric RF discharge for different input powers

Comparison that we have done in previous work has evidenced drawbacks and advantages of both sensors [22]. The main advantage of the calorimetric probe is its low cost, simplicity and sturdiness. It has been shown that this probe provides reliable results in high energy plasma processes as plasma jet, ion beam and magnetron discharge [3, 21, 23]. However, the energy influx evaluation method can cause errors of about 10%. The method also requires a certain acquisition time (seconds to minutes) which can be a problem for detecting low energy contributions or transient energy transfer processes. Thus, the calorimetric probe is a cheap and powerful tool for the measurements of total energy influxes when detection of fast transfer processes is not required.
The main HFM drawbacks are its high cost and fragility. High energy influxes can damage thermopile, but this problem has been solved by positioning a substrate (copper) foil in front of the sensor. The HFM is characterized by a very good time resolution which can even be increased by the ablation of the black coating (zynolithe) and the optimization of the

acquisition system. The HFM is an interesting tool to separate energetic contributions and detect low energy influxes. This will be illustrated in the examples given in the following sections.

3. Energy influxes involved in plasma surface interaction

The different contributions in the energy flux are detailed in a review article [3]. We explain the main contributions, which will be useful for the examples of measurements we present in the next sections.

The thermal power which is transferred at the surface of a material immerged in a plasma is the sum of the following energy fluxes:

- radiation flux Jrad (plasma and reactor wall radiation)
- energy flux due charged particles Jch (electrons and ions)
- energy flux due to neutrals Jn (gas conduction, metastable destruction at the surface (Jmet), adsorbed species (Jads), chemical reactions (Jreact) and rapid neutrals).

The total power Pin is given by :

$$P_{in} = \int_A \left(J_{rad} + J_{ch} + J_n \right) dS \tag{6}$$

A is the surface area of the sample interacting with the plasma.

3.1 Radiation

Heating by radiation can be due to reactor walls, which emit an IR radiation.

A part of the radiation flux J_{rad} corresponds to direct radiation of the plasma by excited states of the different species.

The energy transfer contribution of the reactor walls is usually quite weak in classical reactors [3]. To evaluate the part due to plasma radiation, one can use the following expression [24] :

$$J_{rad,p} = \xi_{ph} j_{ph} E_{ph} \tag{7}$$

ξ_{ph} is the absorption probability of the photon by the surface. It depends on the material.

According to [14], J_{rad} contribution remains of the order of 5 to 10 % of the total energy in a TCP discharge working at 100 W in Argon.

3.2 Electrical charges

In most of cases, charged particles (J_{ch}) represent the most significant contribution in the energy flux. [3]

For positive ions, the kinetic energy acquired in the sheath, the recombination energy lost at the surface and the secondary electron emission have to be considered to evaluate their contribution in the energy transfer. One part only of the ion kinetic energy is transferred to the surface. To estimate the energy of the ions at the surface, their energy distribution function (IEDF) has to be determined. However, the maximum energy flux density can be estimated. It corresponds to the energy flux density, which would be transferred if no energy loss by collisions occurred in the sheath and if the whole ion energy (kinetic energy and recombination energy) was transferred to the surface without reemitting any secondary electrons or sputtered atoms. In the case of a non collisional sheath, ($\lambda_i > d_{sh}$: the mean free path of ions is greater than the sheath thickness), the energy flux is perpendicular to the

sheath. The Bohm criterium can be applied to estimate the incident ion flux, which is equal to the electron flux. The mean energy reaching the surface is equal to $2k_BT_e$, (k_B: Boltzmann constant and T_e : Electron temperature) [24].
Hence, the maximal energy flux due to charged particles is given by [24, 14] :

$$J_{ion} = 0,6\, n_i u_B \left(2kT_e + e(V_P - V_S) + E_{rec}\right) \tag{8}$$

n_i : ion density
u_B : Bohm velocity
T_e : electron temperature
V_P : plasma potential
V_S : surface potential
E_{rec} : recombination energy

3.3 Neutrals

Neutrals can contribute under different manners in the energy transfer from the plasma to the surface. First, they can transfer energy by thermal conduction.
At low pressure, the power density ϕ_{cond} from the plasma to the surface can be evaluated if we know the gas temperature. The « free molecule regime » can be applied if the mean free path of atoms is at least ten times greater than the sample dimensions [9]. In this case, the energy transfer linearly depends on pressure. The following expression (9) can be used to estimate the power density due to neutral conduction [9]:

$$\varphi_{cond} = \sqrt{\frac{2k_B}{\pi m_{Ar}}}\frac{aP}{\sqrt{T_g}}(T_g - T_w) \tag{9}$$

with k_B : Boltzmann constant; m_{Ar} : Argon mass (40 uma) ; a: accommodation coefficient ; P: pressure (Pa) ; T_g: gas temperature in K and T_W: surface temperature in K.
The accommodation coefficient "a" has to be determined for Argon atoms bombarding the surface. The accomodation coefficient represents the atom thermalisation degree with the surface. It is defined by the following expression (10) [9]:

$$a = \frac{E_i - E_r}{E_i - E_w} = \frac{T_i - T_r}{T_i - T_w} \tag{10}$$

E_i, E_r and E_w represent the energy of the incident, reflected and surface atoms respectively. « a » is equal to 1 if atoms completely thermalize with the surface after interaction. According to [14], the accommodation coefficient is equal to 0.86 for argon.
At higher pressure (eg. 10 Pa), the energy flux by conduction of neutrals can become more significant. In this case, formula (9) cannot be applied because the regime is no longer the free molecule regime, but rather in so called « temperature jump regime », which corresponds to an intermediate regime between the free molecule regime and the normal conduction [9].
Metastable neutrals can bring a significant energy when they deexcite at the surface. In fact, the energy of $1s_5$ and $1s_3$ argon metastable levels reaches about 11 eV, which is the order of magnitude of the kinetic energy of the ions impinging the surface when it is not biased.
The power density ϕ^* due to metastables is given by the following expression (11) [5]:

$$\varphi^* = \xi^* N_m \sqrt{\frac{8kT_g}{\pi m_{Ar}}} E_m \tag{11}$$

with ξ^* the deexcitation probability ; N_m the metastable density. E_m the metastable energy (11.74 eV for $1s_3$ and 11.56 eV for $1s_5$ in the case of argon metastables).

ξ^* strongly depends on the surface itself. It can vary from 10^{-5} (for ceramics or oxidized surfaces) to 0.1-1 (catalytic surfaces) [5]. In our estimation, we took a value equal to 1 to have the maximum value of the power density due to metastable recombination.

In deposition processes, physisorption and chimisorption can also bring a significant value in the total energy flux [3]. In some particular cases, sputtered neutrals can get significant energies (e.g. 30 eV [3]) and should be taken into account as it will be shown in section 6. Finally, at higher pressure, in collisional sheath regimes, charge transfer can occur and create rapid neutrals [3].

3.4 Surface reactions

Chemical reactions between radicals of the plasma and the surface can bring energy (exothermal reactions) or consume energy at the surface (endothermal reactions). For example, in the case of silicon etching in plasmas containing fluorin atoms, we obtain the following reaction:

$$Si + 4F \rightarrow SiF4 \tag{12}$$

It is a very exothermal reaction whose enthalpy is -1931 $kJ.mol^{-1}$ [24,25]. Determining the etch rate, one can easily estimate the energy flux due to chemical reactions J_{reac} which is given by :

$$J_{réac} = \frac{\rho_{S_i} v_g H_r}{M_{S_i}} \tag{13}$$

ρ_{Si} is the volumic mass of silicium.
v_g is the etch rate
H_r : is the reaction enthalpy
M_{Si} : is the molar mass of silicium
An example of this contribution is presented in section 5.

4. Energy flux measurements in an Ar inductively coupled plasma

The HFM was directly submitted to an inductively coupled plasma of Argon. In this experiment, no substrate was mounted on the sensor. The HFM was left floating. Data were recorded by a sensitive nanovoltmeter as a function of time. The amount of energy influx due to the different species of the plasma was indirectly evaluated using other diagnostics (Langmuir probe, diode laser absorption, ...) which give plasma parameters such as ion density, electron temperature, gas temperature ...

In figures 10(a) and 10(b), we show respectively the obtained metastable temperature and the $1s_5$ Ar metastable density versus RF power. Measurements were carried out by diode laser absorption experiments. Due to the large lifetime of the metastables, we assume they thermalize with other neutrals. Below 150 W, in capacitive regime, the gas remains at ambient temperature. Then, in inductive mode (P > 100 W), the gas temperature increases from about 400 K up to 600 K versus RF power. The change of regime is also observed in the metastable density curve (figure 10(b)). In capacitive mode, the $1s_5$ metastable density increases versus power and reaches 7.10^9 cm^{-3}. In inductive mode, the $1s_5$ density reaches $9.10^9 cm^{-3}$ at 200 W, then, it decreases versus RF power while electron density rises.

Metastables are mainly destroyed by quenching with electrons especially in inductive mode where electron density significantly increases [4]. At 600 W, the $1s_5$ metastable density is about 3.10^9 cm^{-3}.

To summarize energy balances calculated from plasma diagnostic, we plotted in figure 11 three different curves:

- energy flux directly measured with the HFM
- calculated energy flux due to charged particles (indirect evaluation by Langmuir probe measurements)
- calculated energy flux due to charged particles, gas conduction and metastables (gas temperature and meatastable densities given in figure 10 and energy flux calculated using equations 9,10, 11).

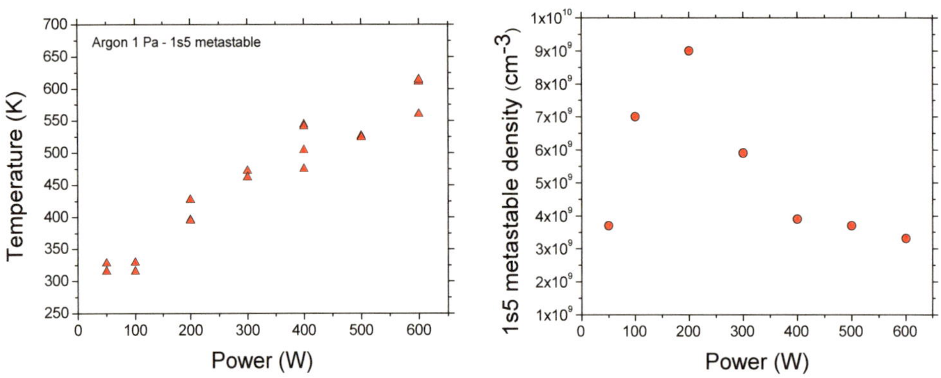

(a) (b)

Fig. 10. (a) $1s_5$ metastable temperature versus RF power,(b)$1s_5$ metastable density versus RF power

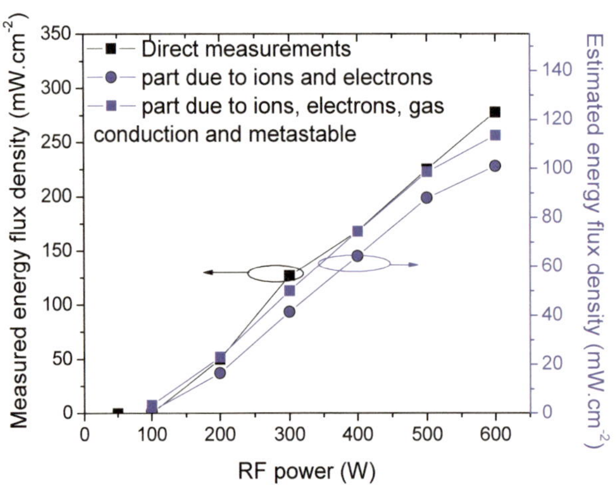

Fig. 11. Power density directly measured or estimated from energy balance versus RF power in an Ar Inductively coupled plasma

We concluded that, in our experimental conditions, most of the energy influx was due to ion bombardment. The contribution due to gas conduction corresponds to about 10 % of the total power density while the energy flux due to metastable de-excitation at the surface was found negligible. From Figure 11, it is seen that the measured heat flux density behaviour vs RF power is in good agreement with the estimations. The values are, nevertheless different, which is attributed to the fact that measurements by Langmuir probe are not very accurate. An error of the order of a factor of two can be typically made in such measurements.

5. Energy flux in a SF_6 plasma interacting with silicon

As seen in section 3, a silicon surface submitted to a SF_6 plasma leads to very exothermal chemical reactions between fluorine radicals and silicon atoms at the surface. A measurement of the energy transfer due to these reactions was carried out by placing a silicon sample on the HFM and by submitting it to a SF_6 inductively coupled plasma (figure 12) [26] .

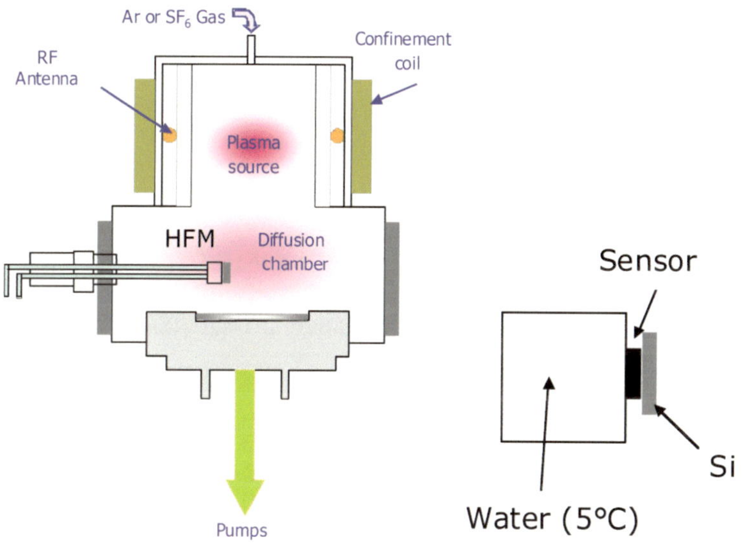

Fig. 12. (a) Schematic of the experiment to evaluate directly the energy flux due to chemical reactions, (b) detail of the sample mounted on the HFM

In figure 13(a), we show the results we have obtained by alternating Ar plasmas and SF_6 plasmas when a silicon sample was mounted on the HFM. Whereas a low energy flux is measured in non-reactive atmosphere (in argon only physical interaction takes place),a high energy flux is obtained in SF_6 plasma due to chemical reactions. The energy flux as a function of the plasma source power is presented in figure 13(b) in different cases. It is clear that low values are obtained in the case of Argon plasma or when the sample is oxidized, which decreases significantly the etch rate.
The energy flux due to chemical reactions is clearly demonstrated by these measurements. The reaction enthalpy was estimated by using the expression (13). We found a rather good agreement between our evaluation (−2200 kJ.mol^{-1}) and the theoretic value (-1931 kJ.mol^{-1}) [26].

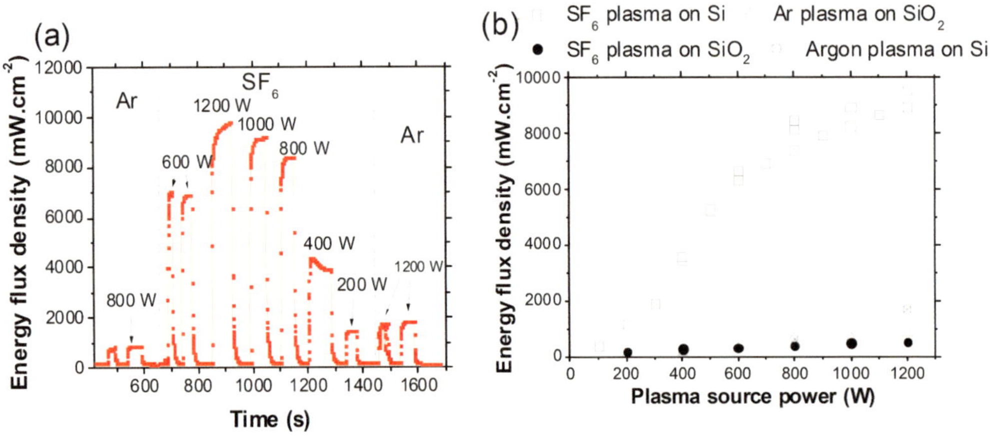

Fig. 13. (a) Time evolution of the HFM signal during an Ar plasma followed by a SF_6 plasma in interaction with a silicon sample for different source powersvs time.
(b) Maximum energy flux density vs the source power density obtained for various plasma/substrate interactions (etching condition: SF6/Si; non-etching conditions: SF6/SiO2, Ar/SiO2 and Ar/Si)

6. Measurements in deposition plasmas

The HFM was used to investigate different kinds of low pressure plasma deposition processes. First results were obtained for cathodic sputtering in an ICP argon plasma. In this experimental configuration, sputtering of the target is initiated by applying a bias voltage to a metal plate, and is independent from the creation of the RF plasma. This allows to separate the energetic contribution of the sputter-deposition process (SDP) from the sputtering plasma ones (see Figure 14)..

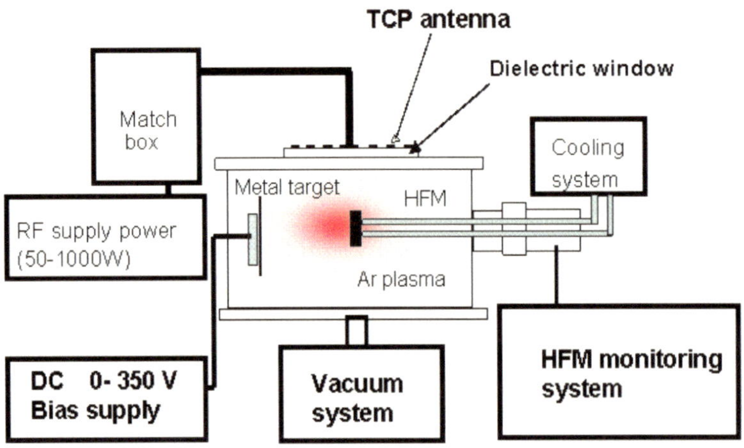

Fig. 14. ICP reactor dedicated to measurements of energy influx in cathodic sputtering deposition process. A plasma is initiated in Ar gas independently from the sputtering process that takes place when the metal target is biased.

Thanks to the good sensitivity of the sensor, the very low contribution of condensing atoms (several mW/cm2) was successfully measured. A typical energy flux time evolution recorded during sputtering of iron is presented in Figure 15. This experiment consists of a sequence of six sputter-deposition steps, lasting 1 min each, with -200V bias voltage of the target. As soon as the plasma is turned on (t ≈ 700 s), the heat flux through the substrate surface increases sharply within 2 s. The plasma contribution (here of about 250mW.cm^{-2}) has been studied in [4] and is due to energy transfer from charged particles, especially Ar ions. After this switching on step, the signal continues to increase until it reaches a steady state (at about 1600 s). This behaviour is attributed to the progressive heating of the reactor, inducing radiative transfer from the walls towards the substrate. This thermal contribution is detected by the HFM in addition to the plasma one. It is thus very easy, with the HFM, to separate this low kinetic contribution from the plasma and deposition ones. In Figure 15 (b), signals corresponding to the sputter-deposition steps are clearly evidenced. The evolution of the signal shape has been explained in reference [4]. What should be noted here is that the sputter-deposition energetic contribution (blue arrows) can easily be determined and that the measurement is reproducible.

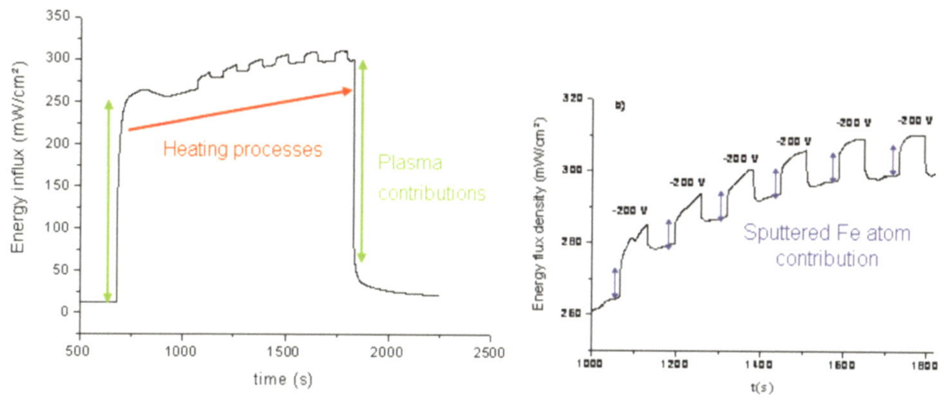

a) (b)

Fig. 15. Fe sputter deposition at −200V target bias, in 0.5 Pa and 400W argon plasma, (a) Energy flux measurements, (b) zoom of the figure showing the 1 min sputter-deposition. The energetic contributions of Ar plasma and sputtered Fe atoms are clearly distinguished.

From this kind of measurements, the energetic contribution of the sputter-deposition process can be studied and followed versus experimental parameters such gas pressure, accelerating voltage and RF Power. An example is given in Figure 16 in the case of Pt sputtering.

A linear evolution is found for the energy brought by the SDP with respect to the Pt target bias voltage. Obviously, as the target voltage becomes more negative, the kinetic energy of Ar^+ attracted by the target increases. This leads to a more efficient sputtering process. The metal atoms sputtered out of the target are thus more numerous. It has been shown that, in our experimental configuration, the mean kinetic energy of sputtered atoms only weakly depends on the energy of the incoming argon ions [15, 28, 29]. The increase of the deposition contribution is thus mainly due to a rise in the number of condensing atoms. Another contribution that can participate to the global SPD energy transfer is the one of the argon ions that are reflected by the target, neutralized and form fast neutrals. It can be predicted from simple calculations given for example in [15]. This contribution is also expected to increase with the negative bias voltage. The behaviour observed in figure 15 was thus expected.

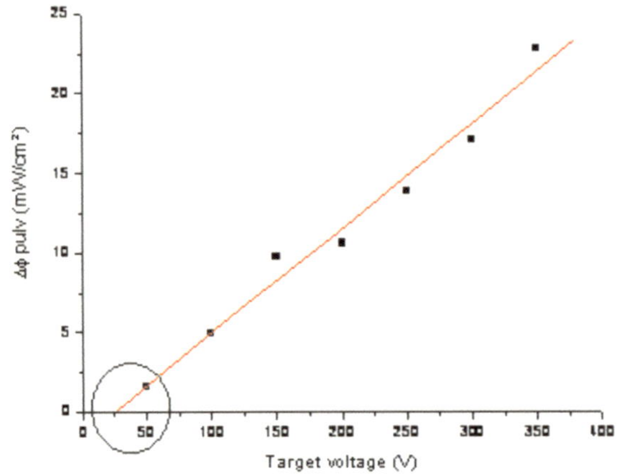

Fig. 16. Energy flux density due to the SDP of Pt versus the target bias voltage in a 0.5 Pa and 400Wargon plasma

In order to study the possible energy brought by fast neutrals, energy balances have been done taking into account the number of Ar ions bombarding the target (calculated from the ionic current), the sputtering and reflection mechanisms [15], and the transport until the substrate. Results are presented in Figure 17 for Pt and Mo.

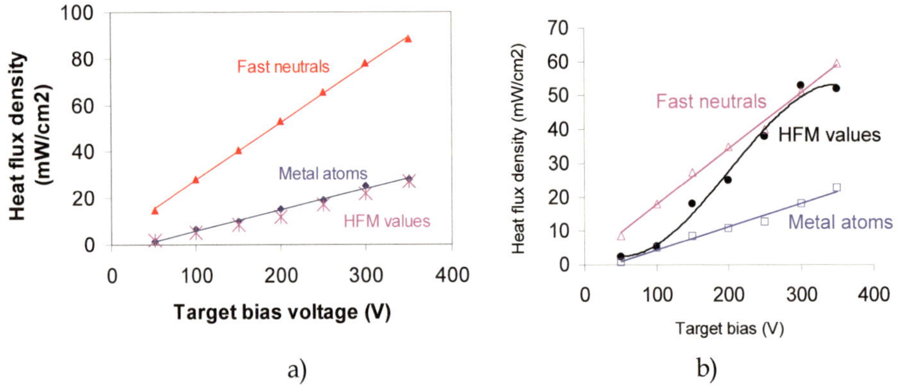

Fig. 17. Evolution of the heat flux density versus the target bias voltage for a) Pt SPD and b) Mo SPD.

The graphs in Fig. 17 clearly show that fast neutrals do not play the same role on the global energy transfer to the substrate depending on the sputtered metal. In the case of Pt, estimated contribution of neutrals is widely higher than the measured heat flux density, showing that the formalism of Drüsedau used in the calculation is not valid in this case. The very good agreement between measurements and calculated metal atom contribution indicates that the only energetic contribution comes from condensing Pt atoms. In the case of Mo, the evolution of HFM values shows that, at low target bias voltages, only metal atoms transfer energy to the substrate, whereas at higher voltages, fast neutrals participate to the global energy flux transfer.

In Figure 16, the target voltage corresponding to $\varphi_{sput} = 0$ is about 25V. Taking into account the plasma potential (≈10 V in our experimental conditions), the corresponding energy of the sputtering argon ions can be estimated to be 35 eV. It is interesting to compare this value with the sputtering threshold of Pt by Ar+ given in the literature (25-30 eV) [30]. The good agreement between both values confirms that the energy transfer from metal sputtered atoms is the main contribution in the SDP of Pt.

By dividing the heat flux density by the estimated metal atom flux at the substrate, it is possible to calculate the energy deposited per atom, which is a parameter known to drive the thin film properties. This parameter is given in Figure 18 versus the target bias voltage.

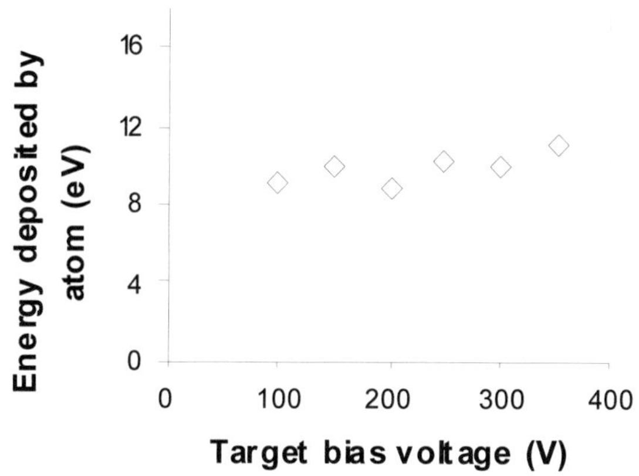

Fig. 18. Evolution of the energy deposited by condensing Pt atoms as a function of the target bias voltage

It is seen from Figure 18 that the energy deposited per atom does not depend on the target bias voltage (in the investigated range). This is in agreement with what is reported in literature [31] and results of our calculations [28]. Indeed, in our ranges of experimental parameters, the kinetic energy of sputtered atoms depends on the gas pressure rather than on the target voltage. The value that can be deduced from the graph is 10.8 eV, which has to be compared to the energy which a Pt atom may release at the substrate surface. Pt atoms carry a kinetic energy estimated to 4.9 eV in the present conditions (0.5 Pa, 10 cm from the target) and transfer their condensation energy, i.e. 5.3 eV, which gives 10.2 eV. The good agreement between calculated and measured values indicates that Pt atoms transfer their whole energy to the surface.

One can wonder if the same energy brought by different species would affects the thin film growth in the same manner. In order to study the role of the energetic vector, deposition of Pt thin films was performed at different locations in the reactor presented in Figure 14. The energetic contributions of both the Ar plasma and the condensing atoms vary. Both contributions are displayed versus the position in the reactor in figure 19. As already reported, in our conditions, the contribution of the plasma is widely higher than the SPD one.

Thin films were deposited at 3 cm and 18 cm from the target. They were analyzed by scanning electron microscopy (SEM) and Rutherford backscattering spectroscopy (RBS). Results are given in figure 20 and Table 4.

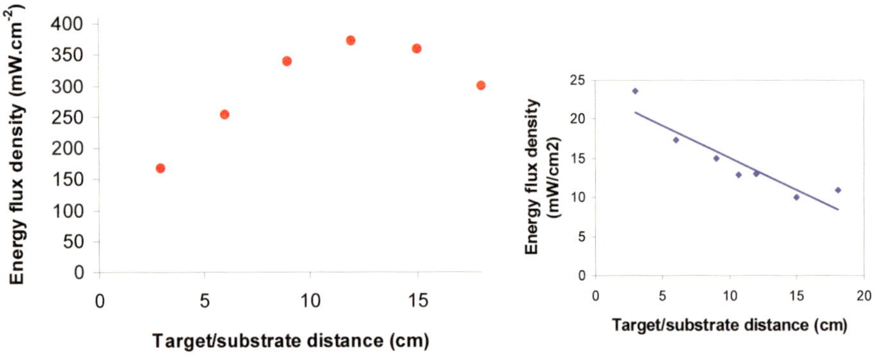

a) (b)

Fig. 19. Evolution of a) plasma and b) SPD energetic contributions versus the distance between the target and the substrate

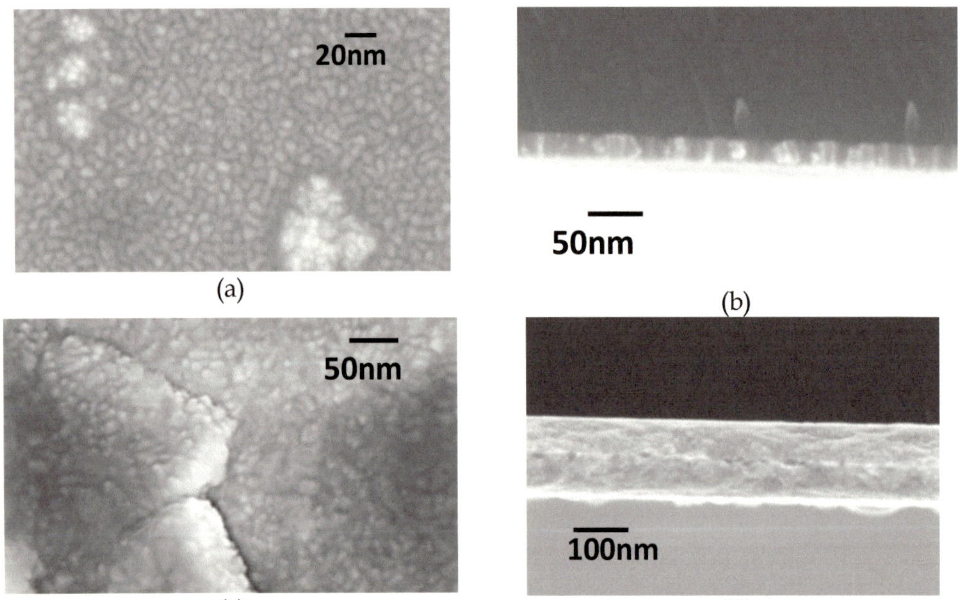

(a) (b)

(c) (d)

Fig. 20. SEM images of Pt thin films synthesized at 0.5Pa, 400W and -200V target bias voltage ; (a) and (b) target/substrate distance of 18 cm, 30 min deposition time ; (c) and (d) target/substrate distance of 3 cm, 11 min 30 s deposition time.

Echantillon	Pt atom number (RBS) (at/cm2)	Thickness calculated from RBS (nm)	Thickness measured on SEM images (nm)	Deposition rate (at/cm2s)
30min, 18cm	1.1×10^{17}	17	31	6.1×10^{13}
11min30s, 3cm	8.5×10^{17}	128	130	1.2×10^{15}

Table 4. Results from SEM images and RBS. The thickness is determined from RBS results by assuming that the film density is the same as the bulk one. The comparison with the value measured on SEM images gives some insight into the film density.

It is seen from the results above that the Pt film synthesized close to the target (target/substrate distance of 3 cm) is denser and more compact that the one deposited 18 cm away. The energetic deposition conditions deduced from figure 18 are given below:

- distance of 3cm $\Delta\varphi_{plasma}$= 160 mW/cm^2 and $\Delta\varphi_{dep}$ = 24 mW/cm^2

$$\Delta\varphi_{global} = 184 \text{ mW/cm}^2$$

- distance of 18cm $\Delta\varphi_{plasma}$= 300 mW/cm^2 and $\Delta\varphi_{dep}$ = 8 mW/cm^2

$$\Delta\varphi_{global} = 308 \text{ mW/cm}^2$$

It is interesting to note that the global transferred energy is higher far from the target because the Ar plasma contains the most energy at this location. However, the obtained films exhibit low density, which is usually attributed to a growth process that takes place when low energy is available at the surface [20]. Indeed, our results show that the energetic contribution that drives the film characteristics is the energy deposited by the condensing atoms, which is higher at 3 cm. The fact that a minor contribution might play the major role had never been reported until now. This indicates that, as usually done, discussing film properties by only taking account the the most abundant species or the one that carries the more energy is not accurate.

7. Conclusions

We have shown that the plasma energy flux probe we have developed from a commercial sensor is a relevant and innovative plasma diagnostic. It is able to give direct measurement of the energy influx from the plasma to a surface. The sensor has to be calibrated following a rigorous procedure so that it can be used in low pressure plasmas. We have compared the values given by this sensor to the ones obtained using an other technique developed by colleagues from Kiel [17] and we found a very good agreement between both probes. The main advantages of this sensor rely on the fact that it has a very good time response (17 µs [5]) and that it does not need further data treatment, but gives directly the measured values in real time. This very good time resolution is obtained because of the very thin films, which form the thermopile. The sensitivity of this new sensor was tested in different experiments such as Argon plasma, silicon etching plasma and sputtering deposition plasmas. In all cases, it gave valuable information and measured values were in good agreement with estimations from indirect methods.

This sensor in its commercially available version is well adapted for measurements onto small samples (typically 1 cm^2). However, we have shown that it is of particular importance to avoid all kinds of heat flux leaks. If a small thermal bridge is obtained by putting an additional mechanical piece on the sample (for example to fix it), then it will impact the measurement. This is why we decided to just stick the different samples to the sensor directly with a thermal paste.

We are now trying to improve the sensor to make it able to measure the energy flux in other conditions. For example, this sensor would be even more of interest for the plasma community if it could be biased. In this case, we would be able to enhance the ion bombardment and evaluate its impact on the energy influx.

This new method in determining the energy influx has opened new research perspectives in the plasma community, especially for materials processing.

8. References

[1] S. Aida and S. Rahmane, Thin Solid Films 288, 83 (1996)

[2] A. Durandet, O. Joubert, J. Pelletier, and M. Pichot, J. Appl. Phys. 67, 3862 (1990)

[3] H. Kersten et Al, Vacuum 63, 385 (2001)

[4] A.L. Thomann et al., Rev. Scient. Instr., 77, 033501 (2006)

[5] http://www.vatell.com

[6] F. P. Incropera, D. P. DeWitt, Introduction to Heat Transfer, 4th ed, Wiley, 93-95,(2002).X.

[7] Zhang , P.Z. Cong and M. Fujii, A Study on Thermal contact resistance at the interface of two solids, International Journal of Thermophysics, 27 (3), 880-895, (2006)

[8] M. J. Persky, Review of black surfaces for space-borne infrared systems, Review of Scientific Instruments, 70 (5), 2193-2217 (1999).

[9] F. M. Devienne, Advances in Heat Transfer, Low Density Heat Transfer, 272-352(1965).

[10] J. A. Thornton J A 1978 Thin Solid Film 54 23

[11] D. J. Ball 1972 J. Appl. Phys. 43 3047

[12] R. Gardon 1953 The Rev. of Sci. Instrum. 24 366

[13] C. Paturaud, G. Farges, M. C. Sainte Catherine and J. Machet 1998 Surf. Coat. Technol. 98 1257

[14] R. Piejak R, V. Godyak, B. Alexandrovich and N. Tishchenko 1998 Plasma Sources Sci. Technol. 7 590

[15] Drüsedau T P, Bock T, John T-M, Klabunde F and Eckstein W 1999 J. Vac. Sci. Technol. A 17(5) 2896

[16] Drüsedau T P, Löhmann M, Klabunde F and John T-M 2000 Surf. Coat. Technol. 133–134 126

[17] Kersten H, Wiese R, Hannemann M, Kapitov A, Scholze F, Neumann H and Hippler R 2005 Surf. Coat. Technol. 200 809

[18] Ellmer K and Mientus R 1999 Surf. Coat. Technol. 116–119 1102

[19] Thornton J A and Lamb J L 1984 Thin Solid Films 119 87

[20] Kersten H, Rohde D, Steffen H, Hippler R, Swinkels G H P M and KroesenGMW2001 Appl. Phys. A 72 531

[21] Wolter M, Stahl M and Kersten H 2009 Vacuum 83 768

[22] Cormier P-A et al 2010 J. Phys. D : Appl.Phys. 43 465201

[23] Lundin D, Stahl M, Kersten H and Helmersson U 2009 J. Phys. D: Appl. Phys. 42 185202

[24] M. A. Lieberman and A. J. Lichtenberg, « Principles of Plasma Discharges and Materials » Processing Wiley, New York (1994)

[25] Ch.Cardinaud, M.C.Peignon, G.Turban, "Si and W etching mechanistic study based on in-situ surface temperature measurements" , International Symposium on Plasma Chemistry 11, Loughborough, UK, August 22-27, 1993, Edited by J. Harry

[26] R. Dussart, A. L. Thomann, L. E. Pichon, L. Bedra, N. Semmar, P. Lefaucheux, J. Mathias, and Y. Tessier, "Direct measurements of the energy flux due to chemical reactions at the surface of a silicon sample interacting with a SF6 plasma" Applied Physics Letters 93, 131502 (2008)

[27] Bedra L, Thomann A L, Semmar N, Dussart R and Mathias J 2010 J. Phys. D: Appl. Phys. 43 065202

[28] Brault P et al 2000 Recent Res. Dev. Vac. Sci. Technol. 2 35

[29] Wendt R, Ellmer K and Wiesemann 1997 J. Appl. Phys. 82 2115

[30] Yamamura Y, Tawara H 1995 NIFS-Data Ser. 23 1

[31] J. A. Thornton, Thin Solid Films 54, 23 (1978).

5

Microsensors for Microreaction and Lab-on-a-chip Applications

Pawel Knapkiewicz and Rafal Walczak
Wroclaw University of Technology
Poland

1. Introduction

Since the first successful applications of the microfluidical devices, measurement of physical, chemical and biochemical parameters of performed reactions and analysis became next challenge and millstone towards successful application of developed instrumentation in many field of science and industry, as well as, deeper understating of micro- and nano-world of fluidics (Ehrfeld at al., 2005). Although, methodology of these measurements was well known from many years, the main problems that occurred were dimensional incompatibility of available macroscopic solutions and sensing problems caused by significant reduction of managed and analyzed volumes. Therefore, microsensors became important part of the microfluidical device enabling real-time and on-chip measurement of measurable parameters like pressure, temperature, conductivity, absorbance or fluorescence.

In this chapter miniature on-chip integrated pressure sensors, discreet corrosion resistant pressure sensor and conductometric flow-through detector will be described in details. Nevertheless, optical microrensors like absorbance NIR and VIS detector, as well as fluorometric detector will be shown. Technology of the sensors utilizes microengineering techniques where silicon and glass play main role as constructional materials. Three-dimensional formation and assembling techniques of silicon and glass allow to fabricate miniature sensors. For each presented microsensor, the fabrication techniques will be described in details. Great attention is also paid for development of the complete measurement system consisting of the microsensors itself but also specialized electronics and information environment for full data management and measurement or analyse result presentation.

2. Miniature sensors and measurement systems for microreaction technology

Pressure and temperature are two the most important parameters of chemical reactions. Steering of those parameters determine chemical reaction course, as well as temperature and pressure inform about actual chemical reaction state. Continues monitoring of temperature and pressure is very important for exothermic, high-speed chemical reactions (Edited by Dietrich, 2009). It can be done relatively easy for standard, macro-scale chemical plants. Commercially available sensors are suitable to large apparatus, can be easy applied and

operate as a part of some automation systems. The microreaction technology stays in opposite to this situation. Chemical reactions are performed in the microscale by the use of microreactors, replacing static chemical reactions by continues-flow reactions. Unfortunately, there is no ready-to-use sensors suitable to the microreactors. Total inner volume of the microreactors is in the range from several µl to few ml, when external dimensions, as well as "dead" volume of available standard sensors are at least ten times larger, in comparison. In consequence, commercially available sensors can not work in tandem with the microreactors.
Let's focus our attention on extremely dangerous, highly exothermic chemical reaction, eg. nitration of organic compounds (Ali at al., 2005; Speight, 2002), where continues monitoring of pressure and temperature inside the microreactor is absolutely required from safety point of view. Moreover, real-time measured data are helpful for conscious steering of chemical reaction, towards high yield (Kralish & Kreisel, 2007).
In this paragraph miniature temperature and pressure sensors, as well as measurement systems dedicated to the microreaction technology will be described. Several requirements of parameters of the sensors and sensors assembling method must be considered:

- temperature sensors operating range: -20°C ÷ +100°C,
- dimensions of the temperature sensors should not exceed typical dimensions of microchannels (typical dimensions are in range of tens µm up to several mm – the average is 1 mm),
- pressure sensors operating range: relative pressure 0 kPa ÷ 400 kPa, overpressure up to 600 kPa,
- "dead" volume of all pressure sensors integrated to microreactor should not exceed 1/10 of total inner volume of microreactor,
- microreactor with sensors work in harsh environment (concentrated acids, organic compounds),
- chemically resistance assembling required.

In the current paragraph hole process, including temperature sensors selection, pressure sensors development, assembling and packaging problems, as well as electronics and software realization, towards complete sensoric system for microreaction technology, will be described in details.

2.1 Temperature sensors

Temperature sensors working in a tandem with microreactor must be characterized by small dimensions, fast response and small measurement error. Sensor can be localised inside microchannel (direct contact to medium) or outside microchannel (indirect contactless measurement).
Direct sensor-to-medium contact gives most precision measurement. Thin film sensors evaporated onto inner walls of microchannel will not survive aggressive chemicals. Moreover, thin film technology is difficult. Discrete temperature sensors can be assembled in the microchannel only by gluing. Both ideas do not fulfil previously listed requirements, what in consequence eliminates this measurement idea and technical realization from the use.
Second method is based on sensor localized outside microchannel. Main requirement is to keep distance between sensor and medium as small as possible to allow to conduct a heat flux with minimum losses.

Several types of commercially available sensors can be investigated. Main parameters of commercially available sensors are collected in Table 1.

Parameter / Type/model	Operating range	Output signal	Tolerance	Housing/Dimensions
Pt100 / Pt500 / Pt1000	-200°C ÷ +860°C (depending on type)	Resistance, linear PTC[1]	≤ 0.3%	SMD[2] 0805, SMD 1206, TO92, SOT223, ceramic or metal tube ø ≥ 1.5 mm, other
Thermistor NTC	-55°C ÷ +150°C (depending on type)	Resistance, non-linear NTC[3]	1% ÷ 10%	SMD 0603, SMD 0805, pill-like ø ≥ 0.8 mm
Thermocouples	-200°C ÷ +1820°C (depending on type)	Voltage, linear	±0.5°C ÷ ±4°C	metal tube ø ≥ 0.15 mm
Transducers with analogue output	-55°C ÷ +155°C (depending on type)	Voltage, linear (usually)	0.5% ÷ 5%	all electronic standards
Transducers with digital output	-55°C ÷ +155°C (depending on type)	digital, serial data transfer	0.5% ÷ 5%	all electronic standards

Table 1. Temperature sensors and their main parameters

The most common are transducers with analogue output. Output voltage signal is linear in relation to the measured temperature and easy to measure. No additional electronic circuits are required to process the signal. Second, most useful are transducers with digital output. The only difference to previous one is, that digital processing by a microcontroller is required. Unfortunately, even the smallest package of temperature transducers (SOT23 electronic standard, plastic body dimensions without electrical connections: 2.9 x 1.3 x 0.85 mm^3) seams to be too large to be directly applied in microreation technology (Fig. 1a).

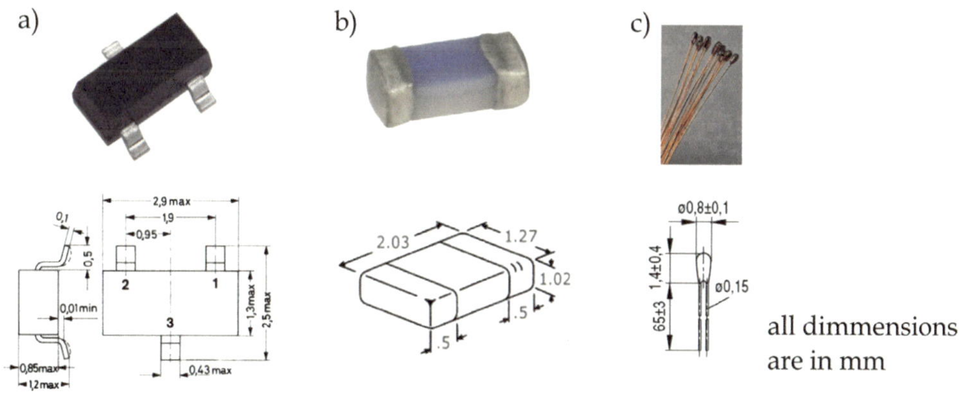

Fig. 1. Examples of miniature temperature sensors: a) temperature transducer in SOT23 package, b) PTC platinum thermoresistor (Pt-series) at 0603 SMD package, c) pill-like NTC thermistor

[1] PTC - Positive Thermal Coefficient
[2] SMD - Surface Mounted Devices
[3] NTC - Negative Thermal Coefficient

Different packages and dimensions of resistance PTC or NTC sensors and thermocouples are available. Platinum temperature sensors (Pt-series) are very accurate and are used as temperature standards. Dimensions of the ceramic or metal tube packages, as well as SMD packages, do not fulfil harsh-environment microreaction requirements (Fig. 1b). NTC thermistors are available in different SMD packages, including miniature pill-like packages distinctive from others (Fig. 1c). The smallest pill-like package has 0.8 mm diameter, what is suitable to mentioned earlier requirements. Thermocouples are usually packaged in the metal tubes. Special constructions are available in tubes of 0.15 mm in diameter. It is the smallest dimension (diameter) of all sensors discussed before. In spite of that, thermocouples need to co-work with sophisticated electronic circuits.
The analysis of commercially available temperature sensors appoints that the miniature pill-like NTC thermistors as the best solution. Small glass package, thin and flexible electrical wires entail small thermal capacity and fast response.
The location of miniature pill-like NTC thermistors onto microreactors made of glass, silicon and glass as cover, ceramics, polymers, was proposed. Sensor should be located inside "blind" hole of 1.0 mm in maximum diameter, fabricated directly above microchannel. The optimal distance of 200 μm between medium and temperature sensor has been found experimentally to be optimal. The schematic cross-section of a microreactor with NTC thermistor and the sensor integration to glass-silicon-glass microreactor is schematically presented on Fig. 2.

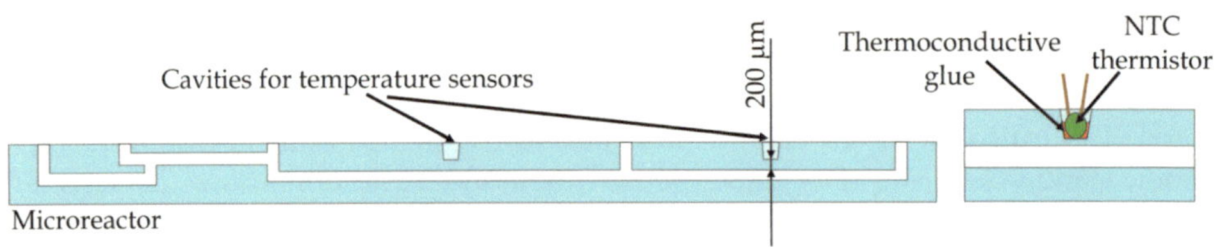

Fig. 2. Indirect temperature measurement: cross-section of a microreactor (left) and NTC thermistor location (right)

Thermal behaviours of the glass-silicon-glass microreactor equipped with miniature pill-like NTC thermistors will be widely discussed later in paragraph 2.3.

2.2 Pressure sensors

As it was mentioned before, required operation range of pressure sensors is from 0 kPa to 400 kPa of relative pressure. 400 kPa is a maximum operating pressure of microreactors made of glass, silicon-glass, ceramics and polymers, in most cases.
Pressure sensor must have direct contact to a measured medium. The "dead" volume of all applied pressure sensor must be as small as possible, in order to do not influence on chemical reaction conducted inside microchannels. The assumption, that "dead" volume of all pressure sensors V_{DV} should not exceed 1/10 of total inner volume of microreactor V_{MR} (1) is appropriate from metrological point of view and minimize negative influence on chemical reaction.

$$1/10\ V_{\mathrm{MR}} > V_{\mathrm{DV}} \tag{1}$$

The "dead" volume of commercially available pressure sensors and transducers (dedicated to harsh environment chemical plants) is counted in mili-litres. Moreover, large size packages disqualify these pressure sensors to be used in the microreaction technology (Fig. 3).

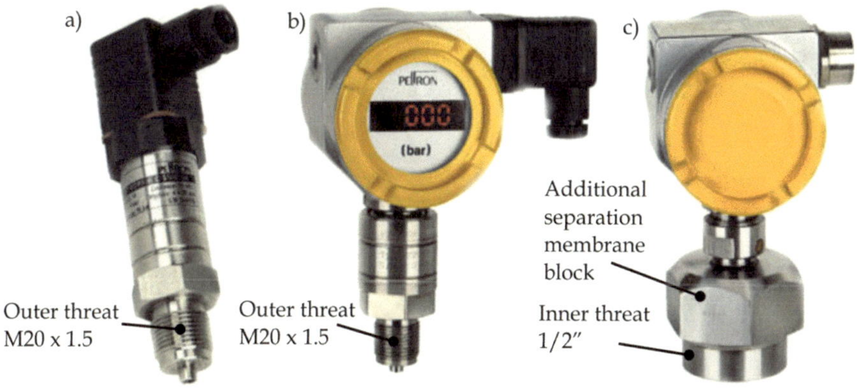

Fig. 3. Examples of commercially available pressure transducers (Peltron, Poland) with stainless steel separation membrane: a) NPX series, b) PXW series with LED display, c) PXC series with additional separation membrane dedicated to harsh environment measurements.

Two variants of pressure measurements inside microchannels have been taken into considerations. The first strategy is to integrate the pressure sensors directly onto microreactor. The second strategy is to use discreet sensors which are independent on the microreactor.

2.2.1 Miniature pressure sensor integrated onto microreactors

Some pre-work tests of chemical resistance of some possible to use assembling methods have been done. Test structures consist of 3" silicon and glass wafers were bathed in H_2SO_4+HNO_3 mixture 1:1 (Table 2).

Assembling method	Gluing - photo-hardened glue	Gluing - epoxy glue	Adhesive bonding - Kapton® foil (DuPont)	Adhesive bonding - Teflon foil (DuPont)	**Silicon-to-glass anodic bonding**
Test: bathed in H_2SO_4+HNO_3 mixture	-	-	-	-	+

Table 2. Resistance against aggressive chemicals of chosen assembling methods

The test shows, that any kind of gluing and adhesive bonding can not be use. Only silicon-to-glass anodic bonding ensures strong, tight and chemically resistant connection. The selected assembling method can by utilized only for glass or silicon-glass microreactors. Moreover, only borosilicate glass (Pyrex-like) can be used (Briand at al., 2004; Knapkiewicz at al., 2007; Dziuban, 2006).

Two ways to integrate pressure sensors directly to the microreactor are possible. Silicon piezoresistive pressure sensor dies are anodically bonded to a top glass cover of the microreactor or sensor dies onto glass posts are integrated to a top glass cover of microreactor trough silicon orifice in two steps of anodic bonding (Fig. 4).

The "dead" volume of two proposed solutions, including "dead" volume of connection channel trough top glass cover, has been estimated assuming 1.1 mm thickness of top glass cover and hole of 0.7 mm in diameter. The "dead" volume is equal to 0.75 μl and 1.65 μl, respectively. The second solution has larger "dead" volume. In spite of that, mechanical stress indication at silicon sensor die (false signal) is lass possible in comparison to the solution with dies bonded directly to the microreactor top cover.

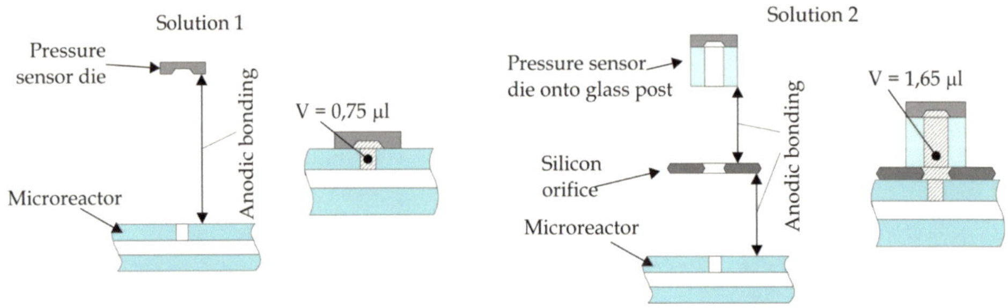

Fig. 4. Two solutions of pressure sensors assembling.

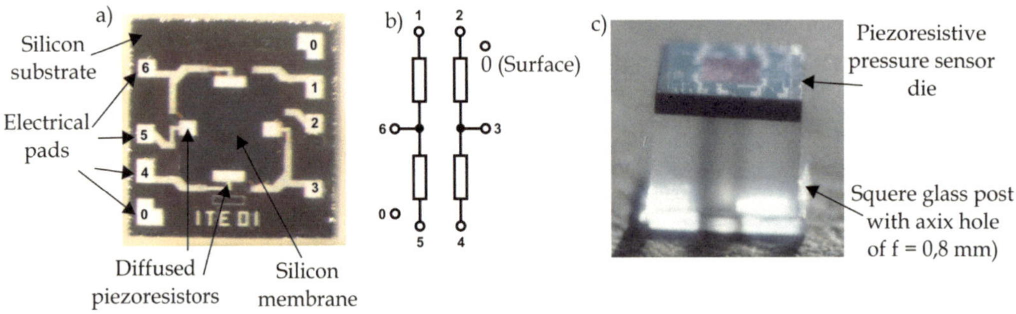

Fig. 5. Silicon piezoresistance pressure sensor (ITE Warsaw): a) top view of the silicon die, b) Wheatston bridge configuration of the piezoresistors, c) silicon die onto glass post

Silicon piezoresistive pressure sensor structures (silicon sensor die on glass post) fabricated at Institute of Electron Technology in Warsaw (ITE Warsaw), has been chosen (Fig. 5). Basic parameters of the sensors are collected in Table 3.

Parameters	Required value	Catalogue value of ITE Warsaw pressure sensors
Nominal pressure	600 kPa	600 kPa
Static overpressure	max 300 kPa	min 400 kPa*
Dynamic overpressure	-	min 750 kPa*
Total error	< 1%	< 0,5%
Work temperature range	- 10°C ÷ 40°C	0°C ÷ 50°C
Response time	< 100 ms	< 1 ms

Table 3. Comparison of requirements and catalogue values of ITE Warsaw pressure sensors

2.2.2 Glass-silicon-glass microreactor with on-chip integrated temperature and pressure sensors

Validation of the concept of temperature and pressure sensors integration was provided by fabrication of the demonstrator of a packaged microreactor made of silicon and Pyrex-like glass (Knapkiewicz at al., 2006). The microreactor has integrated 4 pressure sensors dies and 5 temperature sensors. The silicon pressure sensors dies on 2 mm high Pyrex-like glass post are anodically bonded to the silicon orifice, which is anodically bonded to the glass-silicon-glass microreactor in the next step. Via-holes connecting fluidic channel of the microreactor and pressure sensors, as well as blind-holes for positioning of temperature sensors, are made in proper places of the microreactor (Fig. 6).

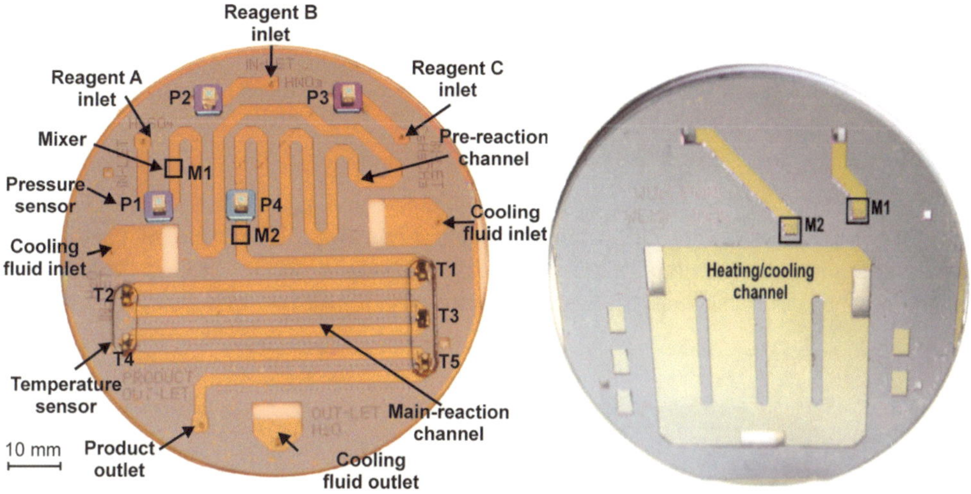

Fig. 6. Silicon-glass microreactor with on-chip integrated pressure and temperature sensors: top (left) and bottom view (right) of assembled microreactor

The microreactor is made of three layers of glass, silicon and glass (Fig. 7). The microfluidical channels (2 mm wide, 120 μm depth), the structure of two micromixers (10 micronozzles of 50 x 50 μm^2 each one) and the cooling chamber are formed by a deep, wet micromachining of (100)-oriented, double-side polished silicon wafer in 40% KOH solution at 80°C. Inner volume of the microreactor has been estimated to be 143 μl. "Dead" volume of the four pressure sensors is 5.7% of the microreactor volume, what is suitable to assumption (1).

After micromachining the microreactor body is oxidized to form 0.3 μm - thick silicon dioxide chemically resistive layer. Following, the body is anodically bonded (450°C, 1.5 kV) to the bottom glass substrate (Borofloat 3.3, Schott, Germany). Next, the top glass with set of pressure sensors is prepared. Firstly, via holes (ø = 0.7 mm) of fluidic inlets, outlets and pressure sensors connections are mechanically drilled in the top glass. Next, cavities (ø = 1.0 mm) for temperature sensors are mechanically drilled. Piezoresistive pressure sensor dies on 2 mm-high glass post (Borofloat 3.3) are anodically bonded to the previously micromachined 5 x 5 mm^2 silicon orifices with via holes (Fig. 7a). The silicon orifices are wet-etched in a separate process. After that, sensor dies with silicon orifices are anodically bonded to the top glass. Such construction of the sensor module eliminates temperature induced stresses and, in consequence, temperature induced errors. It ensures excellent tightness of the fluidic connections. Finally, the top glass with pressure sensors is anodically bonded (450°C, 1 kV) to the microreactor silicon body (Fig. 7b).

The miniature pill-like thermistors (NTC, 5 kΩ, EPCOS) are positioned inside cavities and fixed by the use of thermo-conductive glue. The distance between the sensor and main reaction channel is bellow 200 μm.

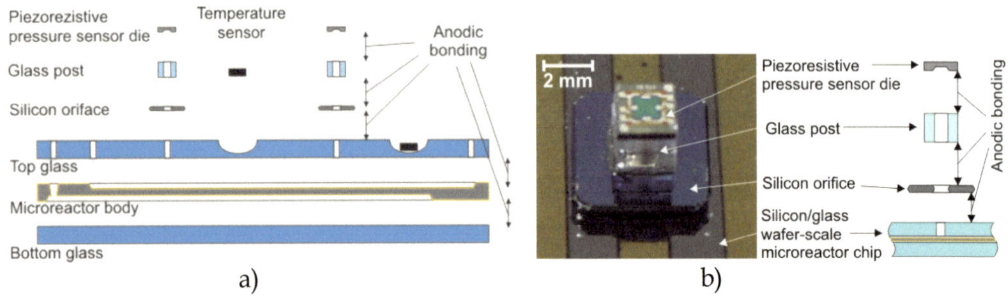

Fig. 7. The microreactor chip: a) cross-section of expanded view, b) details of the pressure senor assembling

Packaging of the microreactor chip is an onerous but crucial problem. Package should fulfil following requirements: mechanically stable fixing of the microreactor chip, protection of the chip and sensors, electrical connections of the sensors and fluidic connections of channels.

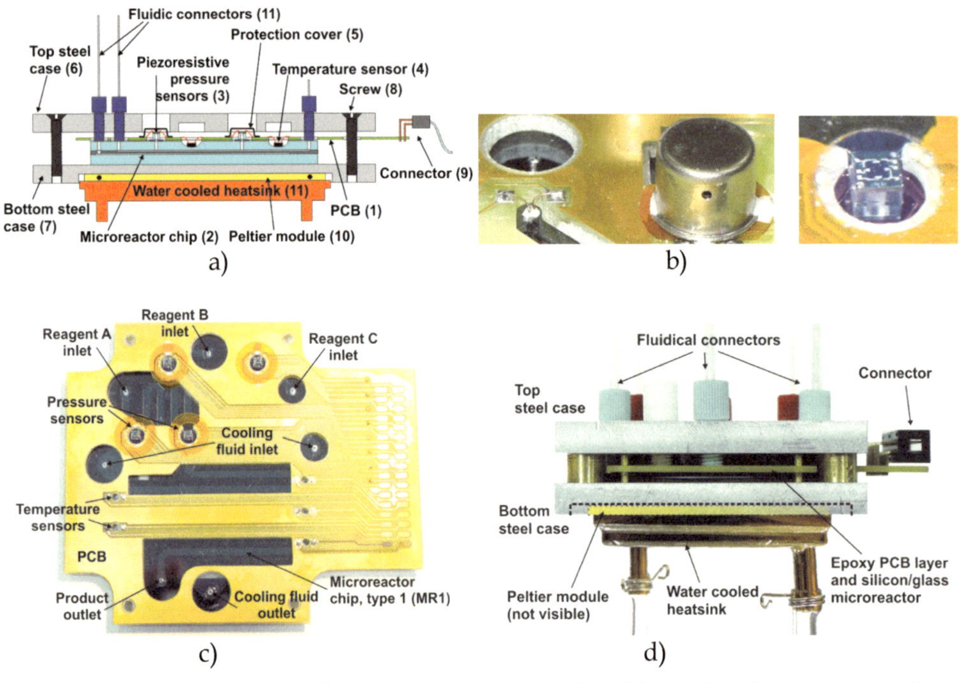

Fig. 8. The microreactor in a package: a) cross-section view, b) details of sensors assembling, c) microreactor chip with PCB board, d) a side view

The cross-sectional view of the packaged microreactor is shown in Fig. 8a. The PCB board (1) with a net of electrical paths is fixed (epoxy glued) to the microreactor chip (2). Pressure sensor dies (3) protrude from PCB board (1), so the wire-bonded electrical connections between dies and PCB can be done. Electrical connections of temperature sensors (4) are soldered with PCB board (1). Each sensor is protected by the metallic cup (5) glued to the PCB board (1). The top (6) and the bottom (7) steel case is positioned and screwed on (8). Output electrical connections are provided by slide-type connector (9). The Peltier module (10) (TM-127-2.0-1.5, Transfer Multisort Electronic, Poland) with heat exchanger is fixed to the back-side of the bottom steel case (7), to increase the cooling efficiency and to steer and/or control general temperature conditions inside the microreactor chip. The heat exchanger is made of water cooled heatsink (11) (CPC, Poland).. Teflon® fluidic connectors (12), are placed in the top case (6).

2.2.3 Miniature, discreet, corrosion-resistant pressure sensor

Independent from the microreactor, corrosion resistant and ultra-low "dead" volume pressure sensor enabling measurement of the pressure inside microreactors made of ceramic, polymers and metal, were strongly required. The new concept of the discrete pressure sensor is based onto an "inverted" principle of packaging and protection of the micromachined silicon pressure sensors die (Fig. 9a). In commonly known designes of corrosion resistive pressure sensors, the die is placed inside a metal case and is protected by a metal separation membrane welded to this case. Inner chamber of the sensor is fulfilled with an oil, working as a hydraulic pressure medium between external environment and thin membrane of the die. Such construction is characterized by the large dead volume, ranging several millilitres, what cannot be tolerated if sensors are applied in the microreactor which dead volume is smaller. The new principle inverses the situation. The metal case is "suspended" on a silicon/glass die, measured media deflects membrane of the die upward. In this solution, there is no need of using of the additional separating metal membrane, dead volume of the sensor is equal to several microlitres and the only wetted surfaces of the sensor are made of corrosion resistant glass and silicon.

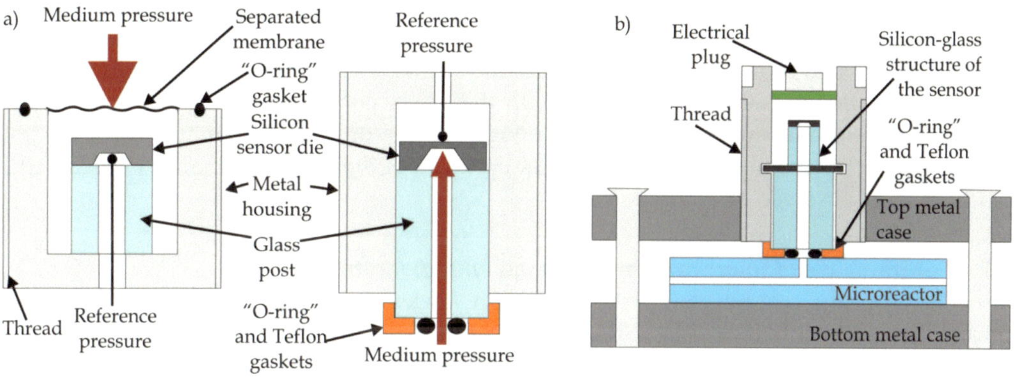

Fig. 9. Principle of the construction of the discrete pressure sensor: a) comparison of typical and novel construction, b) principle of sensor integration with the microreactor

The standard silicon-on-glass pressure sensor chip (silicon die bonded anodically to glass post, ITE Warsaw, Poland) were used. The pressure sensor chip is assembled to a silicon orifice and special glass post afterwards. Assembling was done by the use of the anodic bonding process. The complete multilayer silicon-glass structure is glued into a metal housing with outer thread. Special construction of the metal housing allows to screw the sensor in the top housing plate of the microreactor. Principle of the fluidic connection formation between pressure sensor and microreactor is shown in Fig. 9b. The sensor is screwed into the top metal case, specially designed Teflon® orifice in tandem with Viton® "O - ring" orifice ensure tight fluidic connection between the pressure sensor and the microreactor.

The miniature pressure sensor is chemically robust against concentrated and hot acids (except HF). The distinguishing future is extremely low dead volume - about 8.5 μl for

single sensor. Total dead volume of fixed sensor, including capacity of via-hole made in microreactor connecting the sensor and microfluidical channel, is 9.2 μl.
Metrological parameters, chemical robustness against acids and tightness of set of gaskets of the new pressure sensors were tested. Electro – pneumatic tests have indicated, that the total error of the new sensors is below 1 % for fixed temperature. The temperature sensitivity coefficient, equal to 0.3 %, is significantly higher than noticed for the unpackaged dies (Table 4).

Normal work ratings:

Parameter	Value	Unit
Nominal pressure	600	kPa
Overload pressure max.	750	kPa
Current supply typ.	1,0	mA
Current supply max.	1,5	mA
Voltage supply typ.	5,0	V
Voltage supply max.	7,5	V
Operating temperature max.	50	°C
Operating temperature nim.	-10	°C
Storage temperature max.	85	°C
Storage temperature min.	-10	°C

Metrological parameters:

Parameter	Value	Unit
Full scale output voltage (FSO) *	56,1	mV/V
Offset voltage*	25,5	mV
Bridge resistance *	5	kΩ
Non-linearity *	0,38	%
Pressure histeresis *	0,33	%
Sensitivity coefficient *	9,7	mV/V/100kPa
Temperature offset coefficient **	0,1	%/°C
Temperature sensitivity coefficient **	0,32	%/°C
Total error *	0,85	%

* at 25°C, I_{supply}=1,0 mA, I_{max}600 kPa
** average value over temperature range 0°C to 50°C

Table 4. Basic parameters of discreet pressure sensor

Time-dependent characteristics of the new pressure sensor and reference sensor (Festo, SDE1 series, 600 kPa range) are similar. Chemical resistance of the pressure sensors have been investigated by 100 hours-long test with the use of concentrated sulphuric and nitric acids at ambient temperature and at 60°C. No influence on sensors parameters, leakages, corrosion effects have not been noticed. Leak-tightness of the fluidic connection is excellent. No leaks were observed up to 850 kPa.

2.2.4 Foturan®-glass microreactor equipped with discrete sensors

The developed discreet pressure sensor has been worked in tandem with a Foturan®-glass microreactor (Dietrich at al., 1996; Freitag at al., 2001). The microreactor consists of 12 layers which are photo-structurized in separate processes and following, all of them are bonded together in a fusion bonding process.
Cross-sectional view of the assembled microreactor with sensors is shown in Fig. 10, true pictures of parts and ready-to-work microreactor are shown in Fig. 11 and 12.

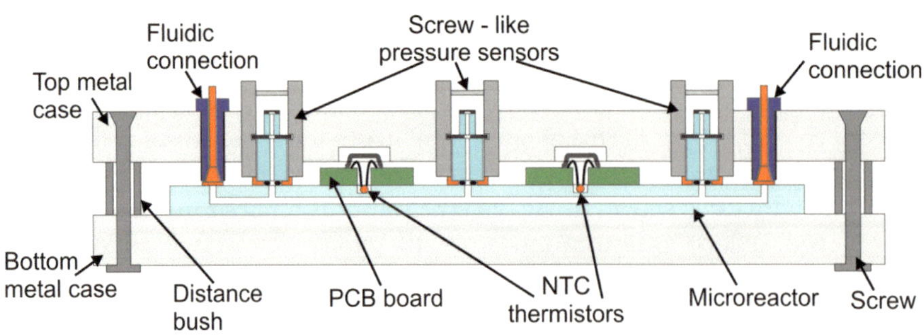

Fig. 10. Cross-sectional view of the packaged microreactor with set of sensors

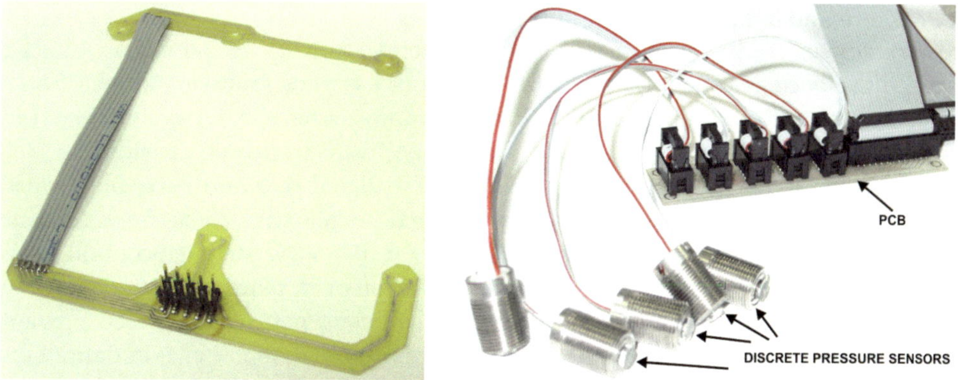

a) b)

Fig. 11. The PCB boards, a) first PCB board with fixed temperature sensors, b) second PCB board for electrical signals collection from five screwedpressure sensors

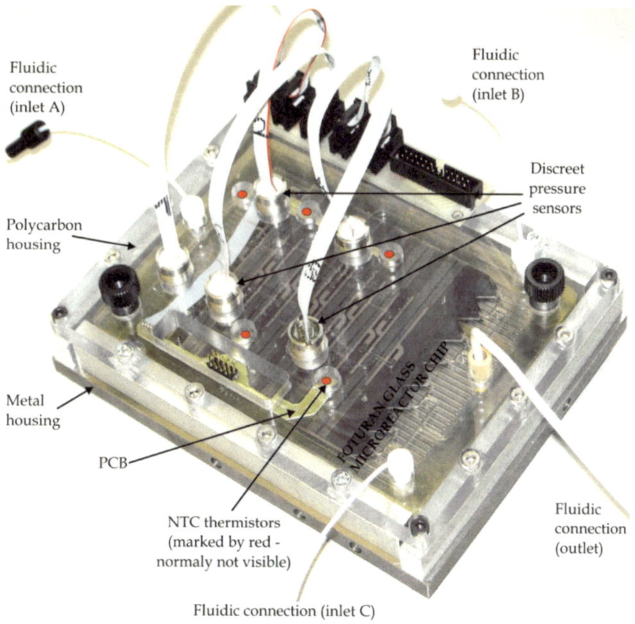

Fig. 12. Foturan®-glass microreactor equipped with 5 discreet pressure sensors and 5 temperature sensors (NTC thermistors)

Foturan® microreactor (Mikroglass GmbH, Meinz, Germany) was positioned between two plates of the case. In the laboratorial version of this package, top plate was made of light transparent carbonate. Pressure sensors were screwed, as described before. Temperature sensors (NTC thermistors, the same as described earlier for silicon/glass microreactor) were placed inside blind-holes made onto the top layer of the microreactor, in a close contact to the fluidic medium flowing inside microchannels and electrically connected , by soldering, to the PCB board with a electrical plug. Standard fluidic connections were screwed. Pressure sensors and fluidic connectors pressed microreactor downward, stabilizing it inside the case.

2.3 Pressure and temperature monitoring systems

No commercial, ready-to-use measurement systems, suitable to previously described sensors, have been available. It extorted to work out electronic/software system from the ground up. Laboratorial and commercial-like monitoring systems, based on data-acquisition card and self designed electronics and software, respectively, have been developed.

The system consist of to main parts: analogue and digital software parts plus software platform. Analogue part is supplying sensors and amplifying signal from sensors to voltage/current signal of useful range (0-3.3 VCD or 0-5 VDC for microcontroller based electronics, 0-10 VDC or 4-20 mA for automation). Moreover, offset correction of pressure sensors output signal is desired. Second, digital (software) part must proceed previously digitalized data, visualized them on graphs (bar-graphs, time-graphs, etc.) in desired units. Nevertheless, data recording is also required.

The block-diagram of the analogue part, driving several pressure and temperature sensors, is shown on Fig. 13.

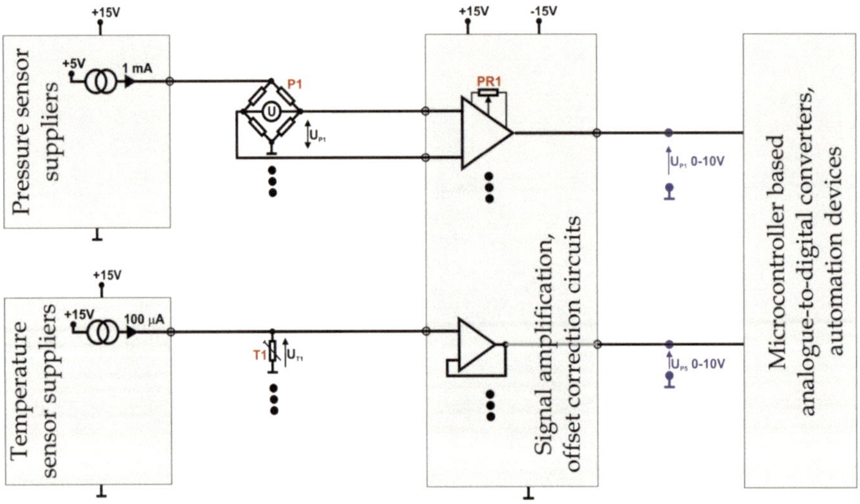

Fig. 13. Block-diagram of the analogue part of the monitoring system

Piezoresistors of pressure sensor are set to the Wheatstone bridge configuration. It is beneficial to supply one diagonal of the Wheatstone bridge by constant, stabilized current of 1.0 mA. Non equilibrium voltage, related to the measured pressure, is collected on the second diagonal of the Wheatstone bridge. Following, corrected (offset) and amplified voltage signal is digitalized.

Constant current supplying of the temperature sensors has been proposed as well. Temperature sensors, especially miniature sensors, must be supplied by low current. High value of supplied current will indicate self-heating of the temperature sensor, what in consequence will change resistance and true temperature can not be measured. Optimal value of supplying current (100 μA) has been found.

The laboratorial version of the measurement system, consists of three PCB boards (PCB of pressure sensors suppliers 5 x 1.0 mA, PCB of temperature sensors suppliers 5 x 100 μA and PCB of instrumentation amplifiers for correction and amplifying signals), data acquisition card (National Instruments). Processed signals are transfer through USB standard from data acquisition card to a PC computer equipped with LabView-based software, where final signals processing, data visualization and data recording are done.

The laboratorial system has been combined with previously described glass-silicon-glass microreactor. The microreactor equipped with 4 pressure sensors and 5 temperature sensors, has been electrically connected to the measurement system. Following, static and dynamic fluidic tests were done. Tests have shown very good work of the system. Probably for the first time ever, true value of the pressure and its changes in time have been measured inside microchannels of the microreactors. System indicated precisely any of pressure instabilities caused by improper work of external pumps or gas bubbles inside channels, as well as blow-up of the microreactor intentionally caused by uncontrolled increase of the inlet pressure (Fig. 14). Temperature and its variation in time has been very precisely measured, the no cooling / cooling effect has been identified as well (Fig. 15).

Commercial-like measurement system, based on the same as previous one block-scheme, has been developed (Knapkiewicz at al., 2008). The electronic part of the system consists of analogue part (sensors supplying, signals amplifying) and self-designed data acquisition card. The system shown on Fig. 16, allows to connect and processed data from 5 of 2nd generation miniature, discreet pressure sensors (Fig. 17). Signals of sensors are pre-processed and sent to the PC computer equipped with specialized software with functions of data processing, presentation and recording. The "electronics" use USB connection for data transfer. Additionally, the system is equipped with analogue outputs useful for any kind of data acquisition systems. The quantity of channels for pressure sensors can be easy enlarge, temperature sensors channels can be added as well.

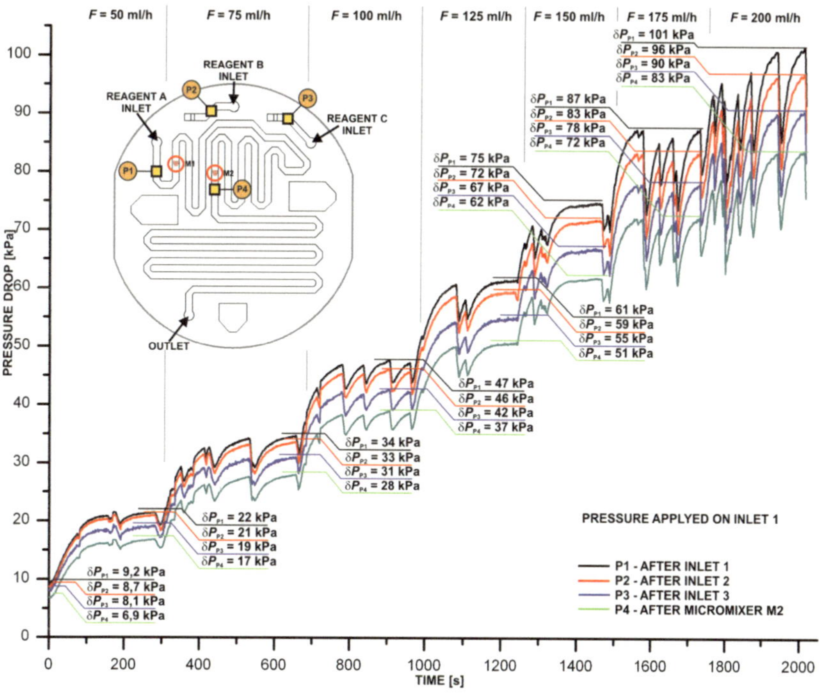

Fig. 14. Pressure drop for different flows of DI water as measured by developed pressure sensors, please note clear detection of pressure instabilities caused by irregular work of the dosing pumps

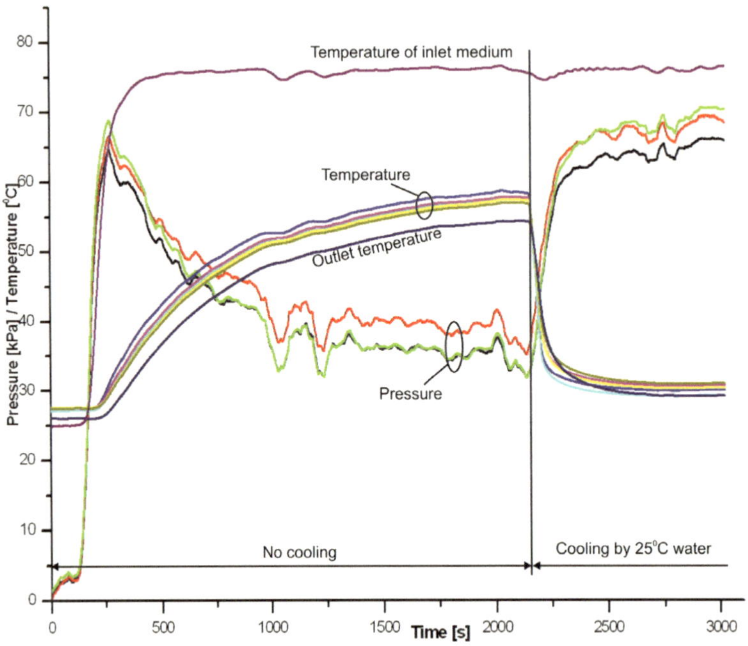

Fig. 15. Pressure and temperature time-related curves as measured by 5 temperature sensors and 4 pressure sensors, please note temperature decreasing while cooling was applied

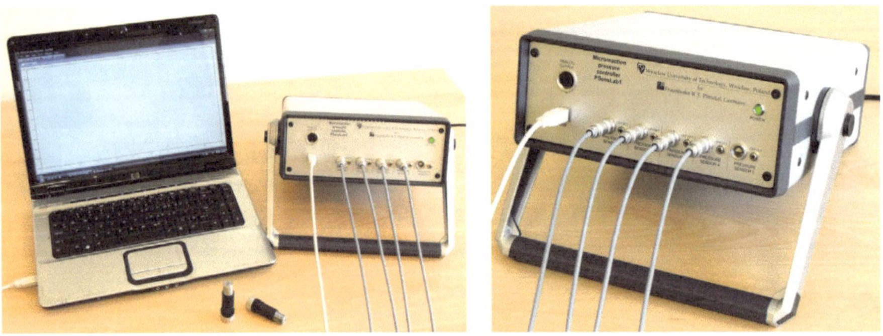

Fig. 16. Commercial-like measurement system driving 5 discreet pressure sensors

The new, discreet pressure sensor and measurements system were used for build the "intelligent" microreactor which were use as key component in "desk-top" microchemical plant for nitration process. The "intelligent" microreactor consists of previously presented Foturan® glass microreactor. Nitration of fluortoluene were conducted in the "desk-top" microchemical plant with simultaneous recording of information about pressure and temperature insight microchannels.

The recorded data were presented as time dependent curves where crucial points of process. Start of flow, initiation of nitration, as well as drastic fluctuations of pressure for high value of reagents flow and corresponding to that fluctuations of temperature involved by pumps (switching of syringe pumps) were observed. Dramatic increase of temperature at outlet of the microreactor being an effect of destabilization of nitration reaction was properly detected. Examples of multi-parametrical measurements of reaction course are shown in Fig. 18.

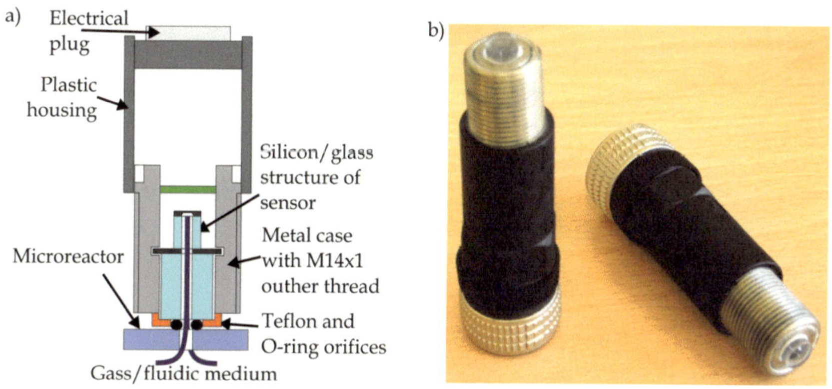

Fig. 17. 2nd generation of discreet pressure sensor: a) cross-sectional view, b) sensors ready to work

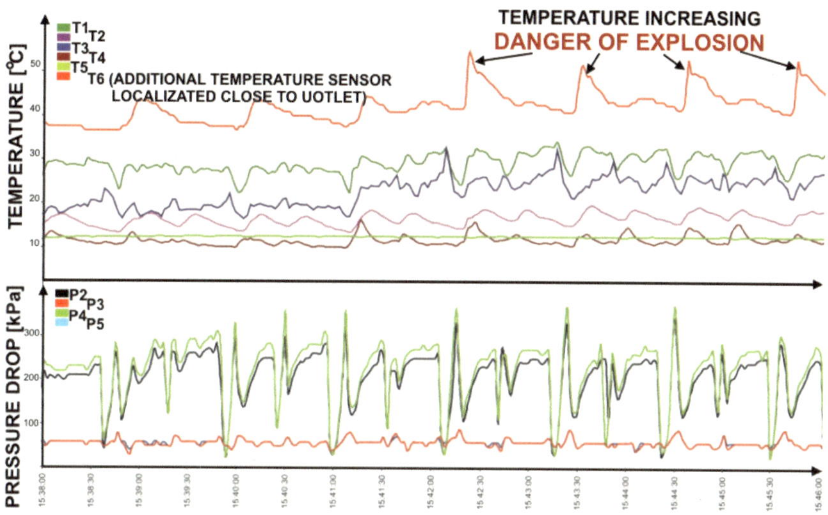

Fig. 18. Pressure and temperature versus time, nitration of fluorotoluene, total flow F = 23,2 ml/min)

2.4 Conductmetric, flow-through detector

Conductometric detection is widely applied in analytical chemistry what leaded aso to wide application of this technique in microfluidical analytical device often called also labs-on-a-chip. This technique utilizes dependency of electrical conductivity on concentration of ions in analyzed sample. It is commonly applied as detection technique in dielectrophoretic separation of organic and nonorganic compounds. Conductometry can be performed in two basic modes: contact and contactless. Various materials can be utilized as the chip body (silicon, glass, polymers, ceramic) but electrodes are usually made of gold, platinum or chromium. The main problem occurring in contact mode of conductometric measurement is possibility of electrochemical reaction observed on the surface of the electrodes. Therefore, technical realizations of contactless mode is developed rapidly. It eliminates reactivity with analyte and enables better electrical isolation of the electrodes in relation to the potential utilized in dielectrophoretic separation of the analyte.

In this subchapter we present en example of discrete silicon-glass microfluidical contact conductometer with various configurations of electrodes. Than, an integrated microdoser with built-in two-electrodes conductivity detector to monitor on-chip volume of dosed analyte is presented.

2.4.1 Discrete microfluidical flow-through conductometric detector

The conductivity microdetector is a silicon-glass structure with galvanic feedback of three metallic electrodes (Fig. 19).

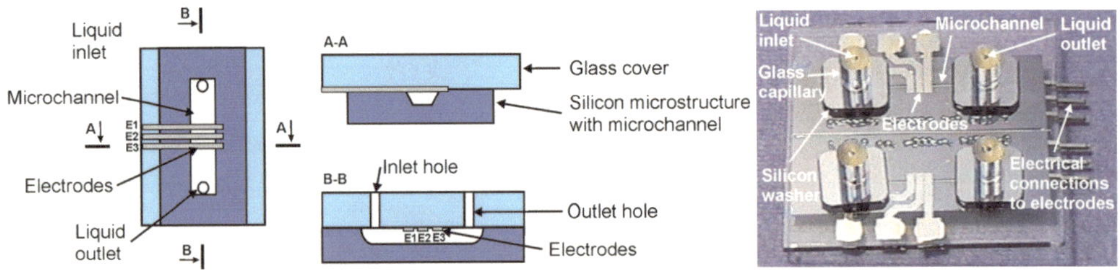

Fig. 19. The conductivity microdetector: a) schematic views of the detector, b) assembled device with double detector structure and the silicon-glass fluidical connections

The silicon body (35 x 35 mm^2) of the microdetector was fabricated on the (100)-oriented, one-side polished, n-type, 5 Ωcm wafer. The microchannel (18 mm long, 300 μm width and 140 μm deep) was anisotropically etched in 40% KOH at 80°C through thermal silicon dioxide (0.5 μm thick) mask layer. Than, SiO_2 was removed in BHF and the wafer was once again thermally oxidized (0.3 μm SiO_2) to form chemically resistance layer. The glass cover of the microdetetor was made of the 1.1 mm thick Borofloat 3.3 substrate. Three metallic electrodes, designed as 1000-μm-width strips, were formed on the selected surface of the glass substrate with different distance between them - 75 μm between E1 and E2 and 120 μm between E2 and E3. The electrodes were made of sputtered Cr/Ni/Au (100/50/150 nm) multilayer. The dead volumes between E1 - E2, E2 - E3 and E1 - E3 electrodes were 87 nl, 89 nl and 135 nl respectively. The input/output holes (ϕ=0.9 mm) for the liquid were mechanically drilled in the substrate. The silicon die and glass cover were washed, hydrofilized and anodically bonded (2 kV, 450°C). In order to minimize the dead volumes of the connections the special silicon-glass connections were applied. The glass capillaries were anodically bonded (1.5 kV, 450°C) through the silicon washers to the microdetector chip. The steel capillaries were glued to the inlet holes of the glass capillaries.

The static characteristics of microdetector were measured at a specially designed test station (Fig. 20). The syringe pump (Prefusor, Germany) ensured constant, pulsation free liquid flow (1.2 ml/h). The microdetector was supplied by sinusoidal-wave generator with constant (0.4 V) amplitude. In the two-electrodes configuration the output voltage was measured at 100 kΩ series loaded resistor by Metex M4650CR (Metex Instruments, Korea) and collected by PC under ScopeView software. In the three-electrodes set up, the outer electrodes (E1, E3) were supplied by the sinusoidal signal and the output signal was collected from the middle electrode (E2).

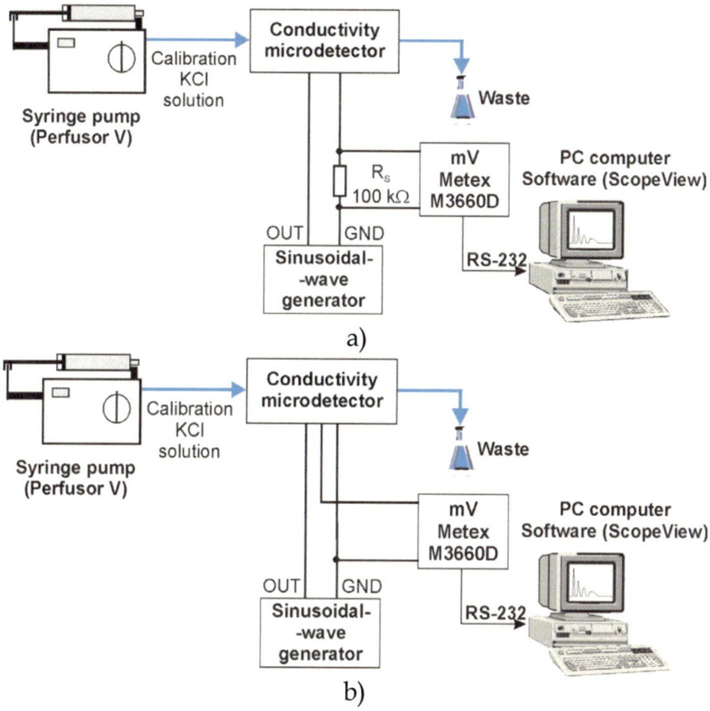

Fig. 20. Schemes of the measurement stations for conductivity microdetector characterization in: a) the two-electrodes configuration, b) the three-electrodes configuration

Tests of the conductivity microdetector were carried out to find the optimal configuration of the electrodes and supplying signal frequency (f_{SUPPLY}). The output signal and sensitivity of the microdetector, defined as change of the output voltage (in mV) in relation to change of the molar concentration of the calibration solution (in mM), have been determined. The highest output signal of the microdetector, for f_{SUPPLY}=1 kHz, was obtained for concentrated KCl solutions and three-electrodes configuration (Fig. 21a). In the range of the low concentrated KCl (0.5 mM ÷ 10 mM) the sensitivity of the microdetector (S_1) was also the highest (S_1=8.38 mV/mM) for three-electrodes configuration. The E1 - E2 configuration had slightly lower sensitivity (S_1=8.28 mV/mM). The E2 - E3 set up had about 60% lower sensitivity than the previous configurations. In the range of more concentrated KCl (10 mM ÷ 100 mM), the highest sensitivity (S_2) was observed for E1 - E2 configuration (S_2=0.4 mV/mM), the lowest - for three-electrodes system (S_2=0.15 mV/mM). It has been concluded, that the two-electrodes configuration (E1 - E2) was the optimal, from the point of view of the micro dosing device, set up of the electrodes. For two-electrodes configuration, the output signal of the microdetector was the highest for above 4 kHz supplying signal frequency and concentrated KCl (Fig. 21b). However, the highest sensitivity (S_1=8.28 mV/mM and S_2=0.4 mV/mM) was observed for f_{SUPPLY}=1 kHz (Fig. 21c). The lowest detection limit was estimated to be about 0.01 µl for 1 mM KCl.
Obtained results were close to the known from the literature typical values of the supplying signal for the conductivity measurements with galvanic feedback (Zemann, 2001). These values ensured high sensitivity of the microdetector in the wide range of the applied KCl

solution concentrations and low, but easily measurable output signal of the microdetector, advantage from the point of view of the life-time of the thin electrodes. It will be also advantage to decrease the distance between electrodes to increase the microdetector sensitivity for the low concentrated KCl solutions.

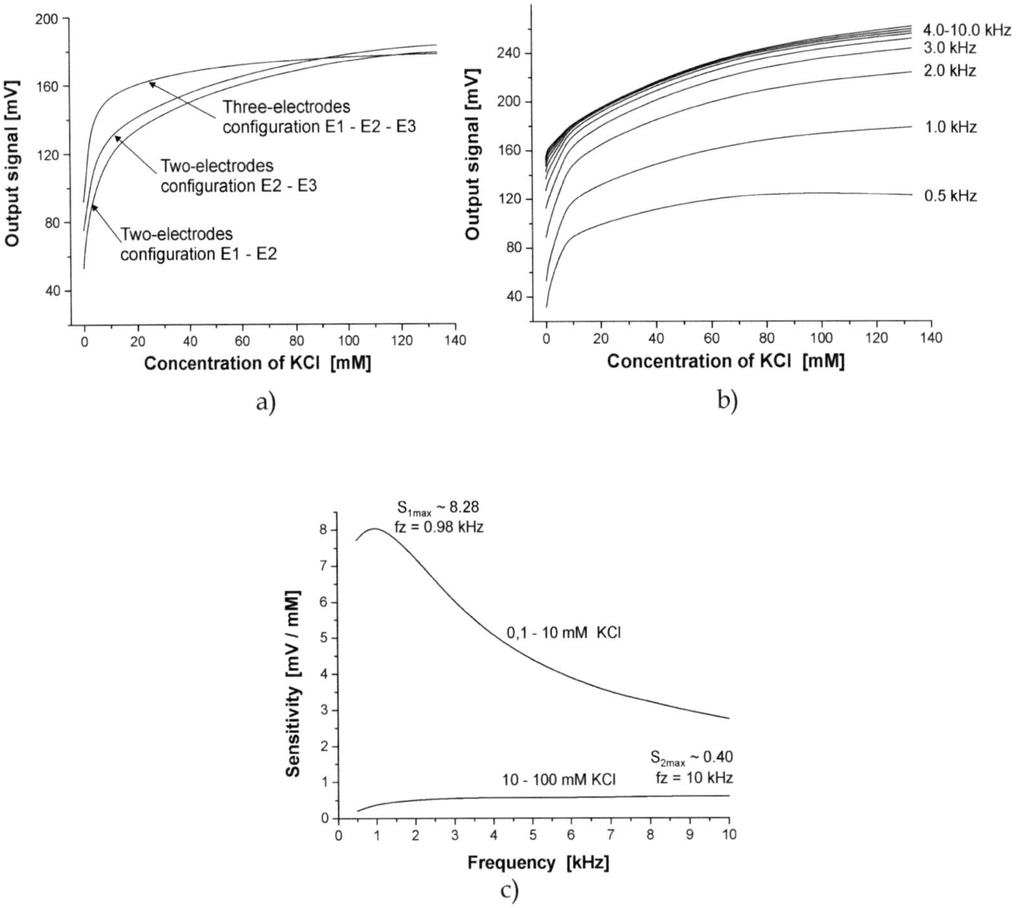

Fig. 21. Conductivity measurements: a) the output signal of the microdetector for different electrodes configuration as a function of the KCl solution concentration, supplying sinusoidal signal with 0.4 V amplitude and 1 kHz frequency, b) the output signal of the microdetector in E1 - E2 electrodes configuration as a function of the KCl solution concentration for different frequency of the supplying sinusoidal signal with 0.4 V amplitude, c) sensitivity of the microdetector in E1 - E2 electrodes configuration as a function of the frequency of the supplying sinusoidal signal with 0.4 V amplitude for different KCl solution concentration, fluids flow 1.2 ml/h

3.4.2 Microdoser with integrated conductometric detector

An idea of a micro dosing device with the pressure driven injection of analyte and on-chip monitoring of the dosed volume is schematically presented on Fig. 22. The on-chip monitoring of the injected volume is ensured by a conductivity microdetector positioned near the outlet.

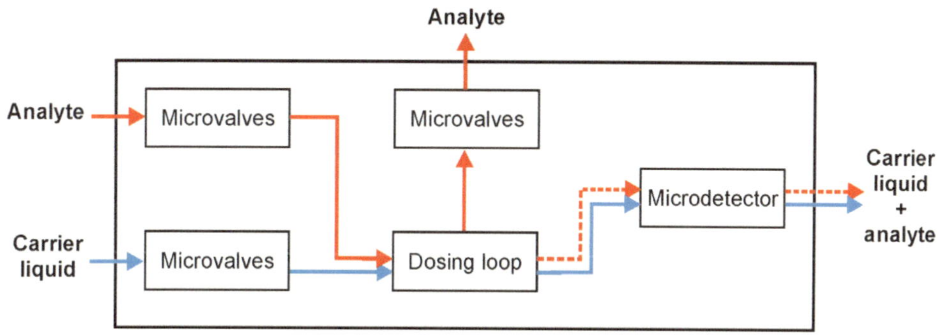

Fig. 22. Scheme of the integrated dosing device with built-in conductometric detector

In a stand-by mode (Fig. 23a), microvalves MV2, MV3 and MV4 were closed while valves MV1, MV5 and MV6 remained open. A carrier liquid (buffer) flowed from inlet 1 to output 3, through the conductivity microdetector. The microdetector measured time-dependent conductivity of the flowing-through samples. The analyte flowed from inlet 5 to outlet 4 and fulfilled the dosing loop. At a dosing mode (Fig. 23b), microvalves MV1, MV5 and MV6 were closed while MV2, MV3 and MV4 were opened. The carrier liquid flowed through the dosing loop and flushed the sample to output 3. The loop was purified and the device was prepared to the next injection. Then, the net of microvalves was switched to the stand-by mode. The fixed maximal volume of the dosed analyte was equal to the volume of the dosing loop (700 nl). Smaller volumes could be dosed by controlling the time of the dosing mode by switching the proper set of microvalves.

Fabrication procedure of the dosing device with Kapton® film as the membrane material of the microvalves was based on microengineering techniques. Structures of the six microvalves and net of 200 μm-deep and 400 μm-wide microchannels were etched in a (100)-oriented n-type 5 Ωcm silicon wafer in 40% KOH at 80°C through thermally grew silicon dioxide (1 μm thick) mask. The after-etch SiO_2 was removed in BHF. Next, micromachined wafer was again thermally oxidized (0.3 μm) to obtain chemically resistance cover. Following, the steering chambers of microvalves (220 μm-deep) were isotropically etched in 40% HF at room temperature in 1.1 mm-thick Borofloat 3.3 through the mask made of the self-adhesive foil (Semiconductor Equipment Corp., USA). The inlet/outlet via-holes (ϕ = 0.9 mm) for gas and fluids were drilled mechanically. Afterwards, two Cr/Ni/Au (50/50/200 nm) electrodes of the conductivity microdetector were deposited and patterned on the processed side of the glass substrate. The Cr/Au (50/150 nm) plates were formed on a chosen side of the Kapton® film to form anti-sticking layer. Next, the via-holes (ϕ = 0.9 mm) were cut in the film. Wafers and film were washed in 30 % H_2O_2 at 80°C, rinsed in DI water and dried in the stream of pure N_2.

The silicon wafer and glass cover, aligned to each other, were sealed under adhesive bonding at about 270°C through Kapton® film (Fig. 24). Fluidical connections were made of cut glass tubes with polished front surface, sealed by use of the UV curable UVO-114 epoxy-glue (EpoTec, Germany). Gas ports were made of medical needles (ϕ=1 mm) glued to glass cover by epoxy-glue. Next, the dosing device was mounted onto PC board. Electrical connections between conductivity sensor and BNC ports were made by wire soldering.

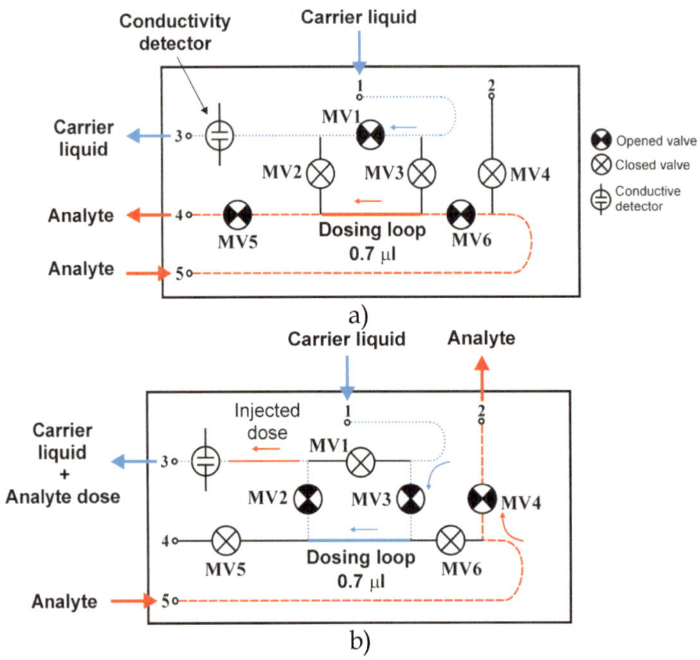

Fig. 23. The dosing device: a) the principle of working of the micro dosing device in the stand-by mode, b) in the dosing mode

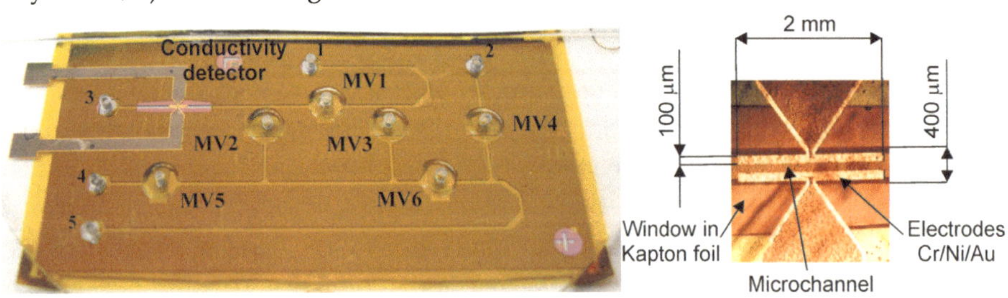

Fig. 24. View of the assembled doser chip (left picture) with enlarged area of conductometric detector (right picture)

The measurement set up for test of the micro dosing device (Fig. 25) contained pneumatic facilities: external electromagnetic valves (EV) (Festo, Germany), normally-open EV1 for MV1, MV5, MV6 and normally-close EV2 for MV2, MV3, MV4, steering pressure P_{STEER} regulator and nitrogen or air gas containers. Flow of the liquid carrier and analyte was ensured by double syringe pump (Perfusor, Germany). The conductivity microdetector was supplied by sinusoidal-wave generator. The output voltage was measured at 100 kΩ series loaded resistor by Metex M4650CR (Metex Instruments, Korea) and collected by PC under ScopeView software. The steering card DAS1402 (Keithley, USA) and LabView 6.0 (National Instruments, USA) software have been used for steering of the external electromagnetic valves. The dosing test were done for the 1 mM KCl injected into DI water. The flow of fluids was 1.2 ml/h. The conductivity detector was supplied with the previously determined optimal sinusoidal-wave signal parameters (f_{SUPPLY}=1 kHz, 0.4 V amplitude).

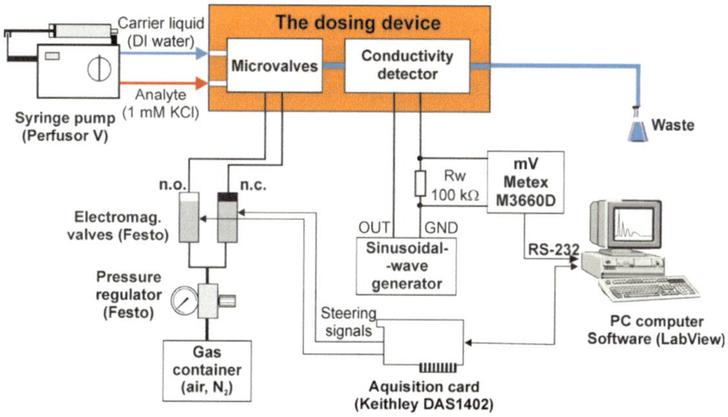

Fig. 25. The measurement set up for the tests of the doser

Thanks to the integrated conductometric detector it has been found, that the dose of the analyte may be adjusted in the range of 100 nl to 700 nl (Fig. 26a) by changing the opening time of the microvalves in the dosing mode. The precise dosing of the analyte with repeatability of the injected volume better than 4 % was obtained for 200 nl to 700 nl volumes. The injected volume had a variation of about 30 % for 100 nl dose. The dose-by-dose test have shown that the dosing device was able to dose the constant (600 nl) volume of the analyte in less that 50 seconds repetition time with the repeatability of the injected volume better than 2 %, what corresponded to 12 nl variation (Fig. 26b).

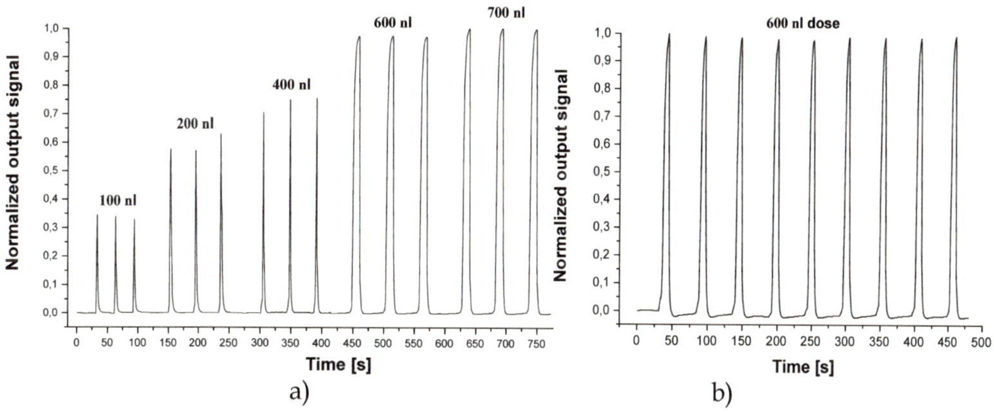

Fig. 26. Normalized output signal of the conductivity microdetctor for: a) various dosing volumes of the analyte obtained by hydrodynamic injection of the volume of the dosing loop (700 nl) or adjusted by the time of the dosing mode (100 nl ÷ 600 nl), b) constant injected volume (600 nl) dosed in 50 seconds repetition time in the multiply-injection mode of the dosing device; dosed analyte: 1mM KCl, carrier: DI water, flow of fluids: 1.2 ml/h, conductivity microdetector supplying parameters: sinusoidal-wave, 1 kHz, 0.4 V

4. Optical microsensors

Among many detection techniques, optical sensing seems to be most widely used. It is because of well known methodology and instrumentation as well as high sensitivity and contactless nature of the measurement. Two optical sensing methods are dominating in microreaction and lab-on-a-chip technologies – spectrophotometry and fluorometry. However, sensing problems arise while analyzed volume is decreased from millilitres to volumes characteristic for lab-on-a-chip techniques - nano- and picoliters. Therefore, usually high sensitivity detectors must be applied. There is a lot of examples of various technical realizations of different microfluidical chips dedicated for both optical methods. In case of spectrophotometry most of these examples are operating in visible light range (VIS), measuring optical properties of non-aggressive liquids (Bargiel at al., 2004). Here we present two unique examples of microfluidical transmittance microsensors for microreaction technique and lab-on-a-chip applications. While fluorometry is applied, typically bulky and expensive epifluorescence-like instrumentation co-working with microfluidical chip is utilized. Here we present a novel methodology and low-cost instrumentation enabling sensitive fluorescence detection induced in various types of labs-on-a-chip.

4.1 Transmittance NIR detector of chemically aggressive liquids

High heat and mass transfer rates in microscale allow the reactions to be performed at higher temperature, providing high yields that are not achievable in conventional reactors. However, real time analytical monitoring of the final product of reaction is necessary to suppress the unwanted by-products. Near-infrared spectroscopy (NIR) appears to be very useful for this purpose due to its capability of a real time, non-invasive monitoring of the chemical processes. The application of NIR spectroscopy in microreactor requires a suitable microsensor characterized by the very low dead-volume and high chemical resistance against for example hot, concentrated nitric and sulphuric acids. Typical spectrophotometric constructions of microfluidical detectors can not be applied in these aggressive conditions mainly due to lack of physical separation between measured liquids and optical fiber. The weakest point of these constructions is a method of optical fibers assembling, utilizing chemically non-resistive glues. It leads to rapid glue destruction and appearance of leakages what can not be accepted due to safety reasons. The unique feature of developed by us microsensor was application of thin silicon wall separating fluidic microchannel from microchannel with optical fibers. It is well known that thin (below 20 μm) silicon layer is transparent for near-infrared light (Fig. 27). Thus, physical separation with simultaneous NIR transmittance was obtained.

The technology of novel optical microsensor utilized standard microengineering techniques. The fluidic channel and alignment grooves were etched by deep reaction ion etching (DRIE) in 380 μm thick, double-side polished, (100)-oriented silicon substrate. DRIE process has been optimized to obtain the vertical side-walls of the channel. Two type of microchannels were formed. The first type was microfluidical channels, the second one – microchannels for positioning of optical fibers. After DRE etching these two types of channels were separated by 20 μm thick silicon wall with perpendicular walls. Photolithographically patterned 100 nm Al mask layer was used to form fluidic inlet/outlet holes form back side of the wafer. The silicon substrate was thermally wet oxidized again to obtain 0.3 μm SiO_2 isolation layer serving as chemically resistive layer. Next, the silicon substrate was anodically bonded (450°C, 1.5 kV) to a Borofloat® 33 glass (Schott, Germany).

High quality bonding process was required to ensure the leakproofness of the channel. Finally, the optical fibers equipped with SMA connections were positioned in the alignment grooves and immobilized by a droplet of UV-curable optical glue UVO-114 (Epo-Tec, Germany) (Fig. 28).

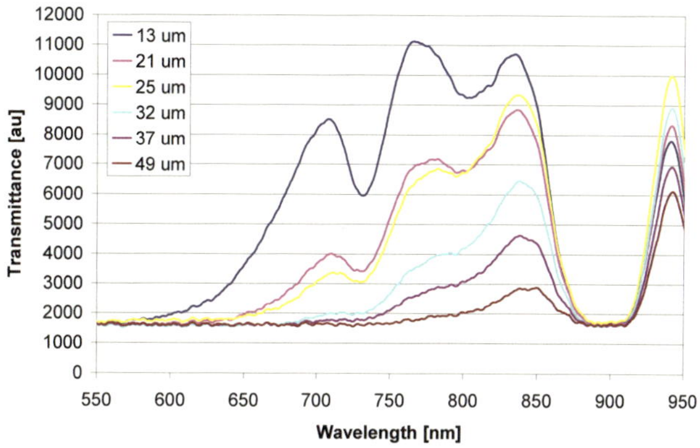

Fig. 27. Transmittance of thin silicon membrane (thickness from 13 μm to 49 μm)

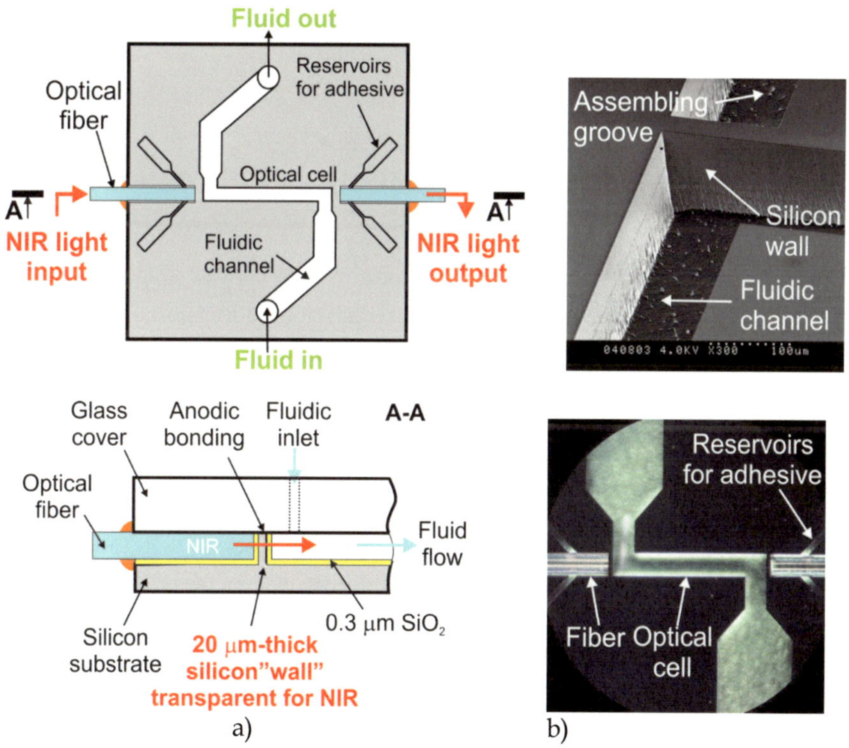

Fig. 28. NIR spectrophotometric microfluidical sensor: a) schematic top view and cross-section of the microsensor, b) SEM picture of the thin silicon wall separating microfludical channels for liquids and optical fiber positioning after DRIE etching (upper picture) and optical microscope picture of the measurement cell with assembled optical fibers (lower picture).

Assembled microfluidical sensor was placed on a PCB carrier and mounted in a metal package with tight and chemically resistive standardized fluidic connections (UpChurch, USA) (Fig. 29a, b).

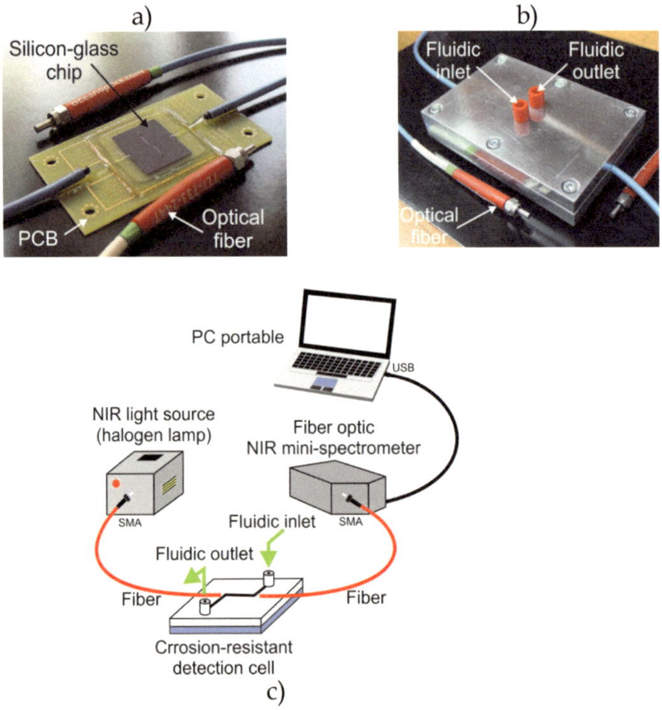

Fig. 29. NIR spectrophotometric microfluidical sensor: a) the chip mounted on a PCB ensuring proper mechanical stiffness and robustness, b) the chip mounted in a metal package ready with fluidic connections, c) scheme of the measurement set-up for NIR spectrophotometric characterization of aggressive liquids by microfluidical silicon-glass sensor

The NIR system was composed of a halogen light source, a silicon-glass corrosion resistant optical cell, and a NIR mini-spectrometer C9406 (Hamamatsu, Japan) (Fig. 29c). The cell with optical path length of 5 mm had detection volume of only 90 nL. The system was controlled by a notebook with suitable software.

The miniature spectrometric system has been tested experimentally by the measuring of NIR spectra of several samples including highly corrosive reactants of nitration reaction. The detection unit worked correctly at wide range of flow rates (0-300 ml/h) what confirmed its mechanical robustness. The 24 h-long test with the measurement cell filled with pure nitric acid followed by sulphuric acid showed corrosion resistance of the detection chip. The spectra of pure nitric and sulphuric acids as well as theirs mixtures with deionized water were successfully obtained (Fig. 30).

In further tests it has been clearly shown that the microsensor recognizes properly different diesel oils and furnace oil as well as gasoline. Concentration of ETOH in Porto red wine has been very well examined. Experimental results confirm the full applicability of the miniature corrosion resistant NIR spectrometric system for use in wide range of applications, e.g. mTAS, microreaction technology, biomedical/medical measurements. The maximal wavelength is limited by the properties of array detector applied in miniature Hamamatsu spectrometer. The use of longer NIR wavelengths (up to 2500 nm) is not limited by the micromachined detection cell.

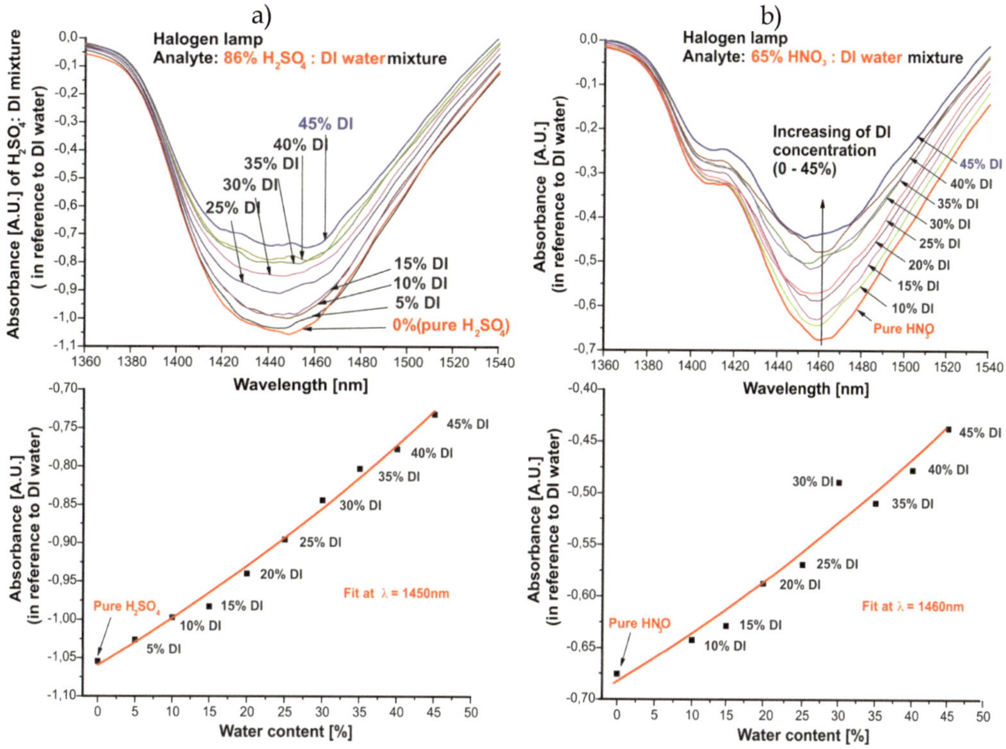

Fig. 30. The spectra of concentrated pure sulphuric (a) and nitric (b) acids as well as theirs mixtures with deionized water, calibration curves describing absorbance versus DI water content are shown below spectra

4.2 Absorbance VIS detector for optical characterization of living oocytes and embryos

Optical characterization of living reproductive cells is an important issue in assisted reproduction techniques. The major goal of these techniques is improvement of in vitro fertilization process towards more successful breeding of farm animals. It is well known that only 5-10% of in vitro fertilized oocytes are viable enough to reach full development competence embryo stage. Assessment of development competence of oocytes and embryos based on lab-on-a-chip system with analyze of the spectral characteristic of the cells, is an important element in research on assisted reproductive technologies. Typical diameter of porcine or bovine oocytes is in 100 μm - 150 μm range, similar dimensions are characteristic for embryos. Due to size and volume incompatibility, spectrophotometric characterization of these cells is impossible in typical measurement cuvette with 10 mm-long optical path and at least a few hundreds microlitres volume. On the other hand, miniaturized spectrometers and light sources co-working with optical fibers as light guiders to and from characterized object are available now. What more, lab-on-a-chip techniques enables fabrication of microchannels with diameter similar to the size of oocytes/embryos and optical fibres (Szczepanska at al., 2009) (Fig. 31a).

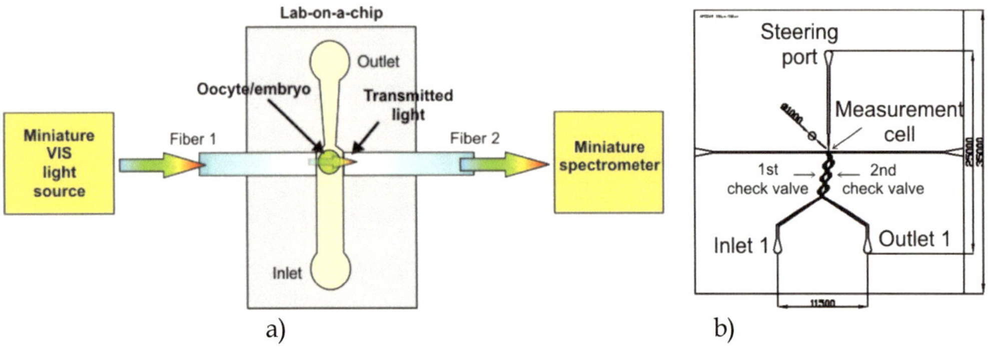

Fig. 31. Scheme of an idea of the spectrophotometric characterization of single oocyte or embryo in a lab-on-a-chip (a) and layout of the real lab-on-a-chip, the dimensions are in μm

Lab-on-a-chip contains measurement cell, net of microchannels and passive valves for steering of fluid and oocyte/embryo flow, and inlet/outlet holes for biological material loading/unloading as well as a steering port for fluid flow management (Fig. 31b). The biological material is introduced into the measurement cell by inlet 1, than passes through set of passive Tesla's valves (1st check valve). Next, characterized cell flows into the measurement cell by sucking of the fluid by a pipette connected to the steering port of the chip. Topology of the measurement cell ensures mechanical immobilization of the cell between two optical fibers. After characterization, the cell is flushed back to the outlet 1 by passing through the second set of Tesla's valves (2nd check valve). Developed configuration enables steering of the fluid flow with examined biological material transport with separation of the inlet and outlet.

The fluidic microchannels and microchannels for optical fibers (all 140 μm deep) were etched simultaneously in DRIE process in the 380 μm - thick monocrystalline silicon wafer (Fig. 32a). After etching, 0,3 μm - thick thermal silicon oxide is was formed to passivate chemically surface of the chip. Next, the wafer was anodically bonded (450^{0}C, 1,5 kV) to a borosilicate glass (Borofloat Schott, Germany) with previously mechanically drilled inlet and outlet via - holes. Following, optical fibers with outer diameter of 120 μm and 100 μm core (Ocean Optics, USA) were mounted. Fronts of the fibers were perfectly aligned each to other thanks to high precision of DRIE etching. Fibers were aligned to the edge of microfluidical channel, ensuring immobilization of the oocyte without its mechanical destruction (Fig. 32c). Fibers were fixed by the use of UV NOA 61 epoxy hard glue (Thorlabs, Sweden). Off-chip ends of both fibers were finished with standard SMA 905 connectors compatible with optical connections of the lamp and the spectrometer. Finally, the chip was positioned in a metal package ensuring stable operation during oocyte/embryo management within the chip (Fig. 32b).

The single oocyte/embryo was introduced into lab-on-a-chip by manual pipeting followed by transport of the cell into the measurement cell thanks to capillary forces. After short measurement (circa 5 seconds) of the optical spectra, the biological material was carefully flushed-back to the outlet by applying pressure into the steering port. Then the cell was captured to a sterile transporting container for further treatment.

The measurement set-up consisted of VIS/NIR light source (a halogen lamp by OceanOptics, USA), developed by us lab-on-a-chip, miniature spectrometer (Ocean Optics, USA) and a personal computer with specialized software (this set-up was similar to presented on figure 31).

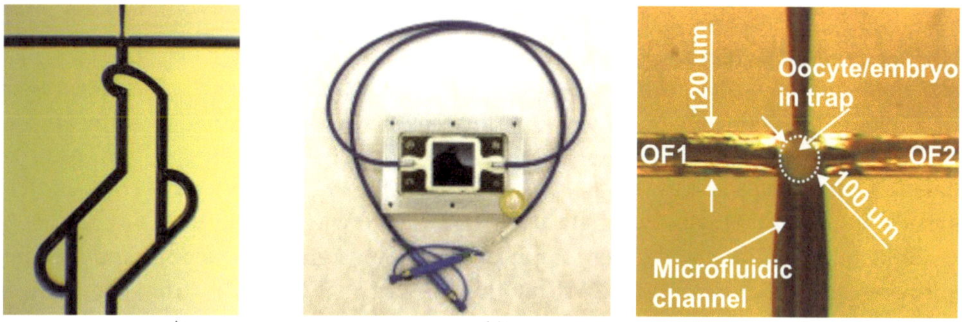

a) b) c)

Fig. 32. Lab-on-a-chip for VIS spectrophotometric characterization of oocytes/embryos: a) enlarged view of the silicon chip after DRIE etching - microchannels for optical fibers, measurement cell in the center and Tesla's valves set are visible, b) packaged lab-on-a-chip ready to work in comparison to Polish 2 zloty coin, c) view of the measurement cell with trapped cell (OF1 and OF2 are optical fibers no 1 and 2)

Totally, over five hundreds of porcine and bovine oocytes, as well as almost one hundred of bovine embryos were optically characterized by novel methodology and lab-on-a-chip. Differentiation of collected spectral characteristics of the cells coming from different classification groups (for example ovarian follicle size or morphological properties) has been observed (Fig. 33). On the base of collected data, set of numerical values, describing subjectively optical properties of examined cell, has been proposed: absorbance level for specific wavelength, absorbance ratio for two wavelengths and wavelength position of the absorbance maxima in VIS region.

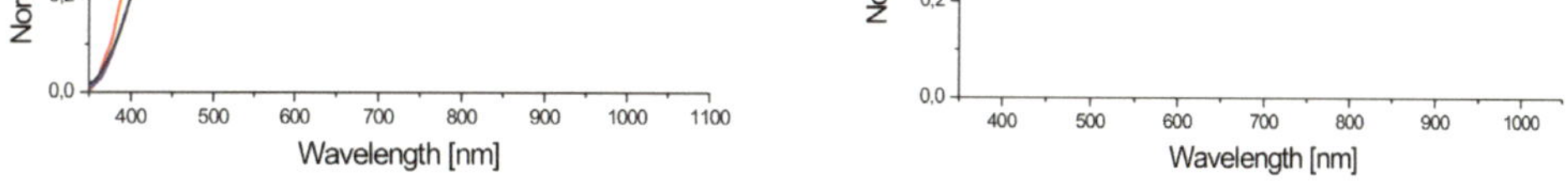

a) b)

Fig. 33. Example of absorbance spectra obtained for: a) porcine oocytes coming from different sizes of ovarian follicles, b) bovine oocytes for different morphological classes

Further experiments confirmed non-destructive nature of spectrophotometric oocyte characterization. Successful in vitro bovine oocyte fertilization after lab-on-a-chip examination has been achieved. This result opens a way towards oocytes selection for artificial fertilization of farm animals oocytes as well as quality assessment of embryos prior to the implantation.

4.3 Fluorometric detector

Labs-on-a-chip dedicated for fluorescence detection of analyte must enable introduction of the fluorescence inducing light and collection of the fluorescence light from an area of interest within the chip. Usually, the chips are whole made of visible light-transparent materials - like glass, PDMS, SU-8, COC or other polymers - or only a top cover of the chip is made of glass, PDMS or other light-transparent materials. Most of the chips are design to co-work with typical apparatus applied for fluorescence induction and readout - epifluorescence microscope (Fig. 34). In this devices, light, usually from arc lamp, light emitting diode (LED) or laser, is restricted to a narrow range of wavelengths that can effectively excitate a fluorochrome and be strongly filtered by the detection channel. The narrow wavelength range is ensured by one or more interference filters and a dichroic mirror. Fluorescence light emitted by the fluorochrome is collected by the microscope objective and guided to a photodetector, passing through filters and dichroic mirror to exclude the excitation light. Common detectors include photomultipliers tube (PMT), semiconductor photodiodes and cooled charge coupled devices (CCD) as sensing matrix in video cameras or lines in spectrometers. Light sources and photodetectors co-work with analog conditioning electronics. The electronic circuits amplify electrical signal with simultaneous reduction of noises. Most of the conditioning electronics is realized by the use of analog circuits. These circuits must ensure high signal to noise ratio (SNR) before the analog signal is digitalized. Therefore, configuration of these analog circuits is sophisticated and only the highest quality elements can be used.

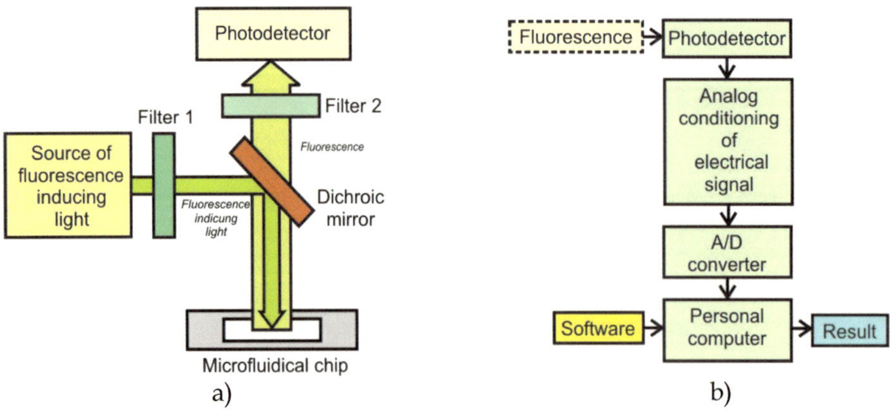

Fig. 34. Fluorescence detection by epifluorescence microscope: a) scheme of the methods, b) path of fluorescence signal conditioning

Although, fluorescence detection is widely used for many years, the configuration of detection apparatus co-working with labs-on-a-chip is based on solutions developed over 20

years ago. Therefore, rapid development of the LOCs must be followed by development of novel methodologies and technical solutions surrounding the chips and leading towards successful application of the microfludical chips in the point-of-care devices.
In the novel concept of the optical instrumentation for fluorescence induction and readout, application of recent developments in optoelectronics and informatics is involved (Fig. 35).

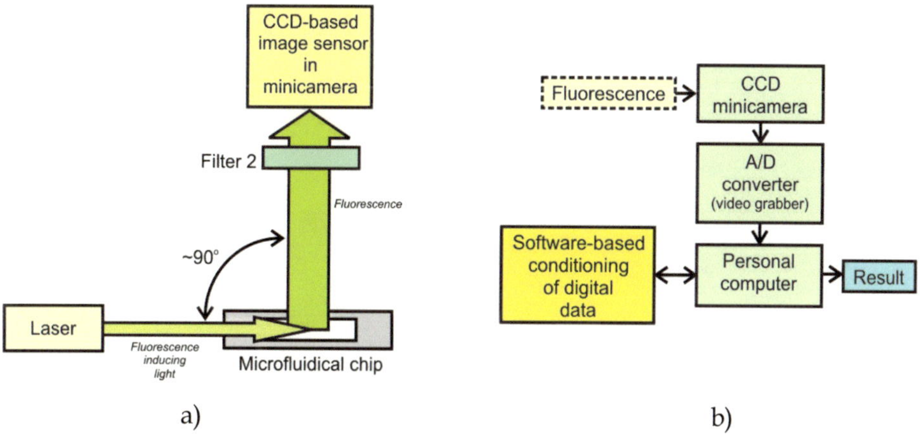

Fig. 35. Fluorescence detection by image sensor and orthogonal configuration of induction/detection channels microscope: a) scheme of the methods, b) path of fluorescence signal conditioning

The fluorescence light is induced by a miniature semiconductor laser. Narrow- spectrum of the laser light eliminates application of emission filter. Collimated laser light is introduced - by edge coupling to a light guiding side wall of the chip - directly into detection area of the chip. The view of the detection area containing fluorescence signal is collected by a CCD image sensor being a part of low-cost minicamera module. The detection unit is positioned perpendicularly in relation to the surface of the chip and laser beam. It enables geometrical separation of the laser excitation light and fluorescence signal without application of dichroic mirror, what significantly simplifies the optical part of the detection unit. Therefore, detection unit consist of only one optical filter passing through the fluorescence light and a miniature black/white CCD camera with the objective. The non-conditioned analog output signal from the minicamera is digitalized by one-channel low-cost frame grabber connected to a personal computer (PC). PC stores images in a memory and specialized software carries out analyze of the captured images to give information on fluorescence intensity. Thus, digital conditioning of the fluorescence signal by the software-base image analyze in spite of analog conditioning is applied in the novel concept. Unique feature of the novel methodology is possible re-analyze of the images which are stored in the computer memory. It is not available in typical instrumentation with "non-imagining" photodetectors (PMT or photodiode) when an operator has no chance for the second analye of carried out experiment.
Novel fluorescence detection methodology and instrumentation co-working with various labs-ona-chip have been successfully applied in many life-science applications –a portable real-time PCR DNA analyzer, a novel portable cocaine detector, a miniature microcytometer for optical characterization of biosamples or on-chip DNA gel electrophoresis.

One of the most interesting and promising applications of LOCs and presented here fluorescence methodology is portable device for detection of food borne pathogens - *Campyloabcer j.* and *Salmonella spp.* by specific amplification of bacteria's extracted DNA with real-time detection of fluorescence - real-time polymerase chain reaction (PCR). This instrumentation has been developed under European project OPTOLABCARD (Ruano-Lopez at al., 2009). The device consists of a disposable real-time PCR chip, a docking station with a specialized chip holder (Fig. 36) and electronics circuits and specialized software for fluorescence signal detection and PCR process temperature profile management.

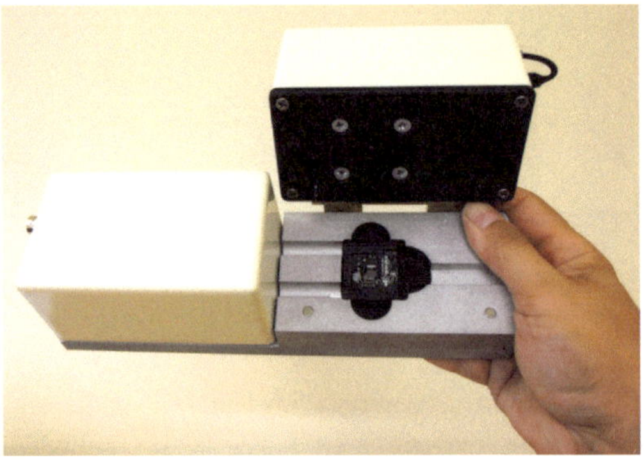

a)

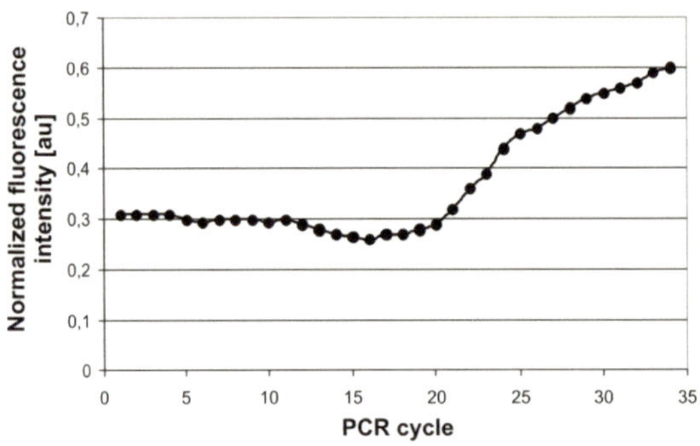

b)

Fig. 36. Portable real-time PCR DNA analyzer utilizing disposable chips: a) view of the docking station with mounted chip holder, b) typical real-time PCR S-curve of *Campylobacter j.* DNA amplification and detection

The disposable glass/SU-8 chip (1 x 1 cm^2) with integrated heater and temperature sensor is placed in a plastic chip holder (2.8 x 2.8 . 0.5 cm^3) with integrated electrical contacts to the chip and some electronics for temperature management. The chip holder has miniature electrical connection to a specialized PCR temperature controller connected to PC. The holder with ready to use chip is positioned in the docking station (15 x 5 x 7 cm^3) in the way ensuring laser light introduction into PCR microchamber and fluorescence light collection. The pre-validation tests of LOC-based system for detection of *Campylobacter j.* were carried out with 48 chicken fecal samples. All the steps - from sample preparation to final result -

were performed in the single chip with 2.5 µl volume of reagents. Red-line fluorochrome (TO-PRO 3) induced by red laser (635 nm, 1 mW) has been applied. The detection unit utilized a long-pass 650 nm interference filter. Typical for real-time PCR fluorescence signal increase during PCR of positive sample has been observed. The ratio of PCR efficiencies between on-chip and on-tube was up to 300%. The sensitivity of on-chip PCR was determined as 0.7-7 ng/ml of template DNA. The real-time PCR process took 30 min - at least 4 times shorter than PCR on-tube.
Similar device but utilizing reusable chip has been developed under Polish national project (Fig. 37). The device was dedicated for rapid detection E. coli in water sample. The chip was made of silicon and glass (Fig. 37b). It was passive chip without integrated heater and temperature sensor. PCR temperature profiling was realized by external in relation to the chip Peltier module-based thermocycler. Due to high chemical resistivity of applied chip materials and assembling technique (anodic bonding) it was possible to clean the chip after PCR by the use of standard sterilization processes (chemical or thermal). Thus, the chip was reusable in contrast to the disposable polymer chips.

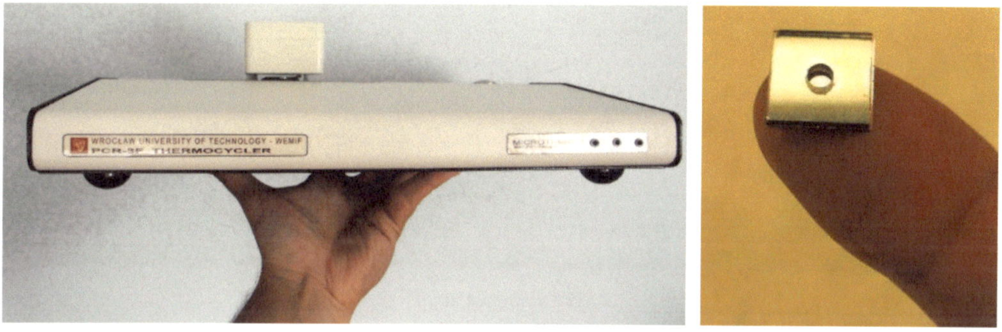

a) b)

Fig. 37. Desktop real-time PCR device co-working with silicon-glass reusable chips: a) view of the instrument, b) 1 cm x 1 cm chip on author's finer

The second interesting application of the miniature semiconductor laser and CCD-based detection unit is a portable cocaine detector developed under European project LABONFOIL (Walczak at al. 2009). The cocaine test is forecasted to be used as prevention test for professional drivers of heavy trucks or buses. The device consist of a disposable wearable cartridge with implemented biological part for cocaine/metabolite detection in a human sweat sample and a hand-held optical reader connected to a computer. The disposable cartridge contains lab-on-a-paper for sweat sample collection and immunochromatography-based cocaine or its metabolite separation and detection. The hand-held reader utilizes semiconductor red laser diode in the excitation channel and 670 nm interference filter co-working with the minicamera in the fluorescence readout channel. The reader is supplied by USB port of a portable computer (Fig. 38). Preliminary tests of the instrumentation confirmed high sensitivity of the optical reader. Lowest detection limit of the cocaine in sweat sample was better than 2 ng/ml, in comparison cut off of cocaine concentration for simple paper-based test with human eye result readout is well above 100 ng/ml.

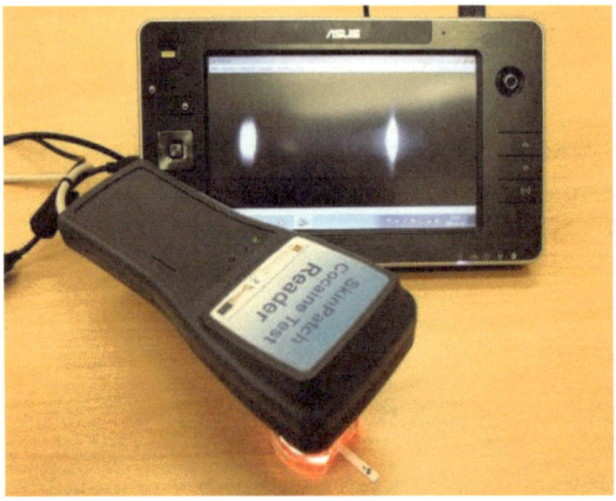

a)

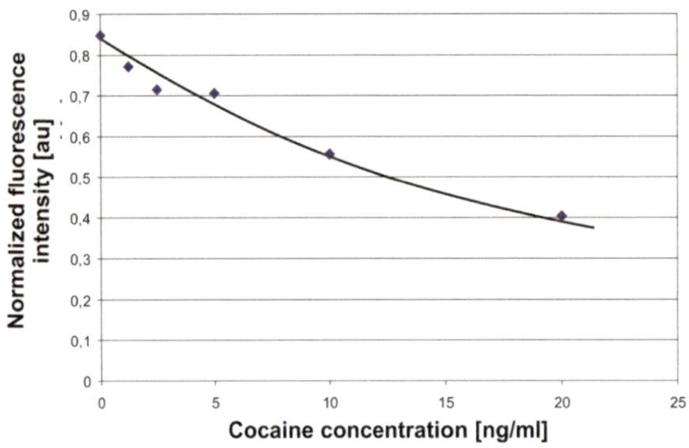

b)

Fig. 38. Fluorometric hand-held reader for cocaine test: a) view of the reader connected to a ultra mobile computer with specialized software, b) normalized fluorescence intensity of control line as function of cocaine concentration in the human sweat sample

5. Conclusion

In this chapter chosen examples of physical, chemical and biochemical microsensors, as the discrete element and as a part of the measurement system have been presented. Main design and fabrication problems of miniature sensors, followed by detailed description of measurement systems and instrumentation have been shown. Afterwards, description of tests with presentation of chosen results were presented. It was also clearly shown, that microengineering technology allows to fabricate microsensors - in some cases this technology is the only useful technique enabling integration of the microsensor with the microfluidical device.

During our activities we are trying to follow a rule: "conscious from-chip-to-instrumentation design" what gives measurable effects of well-fitted and unique parts of the microfluidical system. In our opinion it is only way to develop useful microsensors and instruments for microreaction and lab-on-a-chip applications.

6. Acknowledgment

We would like to direct ours acknowledgments to Sylwester Bargiel from Université de Franche-Comté (Besançon, France), Jan Koszur, Pawel Kowalski and Bogdan Latecki from Institute of Electron Technology (ITE Warsaw, Poland) for close years cooperation. Special thanks are directed to the members of our group: Patrycja Sniadek, Anna Gorecka-Drzazga, Wojciech Kubicki and Jan A. Dziuban.
Most of presented solutions were realized under European projects NEPUMUC (FP6), Optolabcard (FP6), Labonfoil (FP7), as well as Polish projects co-founed by European Union - MNS-DIAG/APOZAR and CiS. We would like to thank to for financing ours activities and persons realizing those projects for grateful cooperation.

7. References

Ali, M.F.; El Ali, B.M.; Speight, J.G. (2005). *Handbook of Industrial Chemistry - Organic Chemicals*, McGraw-Hill, ISBN 0-07-141037-6

Bargiel S.; Górecka-Drzazga A.; Dziuban J. A.; Prokaryn P.; Chudy M.; Dybko A.; Brzózka Z. (2004). Sens. Actuators A, *Nanoliter detectors for flow systems*, No.115, pp. 245-251

Briand, D.; Weber, P.; de Rooij, N.F. (2004). Sensors and Actuators A, *Bonding properties of metals anodically bonded to glass*, No.114, pp. 543 – 549

Dietrich, T.R.; Ehrfeld, W.; Lacher, M.; Krämer, M.; Speit, B. (1996). Microelectronic Engineering, *Fabrication technologies for microsystems utilizing photoetchable glass*, No.30, pp. 497 - 504

Edited by Dietrich, T.R. (2009). *Microchemical Engineering in Practice*, Willey-VCH Verlag GmbH, ISBN 978-0-470-23956-8

Dziuban, J. (2006). *Bonding in Microsystem Technology*, Springer, ISSN 1437-0387, ISBN-10 1-4020-4578-6 (HB), ISBN-13 978-1-4020-4578-3 (HB), ISBN-10 1-4020-4589-1 (e-book), ISBN-13 978-1-4020-4589-9 (e-book)

Ehrfeld, W.; Hessel, V.; Löwe, H. (2005). *Microreactors – New Technology for Modern Chemistry*, Wiley-VCH Verlag GmbH, ISBN 3-527-29590-9

Freitag, A.; Vogel, D.; Scholz, R.; Dietrich, T.R. (2001). Journal of the Association for Laboratory Automation, *Microfluidic devices made of glass*, Vol.6, Issue 4, pp. 45 - 49

Knapkiewicz, P.; Walczak, R.; Dziuban, J.A. (2006). *On integration of silicon/glass micromachined sensors to microfluidical devices - toward intelligent microreactor*, Proceedings of XXth Eurosensors Conference, pp. 40-41, ISBN 91-631-9280-2, ISBN 978-91-631-9280-7, Göteborg, Sweden, September 17-20, 2006

Knapkiewicz, P.; Walczak, R.; Dziuban, J.A. (2007). Optica Applicata, *The method of integration of silicon-micromachined sensors and actuators to microreactor made of Foturan® glass*, Vol. XXXVII, No.1-2, pp. 65 – 72

Knapkiewicz, P.; Dziuban, J.A.; Boskovič, D.; Loebbecke, S.; Freitag, A.; Dietrich, T.R. (2008). *The system for multipoint pressure and temperature measuring in microreactor used for nitration process*, Proceedings of XXIInd Eurosensors Conference, pp. 40-41, ISBN 978-3-00-025218-1, Dresden, Germany, September 7-10, 2008

Kralisch, D.; Kreisel, G. (2007). Chemical Engineering Science, *Assessment of the ecological potential of microreaction technology*, No.62, pp. 1094 - 1100

Ruano-Lopez J.; Agirregabiria M.; Olabarria G.; Verdoy D.; Bang Dang D.; Bu M., Wolff A.; Voigt A.; Dziuban J.; Walczak R.; Berganzo J. (2009). *The SmartBioPhone, a point of care vision under development trough two European projects : OPTOLABCARD and LABONFOIL*, Lab on a Chip 2009, vol. 9, iss. 11, pp. 1495-1499

Speight, J.G. (2002). *Chemical and process design handbook*, McGraw-Hill, ISBN 0-07-137433-7, United States of America

Szczepańska P.; Walczak R.; Dziuban J.; Jackowska M.; Kempisty B.; Jaśkowski J.; Bargiel S. (2009). *Lab-on-chip quality classification of porcine/bovine oocytes*, Procedia Chemistry 2009, vol. 1, iss. 1, pp. 341-344

Walczak R.; Dziuban J.; Szczepańska P.; Scholles M.; Doyle H.; Krüger J.; Ruano-Lopez J. (2009). *Toward portable instrumentation for quantitative cocaine detection with lab-on-a-paper and hybrid optical readout*, Procedia Chemistry 2009, vol. 1, iss. 1, pp. 999-1002

Zemann A. J. (2001). *Conductivity detection in capillary electrophoresis*, Trends Anal. Chem. 20 6+7, pp. 346-354

6

Chemical Microsensors with Ordered Nanostructures

Marina Vorozhtsova, Jana Drbohlavova and Jaromir Hubalek
Brno University of Technology, Faculty of Electrical Engineering and communication
Laboratory of Microsensors and Nanotechnologies (LabSensNano)
Czech Republic

1. Introduction

Current issues solved in Microsensors are focused on finding of new approaches to increase sensitivity with decreasing dimensions at the same time, together with low-cost ability in manufacturing. The chapter deals with non-lithographic techniques of nanostructuring surfaces on sensing layers of microsensors which are promising to improve their parameters, mainly to amplify sensitivity. Especially (bio)chemical sensors for environmental, pharmaceutical and medical applications employ nanostructures in their constructionuse.

2. Template based techniques for ordered nanostructures fabrication

One of the ways to achieve better and better detection characteristics of sensors is the use nanoparticles to modify the surface of the sensitive detection sensor part. A big challenge in fabricating various nanostructures fixed on solid supports is the uniformity and reproducibility in size and spatial distribution. This can be accomplished in several ways. Among proven methods, the lithography is very popular. It allows creating very precise structures and reliefs, but the price for the acquisition of apparatus and the service are very high.

On the other hand, the template based methods are exceptional techniques how to create freestanding, non-oriented and oriented nanostructures like nanotubes, nanorods, nanowires as well as nanodots over large areas on substrates in a fast, cheap and easy reproducible way. The possibility of using these well aligned nanosized structures as sensor arrays makes them very attractive candidates for potential applications in chemical analysis and medicine, especially for biosensing purposes. The template based methods can be applied in the current technology of thick film sensors, but it increasingly penetrates into areas of thin film applications. The other well known applications include electronic (e.g. as embedded capacitors) and optoelectronic devices, for example dye-sensitized solar cells, light emitting diodes and so on.

There is variety of the material which can be used for nanostructures fabrication by the template based methods; from semiconducting oxide (selenide, telluride) to pure metals such as Ni, Au and Pd.

Regarding to above mentioned applications, the microsensor for mentioned applications is build up on (bio)chemical transducer as can be seen on Fig. 1. The transducer contains a sensing layer where nanostructures are very promising formation of the sensing layer to obtain advanced sensing properties.

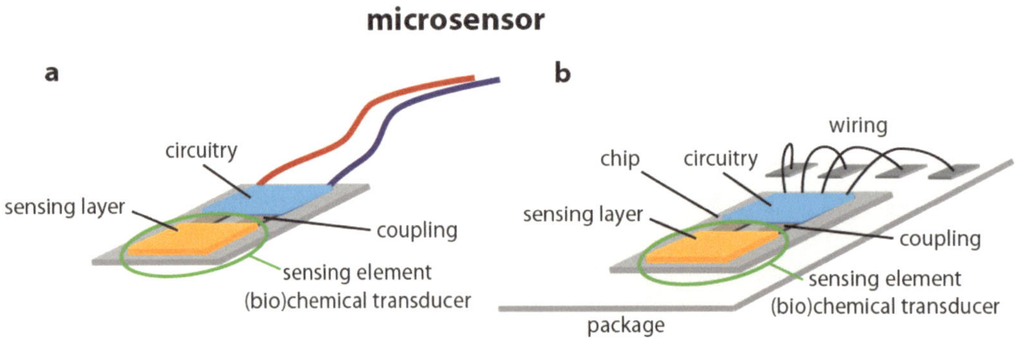

Fig. 1. Microsensor fabricated by hybrid technology (a), microsensor on a single chip (b)

2.1 Template fabrication and properties

Among various materials (activated carbon or carbon nanotubes, polymer gel, radiation track-etched polycarbonate or mica, zeolites, porous silicon prepared by electrochemical etching of silicon wafer and nanochannel array on glass) used as templates, alumina is the most frequently used one (G. Z. Cao & D. W. Liu, 2008). The pores in an anodic aluminum oxide mask are self ordered as a close-packed array of columnar hexagonal cells, each containing a central pore normal to the substrate. Similarly like alumina, the other metals (e.g. titanium) with specific characteristics can be changed by the electrochemical method to porous oxides, so called ceramics, with periodic nanoporous hexagonal structure (Alivov, 2009; Chu, 2005). Aluminum and titanium have a special ability, which was discovered in 1970 by a group of Wood, Sulivan and others. The controlled anodic oxidation means creating of nanoporous structure (Shingubara, 2003). During the 90s, the production process was being improved. The research team of Japanese scientists Madsuda and Fukuda, who developed the production of porous ceramics using the "two-steps" method, which greatly contributed to improving the quality of the resulting structures, was of great importance in nanoporous ceramic research (Masuda & Fukuda, 1995).

During anodization process, the diffusion of metal ions or oxygen through the oxide layer growing in an electric field generated in this layer by applying an external voltage is the steering process. The rate of oxide layers growth during anodization depends exponentially on the electric voltage. The diameter and density of nanopores in template can be tuned in a wide range (from 3 to 500 nm) by controlling the anodization conditions, namely anodization voltage, time, and electrolyte. Kokonou et al. found, that anodization conditions strictly depend on aluminium layer thickness (Kokonou, 2007). In other words, the conditions for thin Al layers differ significantly from those used to grow thicker films. Particularly, much lower electric fields are necessary in the case of aluminium thin films, since the strong electric field causes fast dissolution of the grown alumina. Generally, the lower voltage is applied, the smaller pore size, the higher pore density and the better homogeneity. Concerning the temperature, the lower value is maintained during the anodization process, the slower growth of pores and consequently their better uniformity can be reached. The uniformity of the template can be also influenced by number of anodization steps. The well-ordered nanoporous alumina masks can be obtained by two-step or even by three-step anodization of aluminum layer (Mozalev, 2009; Zhou, 2006).

There is no doubt that the quality of aluminium layer used for template fabrication plays a key role in reaching the uniform growth and regular distribution of nanopores. In other words, the level of film surface roughness has a direct effect on the regularity of pore arrangement (A. P. Li, 1998; Masuda, 1998; A. W. Wang & White, 1995). For sensor applications, a porous mask can be made in two ways. One of them is the production of several tens of micrometers thick membranes of high purity aluminum foil (thickness of 250 μm), and the other one is direct anodization of thin aluminum layer deposited on the substrate. Both of these options bring several problems. When using foil, the first step is electrochemical polishing of the surface followed by annealing step for aluminium recrystallization and to increase homogeneity. The homogeneity of thin aluminum deposited layer depends strongly on the deposition technique. Usually, the aluminium layer is deposited by several PVD techniques like pyrolytic evaporation, magnetron sputtering or ion sputtering on various solid supports, most currently on insulating substrate, like SiO_2 coated silicon wafer, glass, sapphire or on some semiconducting materials (e.g. GaAs, InP). During anodizing it is also necessary to consider the influence of adhesion between the aluminum layer and the underlying substrate (Hrdy & Hubalek, 2005).

As mentioned above, the choice of electrolyte is also a very important factor influencing the anodization process. The most frequently used electrolytes for alumina template fabrication are sulphuric, oxalic or phosphoric acid aqueous solutions at low concentration. The smallest pore size is generally reached using the sulphuric acid which is commonly used under lower constant voltage (in the range of 18 V up to 30 V) compared to oxalic acid (constant voltage from 30 to 60 V). The highest pore size is reached in the case of phosphoric acid. The other electrolytes composition was studied as well. Anitha and colleagues tested the influence of three different fluorine containing electrolytes (aqueous hydrofluoric acid (HF), HF containing dimethyl sulphoxide and HF containing ethylene glycol) on fabrication of TiO_2 nanotubes (Anitha, 2010). The experiments were carried out over a broad voltage range (2-200 V) and HF concentrations (0.1–48 wt%) which resulted in variation of anodization time from 5 s to 70 h. The authors observed that the composition of the electrolyte and its fluorine inhibiting nature has significant impact on nanotube formation as well as on controlling the aspect ratio. In other work, Song et al. examined the influence of defined water additions to an organic anodization electrolyte on wall thickness oscillations in electrochemically grown TiO_2 nanotube arrays (Song & Schmuki, 2010).

Finally, the choice of etching conditions during subsequent chemical etching step after anodization process can induce the type (shape) of prepared nanostructures. For example, the hole array structure can be obtained by the selective removal of silicon oxide from the Si substrate using wet etching in HF solution while the column array structure can be obtained by the selective etching of Si substrate in KOH solution using silicon oxide as a mask (Oide, 2006).

Sometimes, it is necessary to use other post-processing methods such as thinning to obtain very thin nanoporous template. Graham et al. employed the anodized nanoporous alumina to form aluminium electrodes on integrated circuit (IC) which can be applied as biocompatible material enabling a low-cost solution for drug discovery pharmacology, neural interface systems, cell-based biosensors and electrophysiology (Graham, 2010). The porous alumina was electrochemically thinned to reduce the alumina electrode impedance. For applications where a porous electrode surface is either preferred or acceptable, the authors demonstrated that porosity can be manipulated at room temperature by modifying the anodizing electrolyte to include up to 40% polyethylene glycol and reducing the phosphoric acid concentration from 4% (w/v) to 1%. For applications requiring a planar microelectrode surface, a noble metal was electrodeposited into the pores of the alumina film. Limited success was achieved with a pH 7 platinum and a pH 5 gold cyanide bath but good results were demonstrated with a pH 0.5 gold chloride bath which produced planar biocompatible electrodes. A further reduction in impedance was produced by deposition of platinum-black, which may be a necessary additional step for demanding applications such as neuronal recording.

Montero-Moreno studied a barrier layer thinning (BLT) process to decrease the high electrical resistance generated by the barrier layer that isolates the metallic base from the electrodeposition bath (Montero-Moreno, 2009). The authors showed that during thinning, a controllable branched-shaped porous structure of AAO is generated. Finally, the use of stepped techniques to obtain alumina templates with narrower pores than expected in an oxalic acid bath was also analyzed.

3. Vertically ordered nanostructures

Vertically ordered nanostructures fabricated using template-based method can be used for sensor surface modification to enhance its sensitivity or they can create nanosensor array. High surface area and unit impedance behavior is expected from this formation. These nanostructures can be also functionalized with various biomolecules with specific sensitivity for biorecognition transducer construction. The usage of template-based techniques for nanostructures fabrication (especially nanorod, nanowire and nanotube arrays) can be performed either in the solution or with the help of other deposition methods. One of the greatest advantages of template-based synthesis for the growth of nanotubes and nanotube arrays is the independent control of the length, diameter, and the wall thickness of the nanotubes. While the length and diameter of resulted nanotubes are dependent on the templates used for the synthesis, the wall thickness of nanotubes can be readily controlled by the growth duration. Another great advantage of template-based synthesis is the possibility of multilayered hollow nanotube or solid nanocable structures formation [1].

The deposition from solution is known as electrochemical deposition or simply as electrodeposition, which involves oriented diffusion of charged reactive species through a solution when an external electric field is applied, and reduction of the charged species at deposited surface which also serves as an electrode. In industry, electrochemical deposition is widely used in making metallic coatings in a process known as electroplating. In general, this method is only applicable to electrical conductive materials such as metals, alloys, semiconductors, and electrically conductive polymers and oxides. After the initial deposition, the electrode is separated from the depositing solution by the deposit and the electrical current must go through the deposit to allow the deposition process to continue. When deposition is confined inside the pores of template membranes, nanocomposites are produced. If the template membrane is removed, nanorod or nanowire arrays are prepared. However, when the deposition occurs along the wall surface of the pore channels, nanotubes would result (G. Z. Cao & D. W. Liu, 2008). The example of deposition from solution can be the work of Wang, who prepared vanadium pentoxide nanotube arrays (Y.

Wang, 2005). Another example is formation of mesoporous oxides with well defined and ordered porous structure, which can be readily synthesized using surfactant or copolymer micelles as templates through sol–gel processing. Next recent example is Au nanoparticle templated synthesis of poly(N-isopropoylacrylamide) nanogels (G. Z. Cao & D. W. Liu, 2008).

The second mentioned case, i.e. the nanostructures formation using other deposition methods, is for example atomic layer deposition (ALD) as a perfect technique for the synthesis or fabrication of alumina nanotube arrays with well controlled wall thickness and morphology (C. C. Wang, 2007). ALD has also been employed for the fabrication of TiO_2 coated alumina membranes and TiO_2-coated Ni nanowires, and TiO_2 nanotube arrays were readily obtained by dissolving the templates (Kemell, 2007; Y. Wang, 2005).

3.1 Nanodots and nanocolumns

Nowadays, there are lots of papers dealing with ordered nanotubes array fabricated using template methods, but only very few works concerning the application of anodization technique for nanodots preparation. Sometimes, the scientists combine the usage of nanoporous mask and other ways of nanodots deposition, like ion beam evaporation, electron gun evaporation, nanoscale selective epitaxial growth (Y. D. Wang, 2006a), selective anodization using AFM tip, electrodepositon etc.

In order to achieve nanocrystals directly grown by anodization with sizes in the range of 1 to 10 nm which is essential for their quantum effect, the diameter of pores in the template must also be in this range. Beside the pore diameter, the template film thickness is also a crucial parameter which determines the quality of prepared nanostructures. In the case of nanodots, the scientists found that thinner template is more convenient in order to reach the high density of QDs array; particularly the thickness must be in the maximum 3 to 5 times higher than template pore size. Mao et al investigated well reproducible direct preparation of ultrathin template with thickness about 50 nm by utilizing a stop signal, a vivid color appearing at the air-electrolyte interface (Mao, 2009).

From the chemical point of view, several materials can be used for nanodots fabrication. Wang et al. prepared ordered GaN nanodot arrays with an average dot diameter and height of 60 and 100 nm by nonlithographic nanopatterning technique combined with nanoscale selective epitaxial growth for application in high efficiency nitride light emitting diodes (Y. D. Wang, 2006a). The same group of scientists also studied InGaN nanorings, nanodots, and nanoarrows fabrication using GaN layer on a sapphire substrate coated with a thin layer of SiO_2 (around 100 nm) by PECVD and finally coated with evaporated aluminium layer (about 1 μm) (Y. D. Wang, 2006b). The aluminium layer was then anodized to nanoporous template in a two-step anodization process and used to pattern nanopores in SiO_2 transfer layer. The patterned SiO_2 layer was applied as a template for nitride growth by MOCVD. The diameter of the deposited nitride nanostructures varied from 35 to 250 nm and their type was determined by controlling the nitride growth time.

Jung et al. examined the growth of CdTe QDs array (with dot size of 80 nm) on the GaAs substrate by molecular beam epitaxy method using the porous alumina masks (with thickness about 300 nm) for applications in optoelectronic devices in visible range (M. Jung, 2006). Similarly, Alonso-Gonzalez et al. investigated the fabrication of InAs QDs using epitaxial growth on GaAs nanoholes pre-patterned surface (Alonso-Gonzalez, 2006). Their particular approach consists of using the anodic aluminium oxide (AAO) prepared from epitaxially grown aluminium layer on GaAs substrates as a mask for creation of ordered nanoholes in GaAs, which act as preferential nucleation sites for InAs QDs. Liang et al. used highly ordered AAO porous membrane as template for fabrication of hexagonal close-packed nanopore arrays on Si, GaAs, and GaN substrates via reactive ion etching (Liang, 2002).These nanopore structures were then utilized for QDs arrays formation from various metals and semiconductors (ZnO, TiO_2) through evaporation and subsequent etching. Kouklin et al. electrodeposited the hexagonal close-packed array of CdS QDs in a nanoporous AAO template and studied them by capacitance-voltage spectroscopy (Kouklin, 2000). The authors found that these structures are ideally suited for quantum-dot flash memories.

Lim et al. fabricated epoxy nanodots array sensors with sputtered Au electrodes for electrochemical determination of whole blood coagulation time. The authors used the titanium layer instead of aluminium for the replication of nanopatterns into epoxy (Lim, 2009). Even though the ordering of dimples in the titanium is not as good as that in the aluminium, the usage of titanium in this case was found more convenient due to the better hardness of titania and lower surface energy which facilitates the separation of a replica film from the substrate. The fabrication process started by anodization of a titanium foil, which leads to formation of highly ordered nanotubular TiO_2 film. After its removal by epoxy, the hexagonal nanoarrays on the titanium surface are formed similarly to those of the aluminum substrate after the removal of the first anodic oxide. The second epoxy film formation on the titanium substrate mold with hexagonally arrayed dimples. The height of created nanodots was around 28 nm and their diameter was 120 nm.

Nanoporous AAO membrane can also serve as a host material for a variety of magnetic materials which can be embedded for example by electrodeposition, sputtering or infiltration routes. The work of Jung et al. deals with the electrochemical growth of Ni nanodots array using the long-range ordered cylindrical alumina nanopores as a template created on titanium precoated silicon wafer (J. S. Jung, 2008). Such Ni nanodots array is suitable for high density data storage materials.

Another approach of AAO template usage was tested by Yan et al., who prepared the composite of highly ordered nanoporous AAO films loaded with ZnO nanoparticles (10.8 nm) by simple soaking the AAO films in an aqueous solution of zinc acetate followed by annealing at 500 °C (J. L. Yan, 2008). The composite exhibited intense and broad emission spectra in the wavelength range of 350–600 nm.

In other paper, the authors demonstrated an application of ultrathin porous AAO layers (about 50 nm) on Si as templates for $Ba_xSr_{1-x}TiO_3$ nanodot array fabrication (B. Yan, 2007). Furthermore, these aluminum oxide nanotemplates can be employed as lithographic masks to transfer the nanopattern into the silicon substrate.

Yang et al. prepared ordered arrays of Ta_2O_5 nanodots with diameter of 80 nm at the bottom and of 50 nm in height using AAO as a template (C. J. Yang, 2007). The structures were synthesized in a two-step anodization process from TaN (50 nm) and Al (1.5 μm) films deposited successively on p-type Si wafers. Similar approach was applied in the work of Vorozhtsova et al. and Mozalev et al., who used Ta layer as a starting material for Ta_2O_5 nanocrystals or nanocolumns fabrication through AAO template (Mozalev, 2009; Vorozhtsova, 2010). Ta_2O_5 is a material of great interest for fabricating capacitor, semiconducting and photonic devices as well as resistive humidity sensor. This is due to its unique properties such as high dielectric constant, low leakage current density, high index of refraction and low optical propagation losses. Its high dielectric constant and low leakage current density make it popular for a use in the next generation semiconductor electronics.

Li et al. fabricated the hexagonally ordered arrays of ferromagnetic nanodot with narrow size distribution by electron beam evaporation of Fe, Ni of $Fe_{20}Ni_{80}$ into AAO template and subsequent mask lift-off (C. P. Li, 2006). Moreover, exchange-biased bilayer nanodots were fabricated using argon-ion milling instead of the mask lift-off. The authors found that when magnetic dots are fabricated using mask lift-off, the phosphoric acid etching before magnetic materials deposition should be long enough to remove the alumina barrier layer remaining at the bottom of pores. Otherwise, the dots would also be removed during subsequent mask lift-off. On the other hand, with the argon-ion milling, the barrier layer cause any problems. Thus, shorter etching times resulting in smaller pore diameters can be used.

Next frequently used material in QDs array fabrication using AAO template is titanium dioxide. The group of researchers with Chen P. L. intensively studied the uniform nanodot arrays of titanium oxide prepared by electrochemical anodization of Al/Ti film stack on silicon (Chen, 2005), Al/TiN bilayered film on a sapphire (Chen, 2004) or TiN/Al films on silicon substrates (Chen, 2003). In the first case, the authors applied such self-organized nanodot arrays as the electron emission source in field emission triode device. In the second mentioned work, they found that the phase development of the isolated TiO_2 nanodots is very much different from TiO_2 thin films and powders. After high temperature annealing, the nanodots are polycrystalline and consist of a mixed phase of anatase and rutile instead of single rutile phase. While in the third work, the as-prepared nanodots were basically composed of amorphous TiO_x with a hexagonal arrangement and an average diameter of about 60 nm.

Similarly to these papers, Chu et al. reported the TiO_2 nanodots and nanorods (diameter of 20–100 nm, height of 30–260 nm and distance of 50–380 nm) fabricated by combined anodization from superimposed Al/Ti layers sputtered on glass substrates (Chu, 2005a, 2005b). These nanostructures can be applied e. g. as DNA sensors. The authors also tested new anodizing electrolyte, a diluted nitric acid solution, for fabricating uniform, self-organized, ordered nanoporous titania films with parallel cylindrical pores (pore diameter approximate to 30 nm and thickness around 1100 nm). After heating at 600 °C for 2 h, the nanoporous titania films exhibited high photocatalytic activity under UV illumination.

Cha et al. presented a method for fabricating SiO_2 nanodot arrays with average diameter of 68 nm through pattern transfer of self-organized tantalum oxide hard masks onto a Si wafer (Cha, 2004). Tantalum oxide nanopillar arrays are formed at the bottom of anodic aluminum oxide by electrochemical anodization of the Al/Ta films on a Si wafer. Then the tantalum oxide nanopillars were used as hard masks for formation of SiO_2 nanostructures. Ion milling was used for the pattern transfer. The density and diameter of the SiO_2 nanodot arrays could be controlled by varying the anodizing conditions.

3.2 Nanowires and nanotubes created by electrodeposition of metal ions

A variety of metal nanowires, including Ni, Co, Cu and Au with nominal pore diameters between 10 and 200 nm have been synthesized by electrodeposition. For example, Possin prepared various metallic nanowires using radiation track-etched mica (Possin, 1970). Likewise, Williams and Giordano produced silver nanowires with diameters below 10 nm (Williams & Giordano, 1984). Whitney et al. fabricated the arrays of nickel and cobalt nanowires, also using polycarbonate templates (Whitney, 1993). The growth of Au nanorod arrays has also been well studied (G. Z. Cao & D. W. Liu, 2008; J. G. Wang, 2004; H. W. Wang, 2006).

Vertically aligned nanowire arrays realized on interdigited electrodes are very promising for applications where high active surfaces are needed, i.e. sensors and biosensors. Fabrication of nanoparticles takes place in electrolyte by inserting conductive substrate with a porous non-conducting mask. The conductive substrate represents a cathode. After applying a constant current between two electrodes, metal ions are attracted to the cathode. On the cathode, ions diffuse through the open pores to form nanowires, respectively nanotubes. After complete deposition, the porous mask is dissolved usually in NaOH (Matefi-Tempfli, 2009). The step by step anodization process is schematically illustrated on Fig. 2.

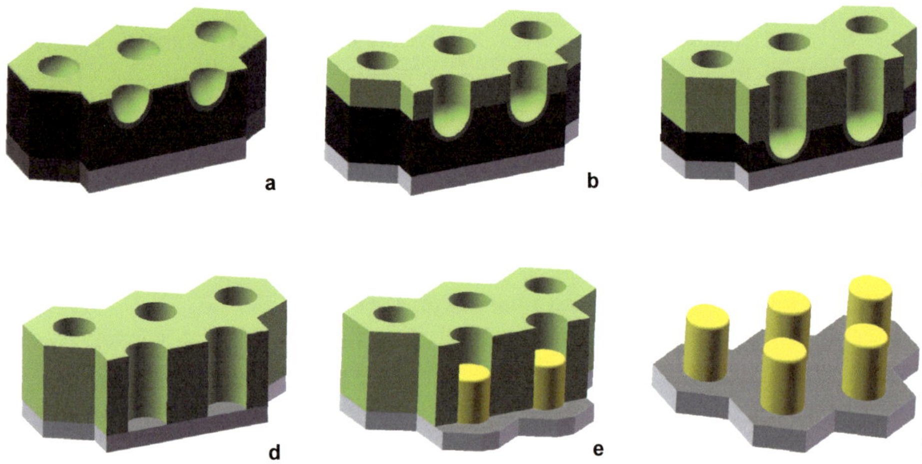

Fig. 2. Nanostructures fabrication process by template-based method: (a) Thin aluminum film with a intermetallic layer. Start of anodization process; (b) Progress of process; (c) Nanoporous template with oxide barrier; (d) Alumina layer with hexagonally arrayed pores created by anodization; (e) Nanostructures which are formed through the template; (f) Nanostructures after the template dissolving

The size of resulting structure is influenced by the geometry of porous mask. The diameter of nanoparticles and their mutual distance are given by the size of individual cells in relation to the pore size, so-called mask porosity. Creating a specific type of nanostructures depends on the mask and electrodeposition conditions. Length of the formed nanostructures is directly proportional to the current density and electrodeposition time. The amount of deposited material is based on Faraday's law. The growth of metallic nanostructures is also affected by temperature and solution concentration, electric potential and pH of the solution. The example of alumina template created in various electrolytes and characterized by SEM can be seen on Fig. 3.

The type of the structure is given by the specific conditions under which the nanostructures are being created. The probability of making nanotubes through a mask grows with the size of pores in the mask. Nanowires almost always occur at very small diameters of nanopores (20 nm). The most frequently used metals for the nanostructure fabrication are nickel, gold, palladium and silver. Some examples of the fabricated nanostructures observed under SEM analysis by Hubalek et al. are presented on following figures: nanotubes on Fig. 4 (left), Fig. 5 (left) and nanowires on Fig. 4 (right), Fig. 6 (left, right).

Fig. 3. Alumina layer with hexagonally arrayed pores (top view): Anodization process in $(COOH)_2$ solution (left) and anodization process in H_2SO_4 solution (right)

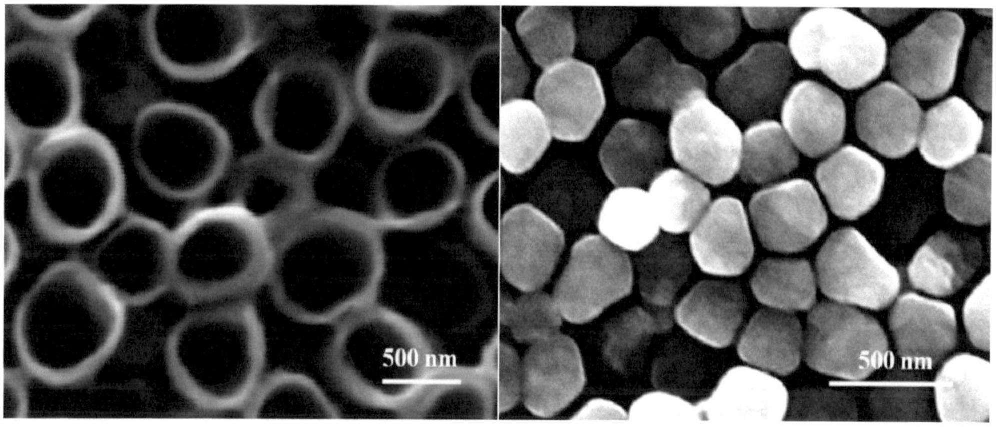

Fig. 4. Nickel nanostructures: Ni nanotubes (left) and Ni nanowires (right)

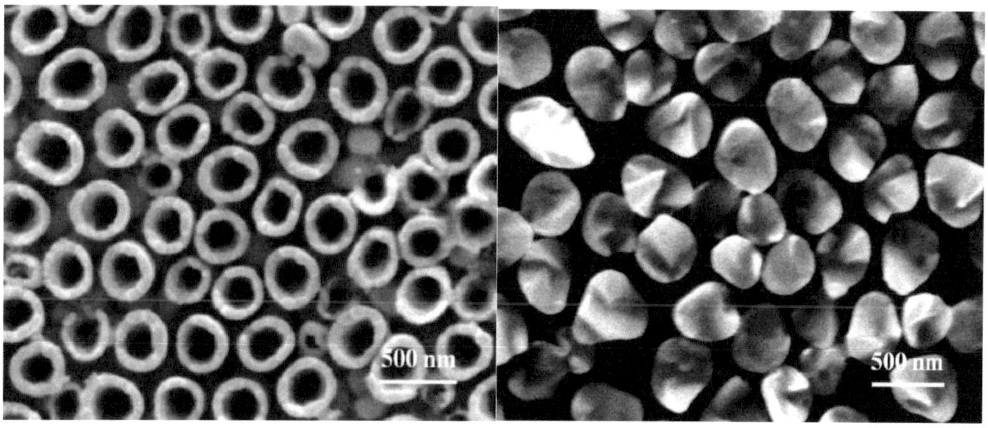

Fig. 5. Gold nanostructures: Au nanotubes (left) and Au nanowires (right)

Fig. 6. Examples of metal nanostructures: Pd long nanowires (left) and Ag nanowires (right)

Very important parameter is the pH of the used electrolyte. At low pH values nanowires are created, while nanotubes grow at high values. It is possible that the formation of nanotubes is somehow connected with the release of gaseous hydrogen at the cathode (the metal layer on the bottom of the pores). This hydrogen release is just affected by the pH of the electrolyte. Among other parameters that were examined, if they have an impact on the type of nanostructures, was the current density. Nevertheless, it was found there was no correlation between the different values of current density and types of nanostructures. Moreover, it is assumed that usage of ultrasound during electrodeposition also affects the type of nanostructure. Even the influence of electrolyte concentration is not precluded (Klosova, 2006a, 2006b).

3.3 Nanotubes created by anodization of metal layers

Most of papers concerning the usage of template based methods for ordered nanotubes fabrication deal with titanium dioxide since this material has a very large surface area and thus it is very attractive for battery, gas sensor, photocatalytic applications, as biomaterial and so on. Li et al. reported a morphologic characterization of the anodic titanium oxide (ATO) films with highly-ordered titania nanotube arrays grown perpendicularly to a titanium foil (L. L. Li, 2010). The ATO film was annealed in an air furnace, forming anatase TiO_2. After $TiCl_4$ post-treatment to enhance the ruthenium-based dye loading on the ATO surface, the optimized ATO arrays were used as a photoanode in a dye-sensitized solar cell. Zhao et al. investigated a novel solid state pH sensor fabricated by anodization of titanium substrate electrode (Zhao, 2010). They found the amorphous TiO_2 nanotubes had better pH response than TiO_2 nanotubes in anatase phase. After being irradiated by ultraviolet light (UV), the potential response of the electrode modified by amorphous TiO_2 nanotube was close to Nernst equation (59 mV/pH). Wang and colleagues employed vertically aligned TiO_2 nanotubes array film produced using electrochemical anodization of Ti foil followed by a nitrogen-doping process in order to form relative humidity sensors (Q. Wang, 2010). These sensors showed resistive and capacitive humidity-sensing properties in the range of 11.3–93.6 %.

Cao et al. discussed a layer-by-layer growth model of anodic TiO_2 nanotube arrays (C. B. Cao, 2010). Many phenomena appeared during the anodization and can be reasonably explained by this model, such as the first sharp slope of current in initial period, current fluctuation, occurring of ridges in adjacent tubes, and the rings broken off from the tube

mouths. Furthermore, key factors which determine the morphology of TNT are discussed which is helpful for the design of nanoarchitectures in related material systems.
Sennik et al. synthesized highly ordered, vertically oriented TiO_2 nanotubes arrays grown by anodic oxidation of titanium foil and investigated their hydrogen sensing properties in the temperature interval of 20–150 °C (Sennik, 2010). The fabricated TiO_2 nanotubes were approximately 1 μm in length and 90 nm in diameter. For the sensor measurements, two platinum pads were used as electrodes on the TiO_2 nanotube arrays. The authors found the sensitivity increased at higher temperature. The sensing mechanism of the TiO_2 nanotube sensor could be explained with chemisorption of H_2 on the highly active nanotube surface.
Similarly like previous authors, Joo et al. examined the formation of TiO_2 nanotube films on pure Ti, Ti-Pt, and Ti-Pd thin films by anodization in a mixture of glycerol and water (1:1 volume ratio) containing 0.5 wt % NH_4F (Joo, 2010). These nanotubes arrays were applied as a resistance-type hydrogen gas sensor. The Pt or Pd particles dispersed in the wall of nanotubes effectively improved the performance of the hydrogen gas sensor perhaps due to the acceleration of hydrogen chemisorption on the wall of the nanotube.
Mun et al. applied TiO_2 nanotube array in label-free sensing of rabbit immunoglobulin G (IgG) using optical interferometry (Mun, 2010). The authors examined the aqueous stability of the TiO_2 nanotube array and compared it with porous silica (SiO_2), which is a more extensively studied thin film optical biosensor. They found that TiO_2 nanotube array is stable in the pH range 2 to 12, whereas the porous SiO_2 sensor displays significant degradation at pH > 8.
Yang et al. studied properties and sensing applications of highly ordered TiO_2 nanotube arrays made by anodic oxidation in fluoride containing electrolytes (L. X. Yang, 2010). The effect of anodization parameters (electrolyte, pH, and voltage) on the titania nanotube size and shape were discussed. The excellent biocompatibility of TiO_2, the high orientation, the large surface area with tunable pore sizes, as well as the high electron transfer rate along with the nanotubes make TiO_2 nanotube array an ideal substrate for the sensor's fabrication and application. The sensors based on the TiO_2 nanotube arrays for sensing hydrogen, oxygen, humidity, glucose and hydrogen peroxide all exhibited low detection limit, high stability, very good reproducibility and high sensitivity.
Tan et al. prepared transparent, well-aligned TiO_2 nanotube arrays on glass substrates via atomic layer deposition of TiO_2 onto free-standing porous anodic alumina template (Tan, 2010). The authors investigated their photocatalytic activity on degradation of aqueous methylene blue solution and solid stearic acid film. They found that Pd functionalized TiO_2 nanotubes revealed the highest photodegradation efficiency thus these nanonstructures might have potential applications in self-cleaning coating, transparent electronics, and solar cells.
Similarly, Liu et al. tested the photocatalytic activity of self-organized, well-crystallized and high aspect-ratio TiO_2 nanotube arrays prepared by anodic oxidation in dimethyl sulfoxide containing 5 wt% HF at 40 V (vs. Pt) (Liu, 2009). The authors compared the TiO_2 nanotubes length influence on photocatalytic degradation of methylic orange. They observed that the photocatalytic efficiency of ultralong nanotubes (19.4 μm) arrays was 1.55 times as high as the short nanotube with length about 500 nm.
Hassan et al. fabricated TiO_2 nanotubes (TiNTs) using electrochemical anodization of titanium in 2-propanol/water containing 0.14 M NH_4F as the supporting electrolyte (Hassan, 2009). The authors studied the effects of the water content, time of anodization and potential on the growth behavior of TiNTs. The TiNTs obtained under optimum conditions had heights of up to 1800 nm and inner diameters of about 90 nm. The ellipsometric characterization confirmed the presence of a thin barrier layer at the nanotubes/metal

interface, which was enriched with the rutile phase, whereas the nanotubes were enriched with the anatase phase. Cho et al. prepared titanium oxide nanotubes via anodization of titanium in various electrolytes: 1 M KH_2PO_4 water solution, glycerine, and ethylene glycol with 0.15 M, 0.17 M or 0.075 M NH_4F (Cho, 2008). The maximum lengths of nanotubes were 3.0 μm in the case of KH_2PO_4 water solution under potential of 25 V, 14 μm in the case of glycerine under potential of 50 V and 164 μm in the ethylene glycol solution under potential of 60 V, respectively. Concerning the TiO_2 nanotubes diameters the smallest one was reached in glycerine (60 nm), then 100 nm using KH_2PO_4 water solution and 150 in ethylene glycol. The nanotubes annealed at 500 °C for 30 min appeared in the anatase phase.
Yoriya et al. described the fabrication of fully separated self-organized titania nanotube arrays by Ti anodization in diethylene glycol containing either HF or NH_4F (Yoriya, 2008). They studied the effect of the fluoride bearing species used in the anodization electrolyte on the tube morphology, degree of tube-to-tube separation, and crystallization.
On contrary to above mentioned papers, Hu et al. described the preparation of highly ordered, vertically oriented TiO_2 nanotube arrays using HF-free aqueous solution (Hu, 2009). The authors investigated the TiO_2 crystalline phase influence on photocurrent generated by an anode consisting of a titanium foil coated by TiO_2 nanotubes and a platinum cathode in an electrochemical cell. It was determined that the anatase crystalline structure converts light into current more efficiently and it is therefore a better photocatalytic material for hydrogen production via photoelectrochemical splitting of water.
Other semiconducting material used for nanotubes fabrication through anodization process was studied by Hahn et al. (Hahn, 2010). Self-organized nanotubular layers of ZrO_2 were electrochemically grown by tailored anodization in an $(NH_4)_2SO_4$ electrolyte containing small amounts of fluoride ions. This semiconducting material is usually used as sensing layer of chemical gas sensors and humidity sensors. Photoluminescence and cathodoluminescence measurements revealed very bright white luminescence of as-grown ZrO_2 nanotubes, hence these nanotubes are suitable for optoelectronic applications.

3.4 Films

Berger et al. demonstrated in their work that highly ordered porous anodic zirconia (PAZ) arrays with cell diameters ranging from 70 to 120 nm can be grown in fluoride containing glycerol electrolytes (Berger, 2008). They showed that this morphology (in contrast to the typically observed nanotubular layers) can simply be obtained by controlling the water content in the electrolyte during the anodization process. It is proposed that the morphology transition from pores to tubes is based on the rate of preferential etching at the hexagonal cell triple points in the oxide.
Zhang prepared the highly ordered TiO_2 thin films by anodic oxidation followed by calcination at various temperatures (300, 400, 500 and 600 °C) (Zhang, 2008). The author investigated the humidity sensing behaviours of prepared samples. The samples calcined at 600 °C showed high sensitivity with nearly two orders change in the resistance and short response and recovery time (< 190 s) during the relative humidity variation from 11 to 95%.
Another method is the deposition of WO_3 thin films on highly ordered nanoporous alumina template. Nanoporous anodic oxide layers were formed by anodizing aluminum films in malonic acid electrolyte. Tungsten trioxide sensing films were deposited on the top of nanoporous alumina layers by rf magnetron sputtering of a metallic target (Fig. 7). The tungsten oxide gas sensing structures supported by nanoporous alumina templates showed high responsiveness to toxic gases, especially to NO_2 (Gorokh, 2006; Khatko, 2009, 2006; Vallejos, 2008).

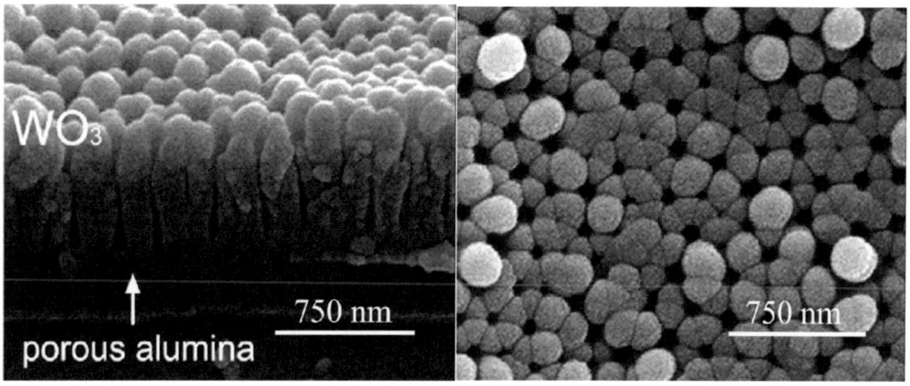

Fig. 7. SEM images of cross-fracture (left) and the surface (right) of alumina films with sputtered WO3

4. Conclusion

Described non-litographic techniques are based on template-assisted method. The template preparation of thin film with highly ordered pores is a suitable way for nanostructured material synthesis since they are cheap, fast and easy reproducible. Due to the special properties arising from their behavior, these highly ordered nanostructures can find various applications in environmental analysis as well as medicine and pharmacy.

In the case of environmental analysis application, the nanostructures are used to modify either the sensing elements from the semiconducting materials of vapor and gas sensors or the electrodes of electrochemical sensors.

Concerning the pharmacy and medicine, quantum dots (QDs) in planar form (so-called lab-on-chip) deposited on various solid surfaces seems to be a new approach of template-based method application. The sensor array created from separately deposited QDs, also called "fluorescence array detector", can be used for in-vitro large-field imaging. This allows the easy detection of many different biomolecules at the same time, since each QD can emit the light at different wavelength. Electrochemical biosensors with functionalized electrodes for rapid detection and mass screening are very promising in near future in cases of pandemic and epidemic. Cultivation of cells on gold nanodots has also high impact in biochemistry research for medicine.

5. Acknowledgment

This work has been supported by Grant Agency of the Academy of Sciencies of the Czech Republic under the contract GAAV KAN208130801 (NANOSEMED) and by Grant Agency of the Czech Republic under the contract GACR 102/08/1546 (NANIMEL).

6. References

Alivov, Y. et al. (2009). Titanium nanotubes grown by titanium anodization. *Journal of Applied Physics*, Vol. 106, No. 3, pp. 5, ISSN 0021-8979

Alonso-Gonzalez, P. et al. (2006). Ordered InAs QDs using prepatterned substrates by monolithically integrated porous alumina. *Journal of Crystal Growth*, Vol. 294, No. 2, pp. 168-173, ISSN 0022-0248

Anitha, V. C. et al. (2010). Electrochemical tuning of titania nanotube morphology in inhibitor electrolytes. *Electrochimica Acta*, Vol. 55, No. 11, pp. 3703-3713, ISSN 0013-4686

Berger, S. et al. (2008). Formation of hexagonally ordered nanoporous anodic zirconia. *Electrochemistry Communications*, Vol. 10, No. 12, pp. 1916-1919, ISSN 1388-2481

Cao, C. B. et al. (2010). Layer-by-Layer Growth Mechanism of TiO_2 Nanotube Arrays. *Journal of the Electrochemical Society*, Vol. 158, No. 1, pp. E8-E11, ISSN 0013-4651

Cao, G. Z. & Liu, D. W. (2008). Template-based synthesis of nanorod, nanowire, and nanotube arrays. *Advances in Colloid and Interface Science*, Vol. 136, No. 1-2, pp. 45-64, ISSN 0001-8686

Gorokh, G. et al. (2006). Anodic formation of low-aspect-ratio porous alumina films for metal-oxide sensor application. *Electrochimica Acta*, Vol. 52, No. 4, pp. 1771-1780, ISSN 0013-4686

Graham, A. H. D. et al. (2010). Nanostructured electrodes for biocompatible CMOS integrated circuits. *Sensors and Actuators B-Chemical*, Vol. 147, No. 2, pp. 697-706, ISSN 0925-4005

Hahn, R. et al. (2010). Bright visible luminescence of self-organized ZrO_2 nanotubes. *Journal of Solid State Electrochemistry*, Vol. 14, No. 2, pp. 285-288, ISSN 1432-8488

Hassan, F. M. B. et al. (2009). Formation of Self-Ordered TiO_2 Nanotubes by Electrochemical Anodization of Titanium in 2-Propanol/NH_4F. *Journal of the Electrochemical Society*, Vol. 156, No. 12, pp. K227-K232, ISSN 0013-4651

Hrdý, R. & Hubálek, J. (2007). Using a Porous Alumina Film as a Mask for Formation of Ordered Nanostructures by Deposition Technique, *Acta Metallurgica Slovaca*, Vol. 13, No. 2, pp. 155-158, ISSN 1335-1532

Hrdy, R. & Hubalek, J. (2005). Self ordered Pore Structure of Anodized Alumina Thin Film on Si Substrate, *Proceedings of Electronic Devices and Systems*, pp. 300-304, ISBN 80-214-2990-9, Brno, Czech Republic, September, 2005

Hu, M. Z. et al. (2009). Synthesis and characterization of anodized titanium-oxide nanotube arrays. *Journal of Materials Science*, Vol. 44, No. 11, pp. 2820-2827, ISSN 0022-2461

Cha, Y. K. et al. (2004). Nonlithographic SiO2 nanodot arrays via template synthesis approach. *Japanese Journal of Applied Physics Part 1-Regular Papers Short Notes & Review Papers*, Vol. 43, No. 8A, pp. 5657-5659, ISSN 0021-4922

Chen, P. L. et al. (2005). Fabrication and field emission characteristics of highly ordered titanium oxide nanodot arrays. *Electrochemical and Solid State Letters*, Vol. 8, No. 10, pp. H83-H86, ISSN 1099-0062

Chen, P. L. et al. (2004). Preparation and phase transformation of highly ordered TiO2 nanodot arrays on sapphire substrates. *Applied Physics Letters*, Vol. 84, No. 19, pp. 3888-3890, ISSN 0003-6951

Chen, P. L. et al. (2003). Self-organized titanium oxide nanodot arrays by electrochemical anodization. *Applied Physics Letters*, Vol. 82, No. 17, pp. 2796-2798, ISSN 0003-6951

Cho, S. J. et al. (2008). Titanium oxide nanotubes anodized in aqueous and non-aqueous electrolytes. *Journal of Ceramic Processing Research*, Vol. 9, No. 5, pp. 449-451, 1229-9162

Chu, S. Z. et al. (2005). Self-organized nanoporous anodic titania films and ordered titania nanodots/nanorods on glass. *Advanced Functional Materials*, Vol. 15, No. 8, pp. 1343-1349, ISSN 1616-301X

Chu, S. Z. et al. (2005). A new electrochemical lithography - Fabrication of self-organized titania nanostructures on glass by combined anodization. *Journal of the Electrochemical Society*, Vol. 152, No. 3, pp. B116-B124, ISSN 0013-4651

Joo, S. et al. (2010). Hydrogen Gas Sensor Using Pt- and Pd-Added Anodic TiO2 Nanotube Films. Journal of the Electrochemical Society, Vol. 157, No. 6, pp. J221-J226, ISSN 0013-4651

Jung, J. S. et al. (2008). Electrodeposited Nickel Nanodots Array on the Silicon Wafer. Bulletin of the Korean Chemical Society, Vol. 29, No. 11, pp. 2169-2171, ISSN 0253-2964

Jung, M. et al. (2006). Fabrication of the uniform CdTe quantum dot array on GaAs substrate utilizing nanoporous alumina masks. *Current Applied Physics*, Vol. 6, No. 6, pp. 1016-1019, 1567-1739

Kemell, M. et al. (2007). Atomic layer deposition of nanostructured TiO_2 photocatalysts via template approach. *Chemistry of Materials*, Vol. 19, No. 7, pp. 1816-1820, ISSN 0897-4756

Khatko, V. et al. (2006). Tungsten trioxide sensing layers on highly ordered nanoporous alumina template. *Sensors and Actuators B-Chemical*, Vol. 118, No. 1-2, pp. 255-262, 0925-4005

Khatko, V. et al. (2009). Micro-machined WO3-based sensors with improved characteristics. *Sensors and Actuators B-Chemical*, Vol. 140, No. 2, pp. 356-362, ISSN 0925-4005

Klosová, K. & Hubálek, J. (2008). Advanced electrodes with nanostructured surfaces for electrochemical microsensors, *Physica Status Solidi*, Vol. 205, No. 6, pp. 1435-1438, ISSN 0031-8965

Klosova, K. et al. (2006). New Microelectrodes for Electrochemical Application with Nanomachined Surface, *Proceedings of the International Conference NANO'06*, pp. 210-214, ISBN 80-214-3331-0, Brno, Czech Republic, November, 2006

Klosova, K. et al. (2006). New Approach to Electrochemical Sensor Electrodes Construction, *Proceedings of Junior Scientist Conference*, pp. 139-140, ISBN 3-902463-05-8, Vienna, Austria, April, 2006

Kokonou, M. et al. (2007). Few nanometer thick anodic porous alumina films on silicon with high density of vertical pores. *Thin Solid Films*, Vol. 515, No. 7-8, pp. 3602-3606, ISSN 0040-6090

Kouklin, N. et al. Capacitance-voltage spectroscopy of self assembled ordered arrays of quantum dots. New York: Ieee, 2000.

Li, A. P. et al. (1998). Hexagonal pore arrays with a 50-420 nm interpore distance formed by self-organization in anodic alumina. *Journal of Applied Physics*, Vol. 84, No. 11, pp. 6023-6026, ISSN 0021-8979

Li, C. P. et al. (2006). Fabrication and structural characterization of highly ordered sub-100-nm planar magnetic nanodot arrays over 1 cm(2) coverage area. *Journal of Applied Physics*, Vol. 100, No. 7, pp. 7, ISSN 0021-8979

Li, L. L. et al. (2010). Morphologic Characterization of Anodic Titania Nanotube Arrays for Dye-Sensitized Solar Cells. *Journal of the Chinese Chemical Society*, Vol. 58, No. 5B, pp. 1147-1150, ISSN 0009-4536

Liang, J. Y. et al. (2002). Two-dimensional lateral superlattices of nanostructures: Nonlithographic formation by anodic membrane template. *Journal of Applied Physics*, Vol. 91, No. 4, pp. 2544-2546, ISSN 0021-8979

Lim, J. H. et al. (2009). Electrochemical determination of whole blood clotting time by using nanodot arrays. *Electrochemistry Communications*, Vol. 11, No. 11, pp. 2141-2144, ISSN 1388-2481

Liu, Y. B. et al. (2009). Comparison of photoelectrochemical properties of TiO2-nanotube-array photoanode prepared by anodization in different electrolyte. *Environmental Chemistry Letters*, Vol. 7, No. 4, pp. 363-368, ISSN 1610-3653

Mao, R. et al. (2009). In situ preparation of an ultra-thin nanomask on a silicon wafer. *Nanotechnology*, Vol. 20, No. 2, pp. 6, ISSN 0957-4484

Masuda, H. & Fukuda, K. (1995). Ordered metal nanohole arrays made by a 2-step replication of honeycomb structures of anodic alumina. *Science*, Vol. 268, No. 5216, pp. 1466-1468, ISSN 0036-8075

Masuda, H. et al. (1998). Self-ordering of cell configuration of anodic porous alumina with large-size pores in phosphoric acid solution. *Japanese Journal of Applied Physics Part 2-Letters*, Vol. 37, No. 11A, pp. L1340-L1342,
Matefi-Tempfli, S. et al. (2009). Nanowires and nanostructures fabrication using template methods: a step forward to real devices combining electrochemical synthesis with lithographic techniques. *Journal of Materials Science-Materials in Electronics*, Vol. 20, No., pp. 249-254, ISSN 0957-4522
Montero-Moreno, J. M. et al. (2009). Production of alumina templates suitable for electrodeposition of nanostructures using stepped techniques. *Electrochimica Acta*, Vol. 54, No. 9, pp. 2529-2535, ISSN 0013-4686
Mozalev, A. et al. (2009). Growth of multioxide planar film with the nanoscale inner structure via anodizing Al/Ta layers on Si. *Electrochimica Acta*, Vol. 54, No. 3, pp. 935-945, ISSN 0013-4686
Mun, K. et al. (2010). A Stable, Label-free Optical Interferometric Biosensor Based on TiO2 Nanotube Arrays. *Acs Nano*, Vol. 4, No. 4, pp. 2070-2076, ISSN 1936-0851
Oide, A. et al. (2006). Fabrication of ordered nanostructure on silicon substrate using localized anodization and chemical etching. *Electrochemistry*, Vol. 74, No. 5, pp. 379-384, ISSN 1344-3542
Possin, G. E. (1970). A method for forming very small diameter wires. Review of Scientific Instruments, Vol. 41, No. 5, pp. 772-&, ISSN 0034-6748
Sennik, E. et al. (2010). Synthesis of highly-ordered TiO_2 nanotubes for a hydrogen sensor. *International Journal of Hydrogen Energy*, Vol. 35, No. 9, pp. 4420-4427, ISSN 0360-3199
Shingubara, S. (2003). Fabrication of nanomaterials using porous alumina templates. *Journal of Nanoparticle Research*, Vol. 5, No. 1-2, pp. 17-30, ISSN 1388-0764
Song, Y. Y. & Schmuki, P. (2010). Modulated TiO_2 nanotube stacks and their use in interference sensors. *Electrochemistry Communications*, Vol. 12, No. 4, pp. 579-582, ISSN 1388-2481
Tan, L. K. et al. (2010). Transparent, Well-Aligned TiO_2 Nanotube Arrays with Controllable Dimensions on Glass Substrates for Photocatalytic Applications. *Acs Applied Materials & Interfaces*, Vol. 2, No. 2, pp. 498-503, ISSN 1944-8244
Vallejos, S. et al. (2008). Micro-machined WO3-based sensors selective to oxidizing gases. *Sensors and Actuators B-Chemical*, Vol. 132, No. 1, pp. 209-215, ISSN 0925-4005
Vorozhtsova, M. et al. (2010). Ta2O5 Nanocrystals Created by Anodization, *Proceedings of X. Workshop of Physical Chemists and Electrochemists*, pp. 259 - 261, ISBN 978-80-7375-396-2, Brno, Czech Republic, June, 2010
Wang, A. W. & White, R. M. (1995). Thin-film anodized aluminum on an acoustic sensor. In: *Ieee Ultrasonics Symposium Proceedings, Vols 1 and 2*, Levy, Schneider, McAvoy, pp. 437-440, IEEE, ISBN 1051-0117, New York
Wang, C. C. et al. (2007). Organic nanowire-templated fabrication of alumina nanotubes by atomic layer deposition. *Nano Letters*, Vol. 7, No. 6, pp. 1566-1569, ISSN 1530-6984
Wang, H. W. et al. (2006). Standing [111] gold nanotube to nanorod arrays via template growth. *Nanotechnology*, Vol. 17, No. 10, pp. 2689-2694, ISSN 0957-4484
Wang, J. G. et al. (2004). Microstructure and interdiffusion of template-synthesized Au/Sn/Au junction nanowires. *Nano Letters*, Vol. 4, No. 7, pp. 1313-1318, ISSN 1530-6984
Wang, Q. et al. (2010). Resistive and capacitive response of nitrogen-doped TiO2 nanotubes film humidity sensor. *Nanotechnology*, Vol. 22, No. 2, pp. 11, ISSN 0957-4484
Wang, Y. et al. (2005). Synthesis and electrochemical properties of vanadium pentoxide nanotube arrays. *Journal of Physical Chemistry B*, Vol. 109, No. 8, pp. 3085-3088, ISSN 1520-6106

Part 3

Optical Microsensors

7

Surface-Enhanced Raman Scattering Sensors based on Hybrid Nanoparticles

Rafael Contreras-Cáceres, Benjamín Sierra-Martín and Antonio Fernández-Barbero
Applied Physics Department, University of Almería
Spain

1. Introduction

Surface-enhanced Raman scattering (SERS) is a powerful vibrational spectroscopic technique that allows ultra-sensitive chemical or biochemical analysis (Kneipp, Kneipp et al. 1999). It works by increasing the Raman signal of analyte molecules located nearby the surface of metallic nanostructures that can undergo localized surface plasmon resonance. Among these nanostructures, gold and silver nanoparticles are the dominant substrates, for both experimental and theoretical perspectives (Kneipp, Wang et al. 1997; Nie and Emery 1997), since they can support plasmon resonance properties able to increase the Raman signal up to 14 or 15 orders of magnitude, high enough to detect single molecules (Nie and Emery 1997; Qian and Nie 2008). Since the first report concerning the enhanced Raman signal of pyridine molecules adsorbed on a roughened silver electrode (Fleischm, Hendra et al. 1974), considerable efforts have been made in understanding the SERS mechanisms (Schatz 1984; Campion and Kambhampati 1998). Nowadays, analytical applications have centred the attention, and research is devoted to optimize the specific conditions for detecting each particular analyte (Porter, Lipert et al. 2008). Interestingly, the enhancement factor is found to depend on the different affinity of the functional groups in the analyte toward gold or silver surfaces because it is the affinity which determines the analyte retention (Pearson 1963; Pearson 1966). To improve the surface-analyte interaction, various approaches have been developed, including the functionalization of nanoparticle surface (Guerrini, Garcia-Ramos et al. 2006; Guerrini, Garcia-Ramos et al. 2008); however, a problem inherent to this alternative is that usually the assembled molecules provide strong SERS signals that overlap and screen those corresponding to the analyte. Another alternative relies on controlling the surface charge of the nanoparticles to promote the electrostatic attraction of the analyte onto the particle surface (Alvarez-Puebla, Arceo et al. 2005; Aroca, Alvarez-Puebla et al. 2005). This approach has been reported to consistently enhances the signal for acids and amines, but it hardly helps in the case of alcohols, ethers, and other oxygen containing groups, as well as for non-functionalized molecules. Thereby, there is a clear need for development of new nanocomposites, based on noble-metals, containing a sensitive material that enables the physical trapping of a wide variety of analyte molecules.

Herein we present the synthesis and applications of novel core-shell nanocomposites comprising Au and Au-Ag bimetallic cores, with spherical or rod-shaped morphology, coated with thermally responsive poly-(N-isopropylacrylamide) (pNIPAM) microgel (Contreras-Caceres, Sanchez-Iglesias et al. 2008). In these systems, whereas the metallic core provides the necessary enhancing properties, the pNIPAM shell, that can swell or collapse as a function of temperature, is used to trap the analyte molecules and get them sufficiently close to the core. These materials present unique optical properties as a consequence of the thermally responsive surface plasmon resonance, which can be ultimately exploited for SERS analysis. Although similar systems have been proposed for applications in catalysis (Lu, Mei et al. 2006), temperature or pH sensing (Kim and Lee 2004), or light-responsive materials (Gorelikov, Field et al. 2004), we report here that the hybrid nanoparticles can function as general sensors for detecting different types of analytes. Apart from the SERS enhancement, these nanocomposites can also be used to modulate the fluorescence intensity of adsorbed chromophores as a function of temperature. It is important to note, that the pNIPAM shell not only enhances the colloidal stability of the system in aqueous solutions, but additionally prevents electromagnetic coupling between metal particles, thus providing highly reproducible SERS signal and intensity, which is crucial for quantitative applications. Through a rational choice of model analytes, we report the applications of these thermoresponsive hybrid materials for Surface Enhanced Raman Scattering and Fluorescence (SERS and SEF, respectively). The nanocomposites are first tested using 1-naphthalenethiol (1NAT) as a model analyte with large affinity for gold, and consecutively against a common dye, Nile Blue A (NBA), whose affinity for gold is lower than of 1NAT. In addition, we present the SERS analysis of 1-napthol, a substance that had remained elusive for SERS since it does not easily adsorb onto conventional silver or gold surfaces and whose detection is decisive because is considered a relevant biomarker (Hansen, Omland et al. 1994; Sun, Shen et al. 2008) and also causes genotoxicity under chronic exposure to humans (Kozumbo, Agarwal et al. 1992; Grancharov, Engelberg et al. 2001). To conclude the report, the SERS efficiency of the different hybrid nanocomposites is compared for a couple of analytes. The wide range of systems investigated, lead us to establish the effect of parameters, such as particle morphology or core composition, on the detection capabilities. Interestingly, sensors based on Au-Ag core coated by the pNIPAM shell are found to provide much higher SERS intensities than their Au-pNIPAM counterparts, not only in the case of spheres but particularly for nanorods.

2. Plasmon resonance and surface-enhanced Raman scattering

Plasmons are quantized collective oscillations of the free electron gas density that occurs between any two materials whose dielectric function changes sign across the interface, for instance metal-dielectric interfaces (Barnes, Dereux et al. 2003). Surface plasmons are those confined to surfaces; they can strongly couple with photons resulting in surface polaritons, which are considered quasi-particles that propagate along the metal surface until its energy decays via absorption into the metal or radiation into the free-space (Zayats, Smolyaninov et al. 2005). Light or electric fields can excite those plasmons, then resulting in surface and localized surface plasmon resonance (SPR and LSPR) in the case of planar and nanometric-sized metallic structures, respectively (Mulvaney 1996). Plasmon oscillation is resonant with the light at a particular frequency. The electric field intensity, the scattering and the adsorption cross-sections are then enhanced. Materials exhibiting surface plasmon properties are used to maximize surface sensitive spectroscopic techniques, such as Raman scattering or fluorescence (Hutter and Fendler 2004). The resonance frequency strongly depends on the size and shape of the metal nanoparticles, as well as, on the metal complex dielectric function and surrounding medium. Noble metals such as copper, silver, and gold exhibit strong visible-light plasmon resonance, whereas other transition metals show only a broad and poorly resolved absorption band in the ultraviolet region (Link and El-Sayed 1999). To understand the optical properties of these metals, it is not only necessary to account for the effect of free-electrons, responsible for plasmon resonance, but also for the interband transitions (Wang, Tam et al. 2005). For instance, copper nanoparticles have strong interband transitions which overlap with the plasmon resonance energies, then leading to a damping effect that minimizes its optical response. Contrarily, in case of gold and silver nanoparticles, both effects are well separated in the spectrum. Therefore, electrons of the conduction band can move freely, showing higher polarizability. This fact, in turn shifts the plasmon resonance to lower frequencies with sharp bandwidth. Since copper is also easily oxidized, gold and silver nanoparticles are more attractive for optics-based applications, specifically silver since it has by far the strongest plasmon resonance. In this case, the higher plasmon energy respect to that of the interband transition results in minimum damping effect (Johnson and Christy 1972). Localized plasmon resonance is responsible for the intense colour of metal nanoparticle dispersions (Bohren and Huffman 1983); the resultant absorption bands are exploited for technical applications like photovoltaic cells (Pillai, Catchpole et al. 2007). In other applications, it is desirable to tune the plasmon resonance depending on the availability of a suitable laser to enhance the optical properties (Willets and Van Duyne 2007; Homola 2008). Although increasing spherical-nanoparticle size causes red-shift due to electromagnetic retardation, the range of frequencies is quite limited (Jain, Huang et al. 2008). Alternatively, LSPR can be tuned by changing the particle morphology, from spherical to rod-shaped. Metal nanorods show typically two resonance peaks corresponding to plasmon oscillations along the short and long axis (Murphy, San et al. 2005). As the aspect ratio, defined as the length-to-width ratio, is increased, the LSPR associated to the long axis is red-shift from visible to near infrared region. The same effect can be achieved by coating a solid sphere with metallic shells (Oldenburg, Averitt et al. 1998); the LSPR frequency decreases as the ratio shell thickness-core size reduces, being the relation almost exponential regardless of the core and shell composition (Jain and El-Sayed 2007).

Surface-enhanced Raman scattering is based on the enhancement of Raman signal induced by plasmonic metal surfaces on nearby molecules (Otto, Mrozek et al. 1992). The extent of enhancement depends on the shape and size of the metal nanoparticles, as these factors influence the ratio of absorption and scattering events (Bao, Mahurin et al. 2003). Large particles allow multipole excitation, which are nonradiative modes; since only dipolar transitions contribute to Raman scattering, the overall efficiency of the enhancement is then reduced. On the other side, too small particles lose their electrical conductance and cannot enhance the field. When the size approaches a few atoms, the definition of plasmon, which involves a large collection of electrons to oscillate together, does not hold (Moskovits 2006). The enhancement factor is maximum for nano-structured metals (10-100 nm) (Tian, Ren et al. 2002), being thus excellent materials for SERS. The exact mechanism accounting for the enhancement effect is still a matter of debate (Qian and Nie 2008). Although several models have been proposed in the literature, nowadays, two mechanisms are accepted (Campion and Kambhampati 1998): electromagnetic and chemical. The first one relies on the excitation of localized surface plasmon on metal surfaces, whereas the second one proposes changes of the molecule electronic structure (Vo-Dinh 1998). The chemical enhancement only applies in specific cases that cannot be explained by the electromagnetic mechanism. For instance, the SERS signal for molecules with almost the same polarizability differs in a factor of 200, even though the electromagnetic model establishes a non selective amplification. Moreover, the transition HOMO-LUMO for many molecules (between the highest occupied and the lowest unoccupied molecular orbital) can occurs at about half the energy in SERS experiments. The formation of charge transfer complexes or chemical bounds between the metal surface and the analyte (necessary in the chemical model) can explain the observed enhancement (Lombardi, Birke et al. 1986). The chemical mechanism is short ranged (0.1-0.5 nm) and strongly dependent on the geometry of bonding and on the molecule energy level. Anyhow, the contribution of this mechanism to SERS is relatively small; its enhancement is estimated to be a factor 10^1-10^3. Instead, the electromagnetic mechanism is the dominant contribution in SERS. It accounts for more situations, even for molecules not adsorbed on the surface (Xu, Aizpurua et al. 2000). According to the electromagnetic model (Fig. 1), the incident electromagnetic field, E_i, excites the surface plasmons and induces oscillating dipoles given by $\mu(t) = \propto E_i(t)$, being $\propto$ the polarizability tensor (Stevenson and Vo-Dinh 1996). The plasmon oscillations must be perpendicular to the surface, otherwise scattering does not occurs (Smith and Dent 2005). The induced polarization generates large local fields on the particle; it magnifies the incident field, thus increasing the Raman signal . On a second step, the emitted Raman field, E_R, can also polarize the metal particle, which thereby acts as an antenna to further intensify the Raman signal. The Raman intensity is proportional to the square field . As a result of the first process, the intensity enhances $(E_i+E_{i,s})^2$, whereas for the second step, it increases $(E_R+E_{R,s})^2$ (Schlucker 2009). The frequency-shift between the incident light and the Raman signal

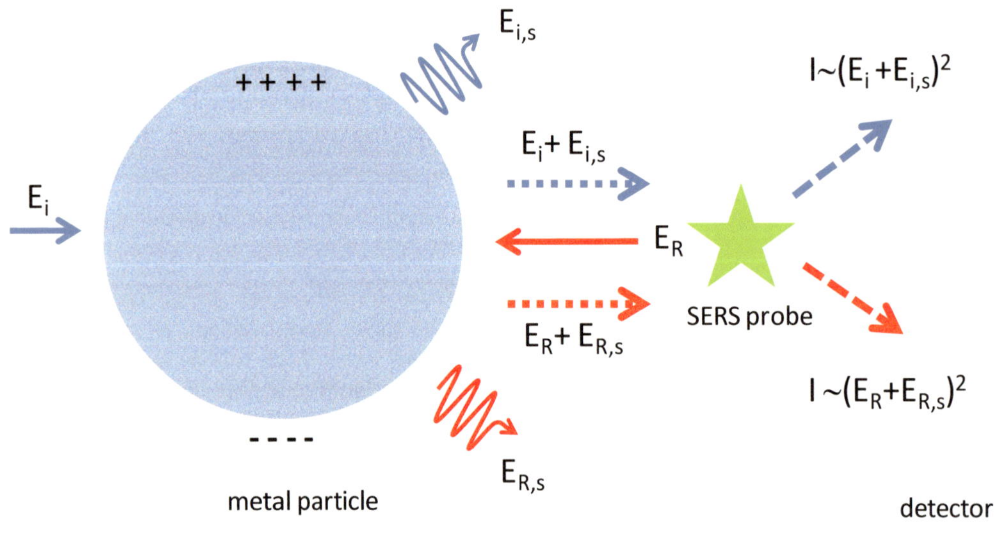

Fig. 1. Schematic representation of the electromagnetic enhancement mechanism in SERS. There are two processes: i) (blue) the incident field, E_i, is enhanced due to the addition of the scattered field, $E_{i,s}$. It arises from the particle polarization, yielding the field $E_i + E_{i,s}$ and exciting Raman modes on the probe molecule; ii) (red) the emitted Raman field, E_R, is also enhanced through the same mechanism. The improved field $E_R + E_{R,s}$ further enhances the Raman intensity, I. Raman signal is enhanced E^2 in each process.

determines the extent of enhancement. For small shifts, both fields can be nearly resonant with the surface plasmon, then establishing a total enhancement of E^4. Gold and silver are mostly used for SERS because their plasmon resonance frequencies fall within the wavelength range used to excite Raman modes, namely visible and near-infrared regions of the spectrum. SERS enhancement also depends on the distance between the nanoparticle centre and the target molecule, R; it falls off as ~ $(r/R)^{12}$, where r is the particle radius (Stiles, Dieringer et al. 2008). This expression indicates that molecule is not required to be just in contact with the surface. Especially useful are those nanostructures capable of controlling R, in order to tune the Raman enhancement in an accurate way (Alvarez-Puebla, Contreras-Caceres et al. 2009).

3. Hybrid nanoparticles: Synthesis and optical properties

3.1 Au-pNIPAM core-shell nanoparticles

Spherical gold nanoparticles are typically prepared in a wide range of size from aqueous solutions of cationic surfactant cetyltrimethylammonium bromide (CTAB) using a seeded growth method, in which ionic precursors are reduced by ascorbic acid in the presence of preformed gold seeds (Jana, Gearheart et al. 2001; Rodriguez-Fernandez, Perez-Juste et al. 2006). The reducing agent role can also be played by other reducing acids, such as salicylic acid, producing similar effect; salicylic acid reduces Au^{III}-CTAB complexes to Au^{I}-CTAB, although the final reduction into Au^0 only happens on the catalytic gold surface (Okamoto and Hachisu 1977). In this chapter we review a novel procedure based on butenoic acid as reducing agent (Contreras-Caceres, Pastoriza-Santos et al.); it is supported on the reducing capability of the vinyl group, present in butenoic acid, toward $HAuCl_4$, used as gold salt precursor (Yavuz, Li et al. 2009). Indeed, the use of butenoic acid is particularly interesting because it provides the particles with vinyl functionality, then making this method useful for direct pNIPAM polymerization on the particle surface and their subsequent encapsulation, avoiding any surface functionalization step. Butenoic acid replaces CTAB molecules from the particle-solvent interface and absorbs on it, thereby promoting polymerization of pNIPAM shells. This may be used on a wide set of preformed particles with interesting morphologies, including nanorods. The polymer shell porosity allows subsequent reduction of metal atoms above the core; this is used as strategy for overgrowing the gold cores encapsulated into pNIPAM, with both spherical and rod-shaped morphologies.

Au-sphere(64nm)@pNIPAM particles. Preparation is based on a seeded growth method using 15 nm Au seed particles, previously prepared under citrate reduction. Briefly, 35 mL of this particle suspension is added to a CTAB aqueous solution (15 mL, 0.03 M). A volume of 4.5 mL of the seed solution is then added to a growth solution prepared under gentle magnetic stirring at 70ºC, with 800 µL of butenoic acid and 50 mL of aqueous solution containing $HAuCl_4$ (1 mM) and CTAB (15 mM). After 10 minutes, the excess of butenoic acid and CTAB is removed by centrifugation (4500 rpm, 30 min). Next, the pellet is re-dispersed into CTAB (50 mL, 4 mM), centrifuged (4500 rpm, 40 min), and the precipitated re-dispersed again into water (10 mL). The resultant dispersion is heated at 70ºC under N_2 flow while the monomers, NIPAM (0.1698 g) and N,N'-methylenebisacrylamide, BIS, (0.0234 g) are added under magnetic stirring. After 15 min, the nitrogen flow is stopped and the polymerization initiated with 2,2'-azobis(2-methylpropionamidine) dihydrochloride, AAPH, (100 µL, 0.1 M). The reaction is extended for 2 h and then cool down to room temperature under stirring. The sample is finally diluted into water (50 mL), centrifuged (4500 rpm, 30 min) and re-dispersed into water (5 fold). The gold core size measured by TEM is 63.9 ± 5.9 nm.

Au-sphere(103nm)@pNIPAM particles. The particles are prepared by growing the gold cores of the particles previously described. Ascorbic acid, AA, (180 µL, 0.1M) is added, under magnetic stirring, to 10 mL of aqueous solution containing Au-spheres(64nm)@pNIPAM (0.25 mM), CTAB (0.1 M) and $HAuCl_4$ (0.9 mM). After 30 min, the growth process is concluded, yielding particles with 103.0 ± 8.4 nm average core size .

Au-nanorod@pNIPAM particles. A gold nanorod suspension is first synthesized following the seed-mediated method reported here (Liu and Guyot-Sionnest 2005). (70.0 ± 7.2) nm long x (17.6 ± 3.1) nm wide gold particles are obtained, which gives aspect ratio AR=4.1. The dispersion (100 mL) is centrifuged (7000 rpm, 1 h) to remove the excess of CTAB and re-dispersed into water (100 mL). Next, butenoic acid (300 µL) is added to this solution at 70ºC. After 1 hour, the dispersion is again centrifuged (6000 rpm, 1 h) to remove any excess of butenoic acid and re-dispersed into water (10 mL). Encapsulation with pNIPAM is performed following the method described for Au(64nm)@pNIPAM particles, differing only on the amount of monomer, NIPAM (0.1924 g) and BIS (0.0265). Once the polymerization is completed, the dispersion is diluted into water (50 mL), centrifuged (5500 rpm, 1 h) and re-dispersed into water (5 fold).

Au-nanorod@Au@pNIPAM particles. The growth of Au-nanorod@pNIPAM particles is carried out as previously described for Au-sphere(103nm)@pNIPAM, but using 0.8 mM $HAuCl_4$ and 160 µL of 0.1 M AA. The final core size is (82.1 ± 7.9) nm x (21.6 ± 2.6) nm, with aspect ratio AR= 3.6. The synthesis using butenoic acid leads to extremely efficient coverage of the gold core (Fig. 2a). TEM image shows that all the particles are perfectly encapsulated, confirming that the presence of vinyl groups on the particle periphery promotes uniform pNIPAM polymerization on the particle surfaces, as previously observed for gold and silica particles (Zha, Zhang et al. 2002; Guerrini, Garcia-Ramos et al. 2006; Karg, Pastoriza-Santos et al. 2006; Guerrini, Garcia-Ramos et al. 2008). The morphology is clearly visible from TEM images. The intense core black-color is due to the higher electron density respect to the organic shell. Further information about the core-shell structure is obtained by atomic force microscopy (AFM), as illustrated by the inset on Fig. 2a. The image is taken in tapping mode on a dried sample, previously prepared on silicon wafer; during dehydration, pNIPAM shell spreads while the gold core stands out from the polymer coverage. The size and plasmonic response of the core-shell nanocomposites may be tuned with temperature in a reversible way. Fig. 2b shows the UV/Vis-NIR spectra, obtained in the range 16ºC (swollen shell) and 56ºC (collapsed shell). As temperature increases, the plasmon band high increases and red-shifts. Both effects are associated to the increase of the microgel shell refractive index during collapse, with enhancement of the Rayleigh scattering (Kozumbo, Agarwal et al. 1992; Grancharov, Engelberg et al. 2001; Karg, Pastoriza-Santos et al. 2007). The localized surface plasmon resonance (LSPR) shifts the peak position with temperature, inversely to the size variation, as shown in Fig. 2c. The pNIPAM shell lower critical solution temperature (LCST) is similar to that reported for pure microgel (Kratz, Hellweg et al. 2001), while the LSPR bands are consistently red-shifted as microgel shell collapses due to the increase on the refractive index around the metal core. As expected, the LSPR position is nicely correlated to the hydrodynamic diameter.

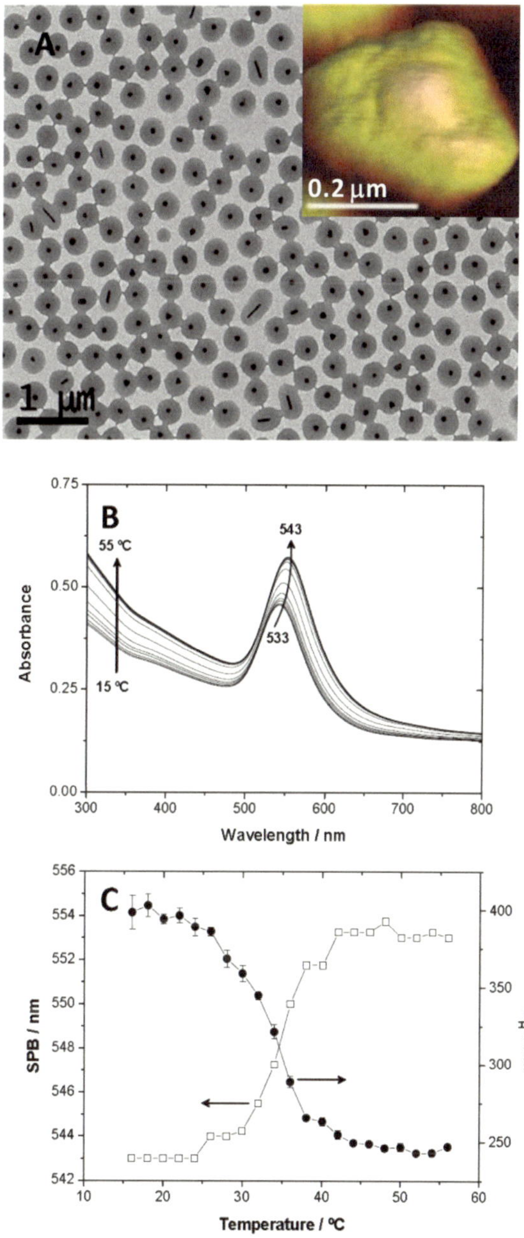

Fig. 2. Characterization of the system Au-sphere(64nm)@pNIPAM. (A) TEM image (microscope JEOL JEM 1010, operating at an acceleration voltage of 100 kV). The inset corresponds to a tapping mode AFM image (Veeco Nanoscope, spring constant k = 20-80 Nm^{-1}; (B) UV-vis spectra at different temperatures (Cary 5000 spectrophotometer); (C) Localized surface plasmon band position (SPB) (squares) and hydrodynamic diameter (circles) temperature dependence. Particle size is measured by dynamic light scattering (Malvern Zetasizer Nano S) using cumulant-type analysis. Reprinted with permission from (Contreras-Caceres, Pastoriza-Santos et al.), Copyright (2010) by Wiley-VSC Verlag GmbH & Co. KGaA

3.2 Bimetallic-pNIPAM core-shell nanoparticles

An advantage of using core-shell gold-pNIPAM particles is the excellent colloidal stability and porosity provided by the shell, especially when the polymer is swollen. They can behave as microreactors where chemical reactions can be performed. In addition, they allow further reduction of metals on the gold core, addressed not only to control the core size (Contreras-Caceres, Pacifico et al. 2009) but also the core nature (Sanchez-Iglesias, Grzelczak et al. 2009). Inasmuch that silver is significantly much efficient for SERS than gold, it is important to incorporate this metal to the nanocomposites. Direct synthesis of size and shape-controlled silver nanoparticles is complicated. An alternative method is to grow silver shells on preformed gold nanoparticles with both spherical or rod-shaped core. The resultant nanocomposites display optical properties close to those of pure silver (Cho, Camargo et al.). The method simply involves the reduction of $AgNO_3$ using ascorbic acid in the presence of the corresponding particle as template.

Au-sphere(64nm)@Ag@pNIPAM particles. Preparation relies on the process for coating pure gold particles with silver (Yang, Lin et al. 2005). A dispersion of preformed Au-sphere(64 nm)@pNIPAM particles (10 mL, 0.25 mM) containing CTAB (50 mM) is set to pH=9.5 with glycine buffer. Next, $AgNO_3$ aqueous solution (655 µL, 15.2 mM) and ascorbic acid (450 µL, 0.1 M) is added. The reduction reaction is extended over 30 minutes, then rendering a bimetallic 100.3 ± 9.2 nm Au@Ag core size.

Au-nanorod@Ag@pNIPAM particles. The growth process is performed as previously described for Au-sphere(64nm)@Ag@pNIPAM, being the only difference the amount of $AgNO_3$ (540 µL, 14.83 mM) and ascorbic acid (360 µL, 0.1 M) aqueous solutions. The final core size for the bimetallic nanorods is (76.7 ± 7.7) nm long x (40.0 ± 5.1) nm wide, with aspect ratio AR=1.9. In previous works NaOH was used to increase the solution pH, leading to the nucleation of small Ag nanoparticles, and consequently to percentage reduction of core-shell particles (Rodriguez-Gonzalez, Burrows et al. 2005). However, for the present method using a glycine buffer, the reduction slows down (Pearson 1963; Pearson 1966), then avoiding the presence of small silver nuclei. This growing method is very efficient, not only for silver growing but also for gold. Fig. 3 (A and C) shows representative TEM images of particles with Au cores (spheres and rods, respectively), whereas Fig. 3 (B and D) illustrates those formed with silver shell. It is noticeable the intensity contrast existing between Au core and Ag shell; it arises from the electron scattering difference between both metals and clearly shows the uniform coating for both geometries. The overall dimensions for the particles, including the pNIPAM shell, are similar before and after the different core growth processes (Contreras-Caceres, Pastoriza-Santos et al.). The different nanocomposites are expected to have different optical properties. Fig. 3 shows pictures corresponding to the aqueous dispersions; it can be observed by simple visual inspection how color changes, indicating shifting of the localized surface plasmon peak.

The optical signatures for the different nanocomposites are recorded by UV-visible spectroscopy and summarized in Fig. 4. It is noticeable (Fig. 4A) that, whereas the growth of Au spherical cores, from 64 to 103 nm, leads to significant redshift due to retardation effects (Rodriguez-Fernandez, Perez-Juste et al. 2006), the coating with Ag (of almost the same thickness), remarkably results in blue-shift and formation of quadrupolar modes, in agreement with previous works (Rodriguez-Gonzalez, Burrows et al. 2005). In the case of rod-shaped cores growing with gold (Fig. 4B), there is no displacement of the localized

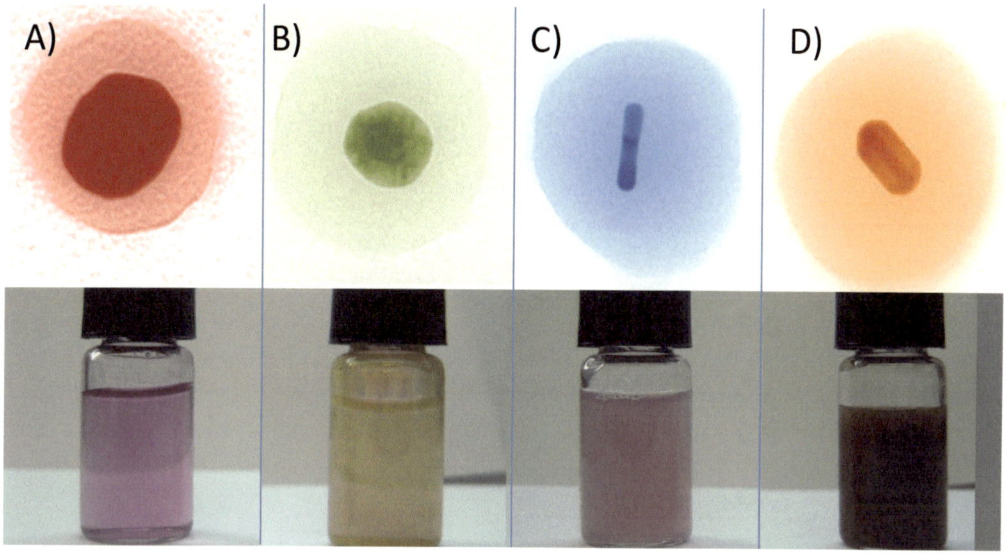

Fig. 3. (Up) Representative TEM images of hybrid nanoparticles with different morphologies; color is retouched to emphasize the core-shell structure. (Down) Pictures of their corresponding aqueous dispersions: (A) Au-sphere(103nm)@pNIPAM; (B) Au(64nm)@Ag(36nm)-sphere@pNIPAM; (C) Au-nanorod@pNIPAM (82.1nm x 21.6 nm); (D) Au-Ag-nanorod@pNIPAM (76.7 nm x 40.0 nm).

plasmon resonance, only the band intensity increases; this demonstrates that the growth process is uniform, with no significant change of the aspect ratio. For silver growth, a dramatic blue-shift of the longitudinal LSPR appears, not only due to the presence of silver on the surface, but also because the growth process occurs preferentially at the rod sides, in agreement with recent reports (Sanchez-Iglesias, Carbo-Argibay et al.). The core-grown nanocomposites preserve the original thermo-responsive properties of the original core-shell particles used as template. Not only the size can be tuned through temperature changes, but also the optical properties are greatly influenced. In general, there is a red-shift of LSPR band as temperature rises above the LCST of 32°C, as a consequence of the polymer shell collapse. The extent of this red-shift depends on the metal core size, shape, and composition. These three factors are known to dominate the LSPR sensitivity toward changes of the local refractive index (Chen, Kou et al. 2008; Sepulveda, Angelome et al. 2009). For instance, within the explored range of temperature, the LSPR shift for Au-sphere(64 nm)@pNIPAM particles is 10 nm, whereas for Au(64 nm)@Ag@pNIPAM increases up to 17 nm. The largest LSPR shift is recorded for Au-nanorod@pNIPAM, about 28 nm, due to the higher electromagnetic field concentration in rods, which in turn leads to a higher sensitivity for local refractive index changes (Chen, Kou et al. 2008).

4. Entrapping target molecules: Surface-enhanced Raman analysis.

Gold-pNIPAM core-shell particles used in this section are synthesized by a slightly different method, consisting of an initial growth of thin polystyrene (PS) on preformed gold nanoparticles (67 nm) coated with CTAB, followed by in situ polymerization of the pNIPAM shell (Contreras-Caceres, Sanchez-Iglesias et al. 2008). Particles with larger metal cores

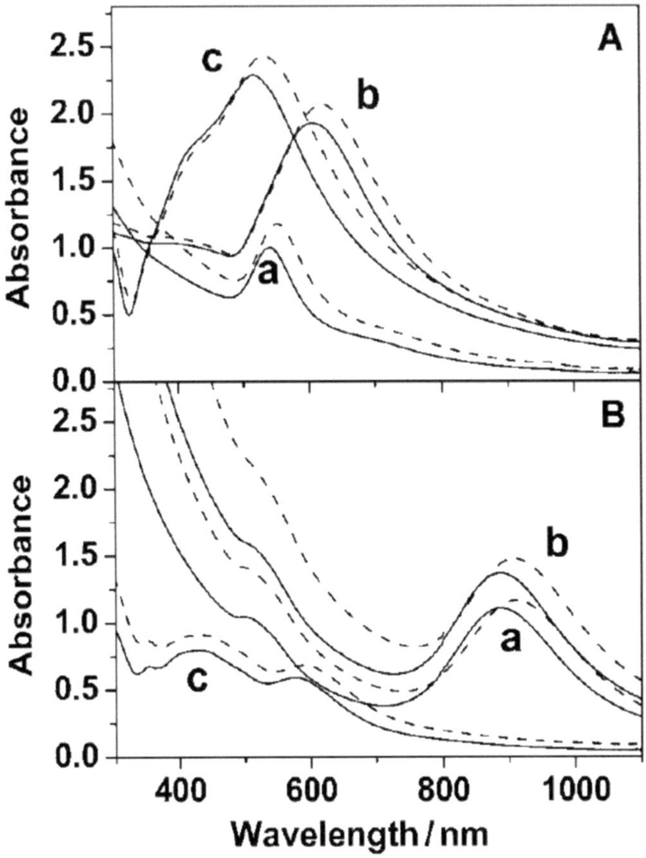

Fig. 4. UV-Vis-NIR spectra for aqueous dispersions of different core-shell metal@pNIPAM nanocomposites. Data are recorded at 22°C (solid lines) and 44°C (dash line), which corresponds to the swollen and collapsed shell states, respectively. (A) Sphere-coated particles: (a) Au-sphere (64 nm); (b) Au-sphere (103 nm); (c) Au@Ag core-shell sphere. (B) Rod-coated particles: (a) Au-nanorod; (b) Au@Au nanorod; (c) Au@Ag nanorod (10 time diluted to avoid scattering effects). Reprinted with permission from (Contreras-Caceres, Pastoriza-Santos et al.), Copyright (2010) by Wiley-VSC Verlag GmbH & Co. KGaA.

(116 nm) are prepared by seeded growth of the 67 nm coated gold cores through the addition of $HAuCl_4$ and ascorbic acid as reducing agent. The SERS spectrum of gold-polystyrene particles is shown in Fig. 5. The peaks correspond to ring C=C stretching (1615 cm^{-1}), CH_2 scissoring (1461 cm^{-1}), ring breathing (1012 cm^{-1}), and radial ring stretching mode (646 cm^{-1}), which are characteristic of polystyrene (Hong, Boerio et al. 1993). Interestingly, as particles polymerize with pNIPAM, the bands disappear showing an effective replacement of PS by pNIPAM, as also observed in Fig. 5, for both selected core sizes Au@pNIPAM (67 and 116 nm). Both spectra fit band to band, being represented by NH bending (1447 cm^{-1}), CN stretching (1210 cm^{-1}), CH_3 rocking (963 cm^{-1}), CH deformation (866 and 841 cm^{-1}), CC rocking (766 cm^{-1}), CNO bending (655 cm^{-1}), and CCO out-of-plane deformation (413 cm^{-1}). There is an important increase in intensity for large gold cores, as result of a considerable enhancement of the optical properties with increasing size, in agreement with previous

reports (Kelly, Coronado et al. 2003; Njoki, Lim et al. 2007). The overall SERS intensity (cross-section) obtained from pNIPAM is low, thus providing an excellent background for analytical applications. pNIPAM shell with thermoresponsive properties allows to entrap analyte molecules and approximate them to the metal core (when the polymer collapses) where Raman enhancement becomes apparent. In addition, shell prevents the electromagnetic metal particle coupling, with highly reproducible SERS signal and intensity. The fluorescence intensity of certain adsorbed chromophores can also be improved in such a way. We present here analytical applications based on SERS, SERRS and surface enhanced fluoresce (SEF), using gold-pNIPAM nanocomposites through a rational selection of analytes. All spectra are taken with a LabRam HR Raman equipment (Horiba-Jobin Yvon), following two kind of experiments. Firstly, the particle dispersion (1mL, $5x10^{-4}$ M in gold) together the analyte (10 µL, 10^{-5}-10^{-6} M) are stabilized at 4°C for 2 h, time enough to reach thermodynamic equilibrium. Next, samples are excited with a 785 nm laser to collect the SERS spectra or with 633 nm laser for SFE and SERRS spectra. Thereafter, the samples are equilibrated at 60°C for 2 h and again at 4°C. After each equilibrium step, spectra are collected under the same experimental conditions. In a second experiment, equilibration steps are repeated, following the inverse temperature sequence, starting at 60°C, cooling to 4°C and heating back to 60°C.

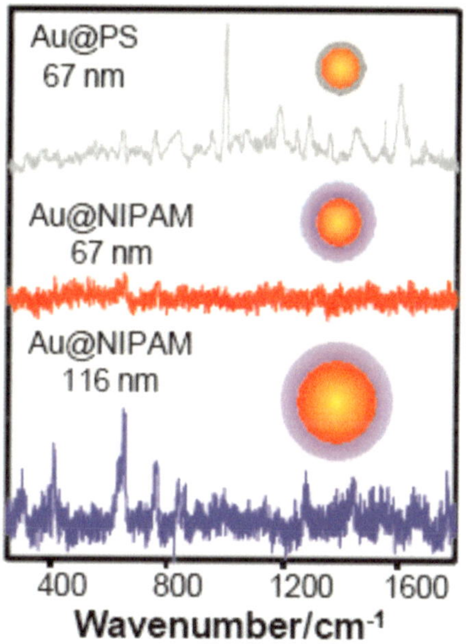

Fig. 5. (From top to bottom) SERS spectra of Au@PS particles, Au(67nm)@pNIPAM (obtained by coating Au@PS) and Au(116nm)@pNIPAM (after in situ growth of gold core). Acquisition time is 50 s. Reprinted with permission from (Contreras-Caceres, Sanchez-Iglesias et al. 2008), Copyright (2008) by Wiley-VSC Verlag GmbH & Co. KGaA.

4.1 Analyte with specific molecular interactions: 1-naphthalenethiol

Raman enhancing properties of Au-pNIPAM nanoparticles are initially tested using 1-naphthalenethiol (1NAT); this is a small molecule with large affinity for gold (through the thiol group). It is considered a model analyte since it easily diffuses across the porous polymer shell. Moreover, its SERS spectrum is well established (Alvarez-Puebla, Dos Santos et al. 2004). As can be seen in Fig. 6a, the Raman spectrum is dominated by the ring stretching (1553, 1503, and 1368 cm^{-1}), CH bending (1197 cm^{-1}), ring breathing (968 and 822 cm^{-1}), ring deformation (792, 664, 539, and 517 cm^{-1}), and CS stretching (389 cm^{-1}). The intensity of the band at 1368 cm^{-1}, corresponding to the ring stretching, is plotted against temperature for both cooling-heating cycles.

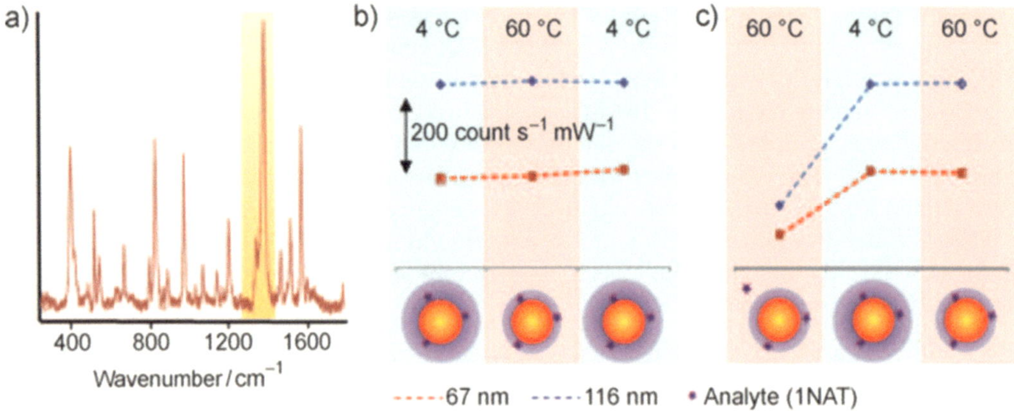

Fig. 6. (a) SERS spectrum of 1-naphtalenethiol dissolved in Au@pNIPAM particle dispersions. Excitation wavelength λ_{ex}= 785 nm. (b, c) Variation of the intensity of the band at 1368 cm^{-1}, ring stretching highlighted in yellow, as a function of gold-core size and temperature in two different cooling-heating cycles: (b) 4–60-4°C; and, (c) 60-4-60°C. Acquisition time is 2s for all experiments. Reprinted with permission from (Contreras-Caceres, Sanchez-Iglesias et al. 2008), Copyright (2008) by Wiley-VSC Verlag GmbH & Co. KGaA.

As the analyte is added to the nanoparticle dispersion at 4°C (Fig. 6b), pNIPAM shells swell, allowing the analyte to diffuse across the polymer to reach the gold-core surface, to which it readily chemisorbs. This results in a high SERS intensity, which remains high after gradually heating up to 60°C and cooling down back to 4°C. Instead, when 1NAT is added to the dispersion at 60°C, SERS signal is substantially lower (Fig. 6c). However, by cooling down temperature to 4°C, the signal surprisingly enhances up to intensities comparable to those of the previous cycle. Furthermore, the high signal remains stable during subsequent temperature changes. The results can be explained by considering the volume transition exhibited by the pNIPAM shell; above 32°C, shell changes from water-swollen to shrunken states, being the process totally reversible (Sierra-Martin, Choi et al. 2005). Thereby, as shells remain collapsed at 60°C, the diffusion of 1NAT through the network is hindered and the gold surface is not longer accessible, giving then low signal. Once the temperature cold down, the analyte adsorbs on the core and retains there regardless the gel swelling state. We suggest that 1NAT forms a covalent bound at the gold surface, which is consistent with the disappearance of the SH stretching peak in the SERS spectra and also with previous reports (Pearson 1963; Pearson 1966). It is interesting to note that the enhancement provided by the larger gold core (116 nm) is considerable higher, partly because of the better match between the excitation wavelength (785 nm) and the plasmon band (Creighton, Blatchford et al.

1979); as gold core size increases, the plasmon band shifts to higher wavelengths (Fig. 4). The enhancement factor calculated for this core-shell system is EF=5.16x10^5, a rather high value if we take into account that 1NAT does not present substantial charge-transfer enhancement (the so-called chemical effect) (McFarland, Young et al. 2005). The polymer shell prevents the electromagnetic coupling between particles, and hence the formation of hot spots. The enhancement factor is estimated by comparing the signal of the analyte with and without hybrid particles; it is given by equation $EF = (I_A V_A / I_B V_B) f$ (Alvarez-Puebla, Dos Santos et al. 2007), where V_A, V_B are the probed volumes, I_A, I_B the respective SERS intensities and f a correction factor that considers the concentration ratio of the target molecule in both experiments. We note that the forthcoming SERS experiments will be developed only with the Au(116nm)@pNIPAM system, provided that it induces the best Raman enhancement.

4.2 Non-interacting analytes

4.2.1 Nile Blue A

A second demonstration of the potential applications of the Au-pNIPAM nanocomposite is developed for a common dye, Nile Blue A (NBA). This molecule is slightly larger than 1NAT. In addition, it contains an amine functional group which diminishes the affinity for gold surfaces respect to 1NAT (Pearson 1963; Pearson 1966). NBA molecules show different spectra, either SERS or SEF/SERRS, depending on the excitation wavelength. Upon excitation with near-IR laser line (785 nm), far away from the electronic absorption band (Alvarez-Puebla, Contreras-Caceres et al. 2009), NBA supported onto the metal core will produce a normal SERS signal. On the other hand, if NBA is excited with a red laser (633 nm), perfectly matching the absorption band, either SERRS or SEF will be produced, depending on the distance to the metal surface. Under these conditions, as the analyte is close enough to the gold, fluorescence can be quenched; however, if the molecule is not as close, it will feel the electromagnetic field enhancement generated by metallic core. Despite SERS and SERRS spectra overlap band to band, their relative intensities are not similar; this is because to the SERRS signal is not only influenced by the surface selection rules (Moskovits and Suh 1984; Moskovits 1985), but also by the resonance effects (Long 2002). Fig. 7 illustrates both SERS and SERRS spectra for NBA molecules immersed into the nanoparticle dispersion; spectra are characterized by the ring stretching (1643, 1492, 1440, 1387, 1351, and 1325 cm^{-1}), CH bending (1258, 1185 cm^{-1}), and the in-plane CCC and NCC (673 cm^{-1}), CCC and CNC (595 cm^{-1}), and CCC (499 cm^{-1}) deformations (Lu, Mei et al. 2006). The bands at 673 and 595 cm^{-1} are significantly more enhanced for SERRS than for SERS, indicating that they correspond to the chromophore (phenoxazine), whereas the electronic resonance tends to enhance scattering bands from chemical groups absorbing the excitation laser line. In addition, SEF spectra is very similar to those obtained for standard fluorescence, with maximum emission at 668 nm (Aslan, Lakowicz et al. 2005). Regarding the temperature influence, as the analyte NBA is added to the particle dispersion at 4°C and excited with NIR laser line (785 5nm), SERS intensity is very weak (Fig. 7a). Unlike the results obtained for 1NAT, where the intensity remains constant with temperature, here the intensity notably increases as shell collapses at 60°C and diminishes again after cooling back to 4°C . When the same sample is excited with a laser operating at 633 nm, the spectrum of the initial, swollen sample shows an intense fluorescence of about 16-fold the normal fluorescence. Instead, as temperature rises to 60°C (collapsed shell), fluorescence quenches and SERRS spectrum is recovered. After subsequent cooling to 4°C, less-intense SERRS spectrum can still be identified on top the strong SEF background.

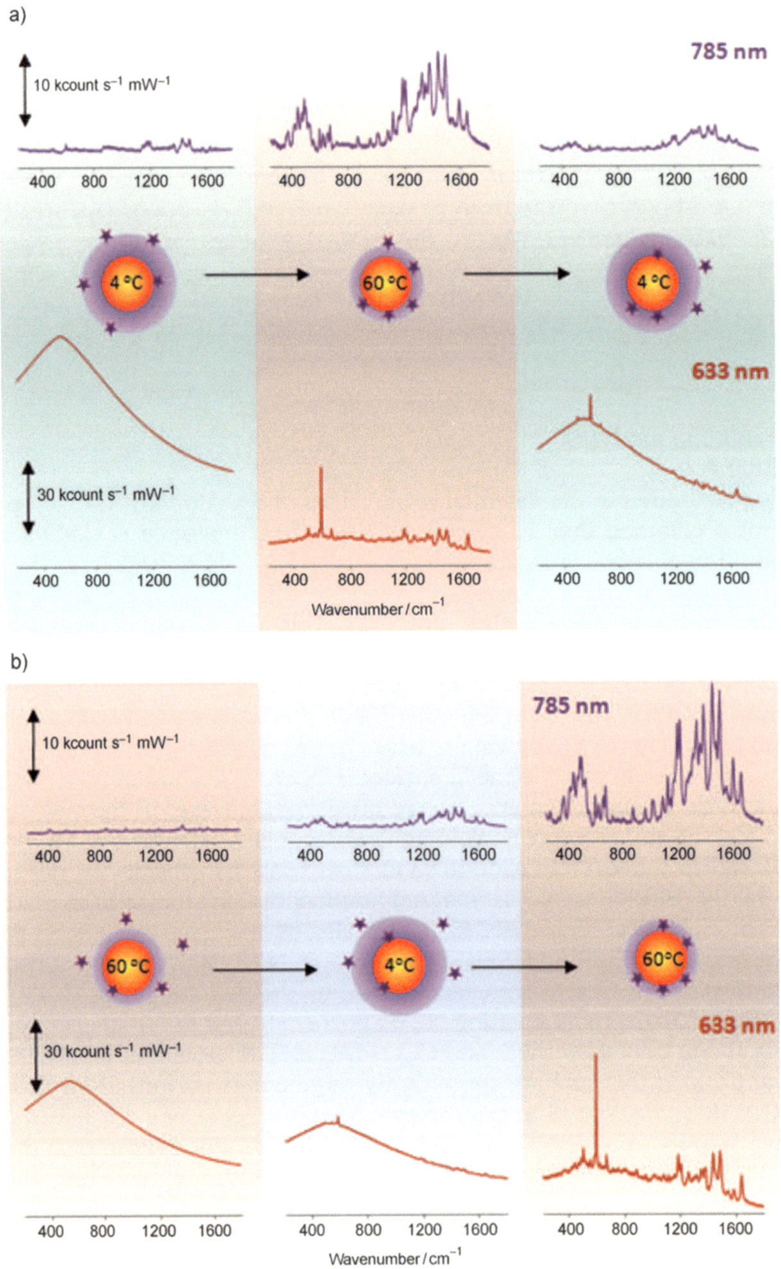

Fig. 7. SERS and SEF/SERRS spectra of Nile Blue A as a function of temperature. The excitation wavelength is λ_{ex}=785 nm (blue trace) and λ_{ex}=633 nm (red trace) for SERS and SEF/SERRS, respectively. Two different cooling-heating cycles are tested: (a) from 4 to 60 to 4°C; and (b) from 60 to 4 to 60°C. The acquisition time is 2s. Reprinted with permission from (Contreras-Caceres, Sanchez-Iglesias et al. 2008), Copyright (2008) by Wiley-VSC Verlag GmbH & Co. KGaA.

The disagreement between results concerning NBA and 1NAT molecules (for the temperature cycle 4-60-4 °C) is attributed to the different affinity between amine and thiol groups for gold; the retention of NBA molecules on gold surface is less stable than for 1NAT, which causes partial release of NBA, thereby contributing to SERS and SERRS weakening (concomitantly to SFE enhancement). Interestingly, for the inverse cycle (60-4-60 °C) (Fig. 7b), strong SEF intensity is recorded at 60°C, which turns upon shell swelling into a weak SERRS signal (4°C) and then to an intense SERRS spectra after final heating up to 60°C. These results are interpreted by considering the shell swelling properties as well as the affinity of the analyte to gold. Due to the low affinity of NBA, even for the particle swollen state, the analyte does not significantly absorb onto gold cores (weak SERS signal at 4°C), but it can be entrapped within the polymer network (strong SEF that completely screens the SERRS signal). When temperature raises up to 60°C, the shells collapse and NBA molecules are entrapped closer to the core, as indicates the notable increase of SERS and SERRS, while SEF signal is quenched. For the second cycle (60-4-60°C), a similar behaviour is found; initially, as particles collapse, only SEF is recorded. Upon particle swelling and subsequent collapse, NBA molecules are retained in close contact to the gold core surfaces. SERS signal then recovers. The entrapping mechanism is closely related to the hydrophilic-hydrophobic transition of pNIPAM microgels and to the microcapillarity effect occurring during particle collapse (Guerrini, Garcia-Ramos et al. 2006; Guerrini, Garcia-Ramos et al. 2008).

4.2.2. 1-naphthol

SERS enhancement for Au-pNIPAM nanocomposite is finally tested for 1-napthol; this molecule does not easily adsorb onto conventional gold or silver surfaces, so its SERS analysis has remain elusive to date. Fig.8 illustrates SERS spectrum, recorded for the first time for 1-naphthol; it is characterized by CH bending (1447 cm^{-1}), ring stretching (1390 cm^{-1}), CCC in-plane deformation (842 cm^{-1}), CH out-of-plane deformation, ring breathing (716 cm^{-1}), ring deformation (655 and 584 cm^{-1}) and, ring twisting (477 cm^{-1}), in close agreement with the Raman assignment previously reported (Lakshminarayan and Knee 1990). SERS signal is properly identified after shell collapse, from 4°C to 60°C; the analyte is first retained within the swollen polymer networks, at 4°C and then, brought into contact with the gold surfaces upon shells collapse. After subsequent cooling, the polymer shells swell again and 1-napthol molecules release the metal surface, resulting in a dramatic loss of SERS signal. The low affinity of hydroxyl groups of the 1-napthol to gold surfaces is clearly shown in the reversibility of the SERS signal along the swell-collapse cycles.

5. Improved SERS detection via bimetallic sensors

In the previous section, we have shown the ability of Au@pNIPAM nanoparticles for entrapping and detecting analytes by means of SERS. Nevertheless, the use of hybrid particles with small cores and the impossibility of those materials to form hot spots due to the physical barrier imposed by the polymer, limits the enhancement and imposes a detection threshold. To overcome this limitation, hybrid materials with different compositions and morphologies are employed. The first alternative involves controlled growth of silver shells onto the gold cores, since it is well known that silver is much more efficient plasmonic material (Zhao, Pinchuk et al. 2008). The second one refers to morphology changes towards rod-shaped cores, with near field concentration areas at the rods edges, (Vesseur, de Waele et al. 2007; Cai, Sainidou et al. 2009). In addition, the molecular affinity of the pNIPAM shell and analyte can be improved by tuning the polymer charge.

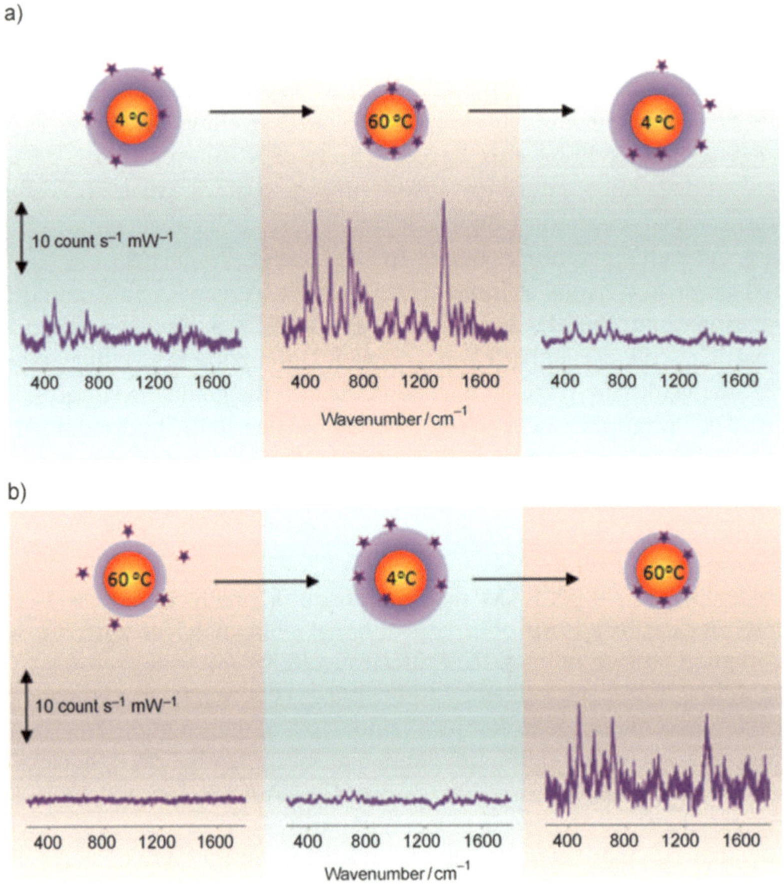

Fig. 8. Variation of SERS intensities of 1-napthol as a function of temperature for two different cooling-heating cycles: (a) from 4 to 60 to 4°C; and (b) from 60 to 4 to 60°C. The excitation wavelength is λ_{ex}=785 nm and the acquisition time 2s. Reprinted with permission from (Contreras-Caceres, Sanchez-Iglesias et al. 2008), Copyright (2008) by Wiley-VSC Verlag GmbH & Co. KGaA.

SERS detection is checked for a couple of analytes, 1-naphthalenethiol and 2-naphthoic (2-NA), using the whole set of hybrid particles prepared in section 3. Experiments are performed with a Renishaw Invia system, equipped with a Peltier charge-coupled device (CCD) detector and a Leica confocal microscope. Spectra are collected in Renishaw extended mode with accumulation times of 10 s using a macrosampling 90° objective adaptor. Samples are prepared by mixing 15 µL of analyte, previously dissolved either in ethanol (for 1-NAT) or aqueous alkaline solution with pH=13 (for 2-NA) at a concentration of 10^{-3} M, to

1.5 mL of the nanoparticle dispersion tested in each case. The mixtures are thermostatizated for 1 h before SERS recording.

5.1. Detection of 1-naphthalenethiol

SERS efficiency for gold and gold-silver nanoparticles coated with pNIPAM is first shown by using 1-naphthalenethiol. The spectra for 1NAT were already recorded in the presence of core-shell gold-pNIPAM particles, as we described in section 4.1. The excitation is now performed at three different lasers lines (inset on Figure 9), to make possible coupling of the respective surface plasmon modes. The intensity of the ring stretching peak (1368 cm^{-1}), at three wavelengths, is compared for all nanocomposites. All systems provide sufficient signal enhancement to allow identification of the analyte, which support the same mechanism proposed for spherical Au-pNIPAM particles irrespective of the core morphology and composition. 1NAT molecules reach the metallic cores by diffusion across the shells and bind onto the surfaces through thiol groups. Nevertheless, it appears substantial differences on signal enhancement; remarkably, for both spheres and rods, SERS intensity is found to increase with core size, although this improvement difference is moderate compared to the signal enhancement coming from the presence of silver. In addition, silver coatings induce spectral changes that lead us to use the green laser line (532 nm) to excite the LSPR; this wavelength is however very inefficient for pure gold particles due to damping effect

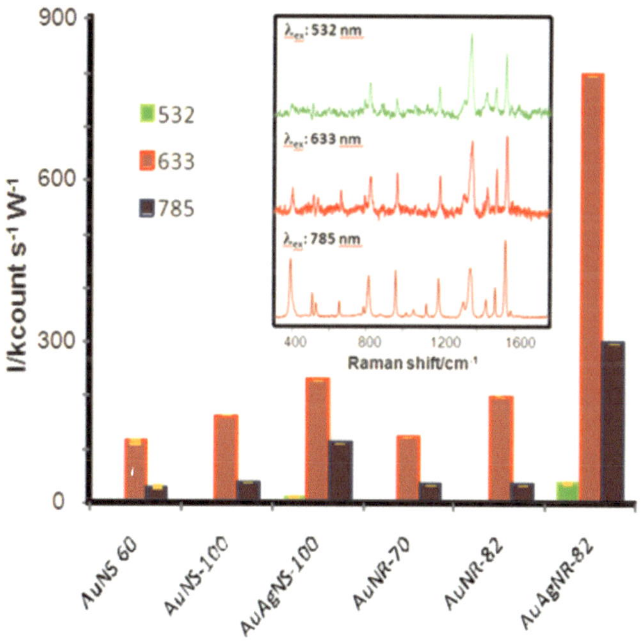

Fig. 9. Comparison of the Raman intensity corresponding to the ring stretching peak (1368 cm^{-1}) of 1NAT in different hybrid nanoparticle dispersions, acquired upon excitation with 532, 633 and 785 nm laser lines. The intensity is normalized with particle concentration in each case. The inset shows the SERS spectra of 1NAT in Au-pNIPAM suspension, recorded also at three different laser lines. Reprinted with permission from (Contreras-Caceres, Pastoriza-Santos et al.), Copyright (2010) by Wiley-VSC Verlag GmbH & Co. KGaA.

provoked by interband transitions. Even more interesting it is the large enhancement caused by the core morphology, with the intensity increasing as the spherical core transforms into rod. Change are particularly noticeable for gold-silver bimetallic nanorods since silver not only improves the optical efficiency, but also blue-shifts the longitudinal surface plasmon resonance, leading to stronger electric field concentration at the ends of the rods (Vesseur, de Waele et al. 2007; Cai, Sainidou et al. 2009). Thereby, both effects contribute to substantial increase of the SERS intensity upon excitation with red laser (633 nm).

5.2 Detection of 2-naphthoic acid

Improved SERS detection with bimetallic core nanoparticles is also shown using ionic analytes; 2-naphthoic acid (2NA). This analyte can be electrostatically entrapped by the polymer mesh in contrast to 1NAT, which chemically binds onto the metal cores through thiol groups. To this end, the analyte is first dissolved in pH=13 alkaline solution and subsequently added to dispersions with different nanoparticles, reaching a final pH=11. At that pH value, 2NA is completely ionized, bearing negative charge; as a result, it refuses to adsorb on standard citrate-stabilized silver particles due to their negative nature. Low Raman signal is then recorded in the presence of citrate-stabilized silver particles (Figure 10, up). Remarkably, Raman signal from NA is greatly enhanced when the analyte spectrum is recorded in the presence of core-shell hybrid particles (Figure 10, down). The well-defined spectra is characterized by the ring stretching at 1632 and 1388 cm^{-1}, CH bending at 1468 cm^{-1}, ring breathing at 1018 cm^{-1}, and CH deformation at 770 cm^{-1} (Alvarez-Puebla and Aroca 2009). The presence of pNIPAM shells is responsible for the signal enhancement; since a

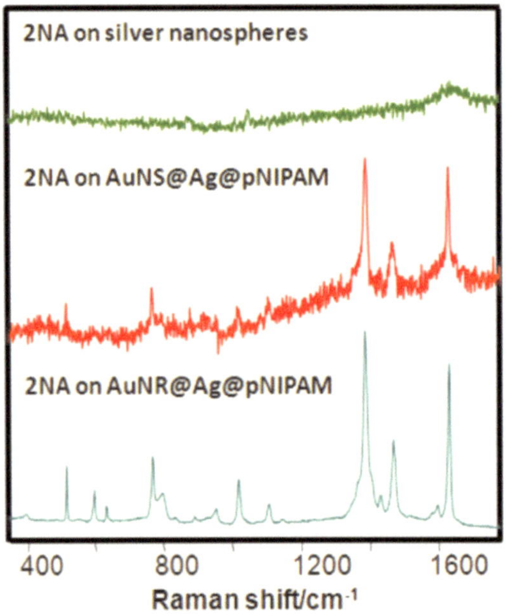

Fig. 10. SERS spectra of 2-naphtoic acid dissolved in an alkaline suspension of the different hybrid nanoparticles, acquired upon excitation with 532 nm laser line. Reprinted with permission from (Contreras-Caceres, Pastoriza-Santos et al.), Copyright (2010) by Wiley-VSC Verlag GmbH & Co. KGaA.

cationic initiator (AAPH) is used during polymerization, it displays a positive surface charge (+25mV), which is also present across the whole polymer network. Positive-charged shells can absorb 2NA anions by electrostatic interaction, then giving rise to significant SERS intensities. Although SERS enhancement is excellent for composites containing Au@Ag spherical cores, the signal is substantially larger for rod-shaped cores, thus confirming that the synergy between composition (silver) and morphology (rods) leads to much efficient SERS detection.

6. Conclusion

In this review we report the design, characterization and application of an advanced optical platform based on gold and gold-silver particles coated with polymeric pNIPAM shells. It allows ultra-sensitive analysis for a wide variety of analytes through Surface-enhanced Raman Spectroscopy. The composite particles combine optical amplification coming from metallic nanoparticles with singular thermoresponsive swelling features pNIPAM gels. Interestingly, the shell leads to entrap. The last one entraps physically analyte molecules and gets them close enough to the core surface, where signal enhancement is maximum. In addition, the shell can absorb charged molecules due to the presence of ionic groups anchored on the network, thus generalizing the use of these systems for SERS sensing. SERS enhancement is checked for different nanocomposite particles, paying special attention to the role of core geometry and metal composition. As expected, bimetallic gold-silver cores provide great SERS signal, especially intense for nanorod morphology.

7. Acknowledgement

Financial support from Spanish Ministerio de Ciencia e Innovación (Project MAT2010-17858), Andalucian Excellence Projects (FQM-02353 and FQM-230-2010), FEDER program and EU-COST Action D43 are acknowledged. The authors thank the great contribution of Prof. Luís Liz-Marzán, Dr. Ramón Álvarez-Puebla and Dr. Jorge Pérez-Juste to the content of this review.

8. References

Alvarez-Puebla, R. A., E. Arceo, et al. (2005). "Role of nanoparticle surface charge in surface-enhanced Raman scattering." Journal of Physical Chemistry B 109(9): 3787-3792.

Alvarez-Puebla, R. A. and R. F. Aroca (2009). "Synthesis of Silver Nanoparticles with Controllable Surface Charge and Their Application to Surface-Enhanced Raman Scattering." Analytical Chemistry 81(6): 2280-2285.

Alvarez-Puebla, R. A., R. Contreras-Caceres, et al. (2009). "Au@pNIPAM Colloids as Molecular Traps for Surface-Enhanced, Spectroscopic, Ultra-Sensitive Analysis." Angewandte Chemie-International Edition 48(1): 138-143.

Alvarez-Puebla, R. A., D. S. Dos Santos, et al. (2004). "Surface-enhanced Raman scattering for ultrasensitive chemical analysis of 1 and 2-naphthalenethiols." Analyst 129(12): 1251-1256.

Alvarez-Puebla, R. A., D. S. Dos Santos, et al. (2007). "SERS detection of environmental pollutants in humic acid-gold nanoparticle composite materials." Analyst 132: 1210-1214.

Aroca, R. F., R. A. Alvarez-Puebla, et al. (2005). "Surface-enhanced Raman scattering on colloidal nanostructures." Advances in Colloid and Interface Science 116(1-3): 45-61.

Aslan, K., J. R. Lakowicz, et al. (2005). "Metal-enhanced fluorescence using anisotropic silver nanostructures: critical progress to date." Analytical and Bioanalytical Chemistry 382(4): 926-933.

Aslan, K., M. Wu, et al. (2007). "Fluorescent core-shell Ag@SiO2 nanocomposites for metal-enhanced fluorescence and single nanoparticle sensing platforms." Journal of the American Chemical Society 129(6): 1524-1525.

Bao, L. L., S. M. Mahurin, et al. (2003). "Study of silver films over silica beads as a surface-enhanced Raman scattering (SERS) substrate for detection of benzoic acid." Journal of Raman Spectroscopy 34(5): 394-398.

Barnes, W. L., A. Dereux, et al. (2003). "Surface plasmon subwavelength optics." Nature 424(6950): 824-830.

Bell, S. E. J. and N. M. S. Sirimuthu (2008). "Quantitative surface-enhanced Raman spectroscopy." Chemical Society Reviews 37(5): 1012-1024.

Bohren, C. F. and D. R. Huffman (1983). Absorption and Scattering of Light by Small Particles. New York, Wiley.

Braeckmans, K., S. C. De Smedt, et al. (2002). "Encoding microcarriers: Present and future technologies." Nature Reviews Drug Discovery 1(6): 447-456.

Brown, L. O. and S. K. Doorn (2008). "Optimization of the preparation of glass-coated, dye-tagged metal nanoparticles as SERS substrates." Langmuir 24(5): 2178-2185.

Cai, W., R. Sainidou, et al. (2009). "Efficient Generation of Propagating Plasmons by Electron Beams." Nano Letters 9(3): 1176-1181.

Campion, A. and P. Kambhampati (1998). "Surface-enhanced Raman scattering." Chemical Society Reviews 27(4): 241-250.

Contreras-Caceres, R., J. Pacifico, et al. (2009). "Au@pNIPAM Thermosensitive Nanostructures: Control over Shell Cross-linking, Overall Dimensions, and Core Growth." Advanced Functional Materials 19(19): 3070-3076.

Contreras-Caceres, R., I. Pastoriza-Santos, et al. "Growing Au/Ag Nanoparticles within Microgel Colloids for Improved Surface-Enhanced Raman Scattering Detection." Chemistry-a European Journal 16(31): 9462-9467.

Contreras-Caceres, R., A. Sanchez-Iglesias, et al. (2008). "Encapsulation and growth of gold nanoparticles in thermoresponsive microgels." Advanced Materials 20(9): 1666-1670.

Creighton, J. A., C. G. Blatchford, et al. (1979). "Plasma resonance enhancement of raman-scattering by pyridine adsorbed on silver or gold sol particles of size comparable to the excitation wavelength." Journal of the Chemical Society-Faraday Transactions Ii 75: 790-798.

Chen, H. J., X. S. Kou, et al. (2008). "Shape- and size-dependent refractive index sensitivity of gold nanoparticles." Langmuir 24(10): 5233-5237.

Cho, E. C., P. H. C. Camargo, et al. "Synthesis and Characterization of Noble-Metal Nanostructures Containing Gold Nanorods in the Center." Advanced Materials 22(6): 744-748.

Fleischm.M, P. J. Hendra, et al. (1974). "Raman-Spectra of Pyridine Adsorbed at a Silver Electrode." Chemical Physics Letters 26(2): 163-166.

Gorelikov, I., L. M. Field, et al. (2004). "Hybrid microgels photoresponsive in the near-infrared spectral range." Journal of the American Chemical Society 126(49): 15938-15939.

Grancharov, K., H. Engelberg, et al. (2001). "Inhibition of UDP-glucuronosyltransferases in rat liver microsomes by natural mutagens and carcinogens." Archives of Toxicology 75(10): 609-612.

Guerrini, L., J. V. Garcia-Ramos, et al. (2006). "Functionalization of Ag nanoparticles with dithiocarbamate calix[4]arene as an effective supramolecular host for the surface-enhanced Raman scattering detection of polycyclic aromatic hydrocarbons." Langmuir 22(26): 10924-10926.

Guerrini, L., J. V. Garcia-Ramos, et al. (2008). "Building highly selective hot spots in Ag nanoparticles using bifunctional viologens: Application to the SERS detection of PAHs." Journal of Physical Chemistry C 112(20): 7527-7530.

Hansen, A. M., O. Omland, et al. (1994). "Correlation between work process-related exposure to polycyclic aromatic-hydrocarbons and urinary levels of alpha-naphthol, beta-naphthylamine and 1-hydroxypyrene in iron foundry workers." International Archives of Occupational and Environmental Health 65(6): 385-394.

Homola, J. (2008). "Surface plasmon resonance sensors for detection of chemical and biological species." Chemical Reviews 108(2): 462-493.

Hong, P. P., F. J. Boerio, et al. (1993). "Surface segregation in blends of polystyrene and deuterated polystyrene." Macromolecules 26(6): 1460-1464.

Hutter, E. and J. H. Fendler (2004). "Exploitation of localized surface plasmon resonance." Advanced Materials 16(19): 1685-1706.

Jain, P. K. and M. A. El-Sayed (2007). "Universal scaling of plasmon coupling in metal nanostructures: Extension from particle pairs to nanoshells." Nano Letters 7(9): 2854-2858.

Jain, P. K., X. H. Huang, et al. (2008). "Noble Metals on the Nanoscale: Optical and Photothermal Properties and Some Applications in Imaging, Sensing, Biology, and Medicine." Accounts of Chemical Research 41(12): 1578-1586.

Jana, N. R., L. Gearheart, et al. (2001). "Seeding growth for size control of 5-40 nm diameter gold nanoparticles." Langmuir 17(22): 6782-6786.

Johnson, P. B. and R. W. Christy (1972). "Optical-Constants of Noble-Metals." Physical Review B 6(12): 4370-4379.

Karg, M., I. Pastoriza-Santos, et al. (2006). "A versatile approach for the preparation of thermosensitive PNIPAM core-shell microgels with nanoparticle cores." Chemphyschem 7(11): 2298-2301.

Karg, M., I. Pastoriza-Santos, et al. (2007). "Nanorod-coated PNIPAM microgels: Thermoresponsive optical properties." Small 3: 1222-1229.

Kelly, K. L., E. Coronado, et al. (2003). "The optical properties of metal nanoparticles: The influence of size, shape, and dielectric environment." Journal of Physical Chemistry B 107(3): 668-677.

Kim, J. H. and T. R. Lee (2004). "Thermo- and pH-responsive hydrogel-coated gold nanoparticles." Chemistry of Materials 16(19): 3647-3651.

Kneipp, J., H. Kneipp, et al. (2006). "In vivo molecular probing of cellular compartments with gold nanoparticles and nanoaggregates." Nano Letters 6(10): 2225-2231.

Kneipp, K., H. Kneipp, et al. (1999). "Ultrasensitive chemical analysis by Raman spectroscopy." Chemical Reviews 99(10): 2957-2975.

Kneipp, K., Y. Wang, et al. (1997). "Single molecule detection using surface-enhanced Raman scattering (SERS)." Physical Review Letters 78(9): 1667-1670.

Kozumbo, W. J., S. Agarwal, et al. (1992). "Breakage and binding of dna by reaction-products of hypochlorous acid with aniline, 1-naphthylamine, or 1-naphthol." Toxicology and Applied Pharmacology 115(1): 107-115.

Kratz, K., T. Hellweg, et al. (2001). "Structural changes in PNIPAM microgel particles as seen by SANS, DLS, and EM techniques." Polymer 42(15): 6631-6639.

Lakshminarayan, C. and J. L. Knee (1990). "Spectroscopy and dynamics of the s1 state of jet-cooled 1-naphthol." Journal of Physical Chemistry 94(6): 2637-2643.

Lee, J., A. O. Govorov, et al. (2005). "Nanoparticle assemblies with molecular springs: A nanoscale thermometer." Angewandte Chemie-International Edition 44(45): 7439-7442.

Link, S. and M. A. El-Sayed (1999). "Spectral properties and relaxation dynamics of surface plasmon electronic oscillations in gold and silver nanodots and nanorods." Journal of Physical Chemistry B 103(40): 8410-8426.

Liu, M. Z. and P. Guyot-Sionnest (2005). "Mechanism of silver(I)-assisted growth of gold nanorods and bipyramids." Journal of Physical Chemistry B 109(47): 22192-22200.

Lombardi, J. R., R. L. Birke, et al. (1986). "Charge-Transfer Theory of Surface Enhanced Raman-Spectroscopy - Herzberg-Teller Contributions." Journal of Chemical Physics 84(8): 4174-4180.

Long, D. A., Ed. (2002). The Raman Effect: A Unified Treatment of the Theory of Raman Scattering by Molecules. Chichester, Wiley.

Lu, Y., Y. Mei, et al. (2006). "Thermosensitive core-shell particles as carriers for Ag nanoparticles: Modulating the catalytic activity by a phase transition in networks." Angewandte Chemie-International Edition 45(5): 813-816.

McFarland, A. D., M. A. Young, et al. (2005). "Wavelength-scanned surface-enhanced Raman excitation spectroscopy." Journal of Physical Chemistry B 109(22): 11279-11285.

Moskovits, M. (1985). "Surface-enhanced spectroscopy." Reviews of Modern Physics 57(3): 783-826.

Moskovits, M. (2006). Surface-enhanced Raman spectroscopy: a brief perspective. Surface-Enhanced Raman Scattering: Physics and Applications. 103: 1-17.

Moskovits, M. and J. S. Suh (1984). "Surface selection-rules for surface-enhanced raman-spectroscopy - calculations and application to the surface-enhanced Raman-spectrum of phthalazine on silver." Journal of Physical Chemistry 88(23): 5526-5530.

Mulvaney, P. (1996). "Surface plasmon spectroscopy of nanosized metal particles." Langmuir 12(3): 788-800.

Murphy, C. J., T. K. San, et al. (2005). "Anisotropic metal nanoparticles: Synthesis, assembly, and optical applications." Journal of Physical Chemistry B 109(29): 13857-13870.

Nie, S. M. and S. R. Emery (1997). "Probing single molecules and single nanoparticles by surface-enhanced Raman scattering." Science 275(5303): 1102-1106.

Njoki, P. N., I. I. S. Lim, et al. (2007). "Size correlation of optical and spectroscopic properties for gold nanoparticles." Journal of Physical Chemistry C 111: 14664-14669.

Okamoto, S. and S. Hachisu (1977). "Ordered Structure in Monodisperse Gold Sol." Journal of Colloid and Interface Science 62(1): 172-181.

Oldenburg, S. J., R. D. Averitt, et al. (1998). "Nanoengineering of optical resonances." Chemical Physics Letters 288(2-4): 243-247.

Otto, A., I. Mrozek, et al. (1992). "Surface-Enhanced Raman-Scattering." Journal of Physics-Condensed Matter 4(5): 1143-1212.

Pearson, R. G. (1963). "Hard and soft acids and bases." Journal of the American Chemical Society 85(22): 3533-3539"

Pearson, R. G. (1966). "Acids and bases." Science 151(3707): 172-177.

Penn, S. G., L. He, et al. (2003). "Nanoparticles for bioanalysis." Current Opinion in Chemical Biology 7(5): 609-615.

Pillai, S., K. R. Catchpole, et al. (2007). "Surface plasmon enhanced silicon solar cells." Journal of Applied Physics 101(9): 93105-93112.

Porter, M. D., R. J. Lipert, et al. (2008). "SERS as a bioassay platform: fundamentals, design, and applications." Chemical Society Reviews 37(5): 1001-1011.

Qian, X. M. and S. M. Nie (2008). "Single-molecule and single-nanoparticle SERS: from fundamental mechanisms to biomedical applications." Chemical Society Reviews 37(5): 912-920.

Rodriguez-Fernandez, J., J. Perez-Juste, et al. (2006). "Seeded growth of submicron Au colloids with quadrupole plasmon resonance modes." Langmuir 22(16): 7007-7010.

Rodriguez-Gonzalez, B., A. Burrows, et al. (2005). "Multishell bimetallic AuAg nanoparticles: synthesis, structure and optical properties." Journal of Materials Chemistry 15(17): 1755-1759.

Sanchez-Iglesias, A., E. Carbo-Argibay, et al. "Rapid Epitaxial Growth of Ag on Au Nanoparticles: From Au Nanorods to Core-Shell Au@Ag Octahedrons." Chemistry-a European Journal 16(19): 5558-5563.

Sanchez-Iglesias, A., M. Grzelczak, et al. (2009). "Synthesis of Multifunctional Composite Microgels via In Situ Ni Growth on pNIPAM-Coated Au Nanoparticles." Acs Nano 3(10): 3184-3190.

Schatz, G. C. (1984). "Theoretical-Studies of Surface Enhanced Raman-Scattering." Accounts of Chemical Research 17(10): 370-376.

Schlucker, S. (2009). "SERS Microscopy: Nanoparticle Probes and Biomedical Applications." Chemphyschem 10(9-10): 1344-1354.

Sepulveda, B., P. C. Angelome, et al. (2009). "LSPR-based nanobiosensors." Nano Today 4(3): 244-251.

Sierra-Martin, B., Y. Choi, et al. (2005). "Microscopic signature of a microgel volume phase transition." Macromolecules 38(26): 10782-10787.

Smith, E. and G. Dent (2005). Modern Raman Spectroscopy: A Practical Approach., John Wiley and Sons.

Stevenson, C. L. and T. Vo-Dinh (1996). Modern Techniques in Raman Spectroscopy New York, Wiley.

Stiles, P. L., J. A. Dieringer, et al. (2008). "Surface-Enhanced Raman Spectroscopy." Annual Review of Analytical Chemistry 1: 601-626.

Sun, H., O. X. Shen, et al. (2008). "Carbaryl, 1-naphthol and 2-naphthol inhibit the beta-1 thyroid hormone receptor-mediated transcription in vitro." Toxicology 249(2-3): 238-242.

Tian, Z. Q., B. Ren, et al. (2002). "Surface-enhanced Raman scattering: From noble to transition metals and from rough surfaces to ordered nanostructures." Journal of Physical Chemistry B 106(37): 9463-9483.

Vesseur, E. J. R., R. de Waele, et al. (2007). "Direct observation of plasmonic modes in Au nanowires using high-resolution cathodoluminescence Spectroscopy." Nano Letters 7(9): 2843-2846.

Vesseur, E. J. R., R. de Waele, et al. (2007). "Direct observation of plasmonic modes in Au nanowires using high-resolution cathodoluminescence Spectroscopy." Nano Letters 7: 2843-2846.

Vo-Dinh, T. (1998). "Surface-enhanced Raman spectroscopy using metallic nanostructures." Trac-Trends in Analytical Chemistry 17(8-9): 557-582.

Wang, H., F. Tam, et al. (2005). "Cu nanoshells: Effects of interband transitions on the nanoparticle plasmon resonance." Journal of Physical Chemistry B 109(39): 18218-18222.

Watson, D. A., L. O. Brown, et al. (2008). "A flow cytometer for the measurement of Raman spectra." Cytometry Part A 73A(2): 119-128.

Willets, K. A. and R. P. Van Duyne (2007). "Localized surface plasmon resonance spectroscopy and sensing." Annual Review of Physical Chemistry 58: 267-297.

Xu, H. X., J. Aizpurua, et al. (2000). "Electromagnetic contributions to single-molecule sensitivity in surface-enhanced Raman scattering." Physical Review E 62(3): 4318-4324.

Yang, Z. S., Y. W. Lin, et al. (2005). "Impacts that pH and metal ion concentration have on the synthesis of bimetallic and trimetallic nanorods from gold seeds." Journal of Materials Chemistry 15(25): 2450-2454.

Yavuz, M. S., W. Y. Li, et al. (2009). "Facile Synthesis of Gold Icosahedra in an Aqueous Solution by Reacting HAuCl4 with N-Vinyl Pyrrolidone." Chemistry-a European Journal 15(47): 13181-13187.

Zayats, A. V., Smolyaninov, II, et al. (2005). "Nano-optics of surface plasmon polaritons." Physics Reports-Review Section of Physics Letters 408(3-4): 131-314.

Zha, L. S., Y. Zhang, et al. (2002). "Monodisperse temperature-sensitive microcontainers." Advanced Materials 14(15): 1090-1092.

Zhao, J., A. O. Pinchuk, et al. (2008). "Methods for Describing the Electromagnetic Properties of Silver and Gold Nanoparticles." Accounts of Chemical Research 41(12): 1710-1720.

8

Optical Fiber Microsensor of Semidrop

Esteban Molina-Flores, R. B. López-Flores, Daniel Molina-Flores,
José A. Dávila-Píntle, Germán A. Muñoz-Hernández
Carlos A. Gracios-Marín and Enrique Morales-Rodríguez
Benemérita Universidad Autónoma de Puebla, FCE, Optoelectrónica
México

1. Introduction

The optical fiber systems until now have turned out to be an ample resource that offers alternatives for tasks of detection and transduction of the energy forms, as a classification of these optical fiber transducers, in this task we presented a sensor that belongs to those of the refractometric type. Since 1995, researchers Khotiaintsev Sergei, Victor De Leon Paredes, Esteban Molina Flores, Alexandre Zemliak made a theoretical modeling analysis of a variety of refractometric sensors based on a pair of fiber optic ends, all useful for measurement of refractive indices. These sensors had an oblique disposition, and these researches are conducted the analysis of best response on the degree of misalignment of two fibers, and to the resulting form of its hemispherical surface of contact with surrounding means. These researches always were within the theoretical regime and for monochromatic irradiation, and never had a tangible evidence of an experimental result (Jones & Zimmer, 1978; Molina-Flores, 1995; Khotiaintsev et al, 1996, 2000, 2002). In 2005 the researcher Esteban Molina Flores, main adviser of this work, proposed and managed to realize the micromachining, one of this microsensorial element with technology of electrical arc, in his particular parallel fiber version, and with surface of semidrop, obtaining theoretical and experimental results in his characterization in the infrared region (1550 nm) for this refractometric microsensor. Like integral part of the research line, the refractometric sensors developed by researcher Esteban Molina Flores (Molina-Flores, 2006), in this occasion through this project, theoretical and experimental results from this microsensor operating in the infrared region are presented, whose particular applying are to solve the problem of the detection, identification and measurement of liquid levels in a cistern tank, by means of the identification of the interface of the immiscible liquids: water and oil engine, and of the interface air-oil engine. The objectives of this work are divided in two kinds of objectives: General objective: realize the characterization of a refractometric microsensor in the infrared region. Particular objectives: a) Make our own sensor of semidrop, b) Establish the conditions of the installation and application of the microsensor with which its efficient operation is guaranteed, c) To realize the theoretical proposal of a design of application of the microsensor of semidrop for the detection of a pair of immiscible liquids. The order in which the sections were considered, obeys to the intention to show the reader the development stages through which it passed the microsensor until the publication of the results in indexed magazine.

2. Manufacturing of the microsensor

In this project the methodology of manufacture of a refractometric microsensor of fiber is presented, which is elaborated from a pair of adjacent fiber tips of parallel disposition, which are fused on a specific form in one of its ends to form a semidrop that serves like detection or refractometric interaction surface with the liquid in submission, see Fig. 1.

2.1 Optical fiber microsensor micromachining

The technology implemented in this project is based on heat source by electrical arc.

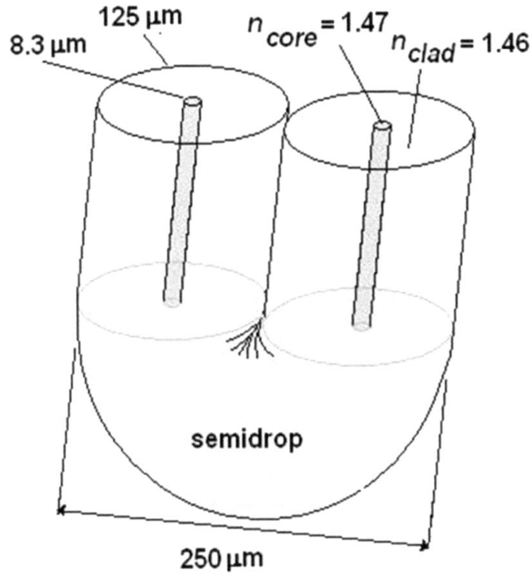

Fig. 1. Disposition of fibers and the fused region is the formed semidrop

The physical manufacture of the microsensor is repeatable every time respecting the characteristics of electric power applied to melt the glass fiber to 1900 ° C, and the specific moments and points in which they are applied time and time again to obtain the form of the semidrop, and to guarantee therefore the awaited optical response in this microsensor. This sensor can be excited with a visible or infrared irradiation source, behaving the fibers like fibers monomodal or multimodal, respectively. The preparation of the optical fiber ends is of the same way in which the optical fiber ends are prepared to realize a common and current splice of low loss. This procedure consists of removing two centimeters of polymer in each fiber end, and realizing an orthogonal cut to the fiber axis, right in its end. After being carefully cleaned with a clean cloth and acetone each of these ends, it can be verified through the camera of the optical fiber splicer machine. Once prepared fibers, they are placed in parallel way, within the fusion chamber of the splicer machine, so that the electrical arc is applied to them, as is shown en Fig. 2.

2.2 Installation of the pair of optical fiber ends in the electrical arc

As experimental installation was used the chamber of fusion of the optical fiber splicer, Fitel mark, whose loss record per splice is of 0,01 dB to monomode fiber, each splice can be realized in a lapse of 13 seconds approximately. The optical fiber splicer is a machine that radiates electrical arcs whose electric power are of 20 W to splice fiber monomode, and 40 W to splice fiber multimode. In this application microsensor was made with monomode optical fibers. After cleaning carefully with acetone the optical fiber ends, these are installed in parallel in the fusion chamber of the splicer machine, in such a way that the pair of fiber ends is tied and right where the electrical arc will form. This task is not easy the first times, but with a little practice and care, it is routine and effective. Fortunately the system of the splicer has control of axial micropositioning with which the point can be chosen where it is desired to apply the discharge in pair of fiber ends.

2.3 Formation of the semidrop: formation stages

In order to obtain the required optical results with the microsensor, it is very important to obtain the form or model of the semidrop. As all artisan work is necessary to have a serious dominion and knowledge on the operation of the splicer machine, to have absolute control on the final physical form that we wished to give our microsensor. In order to explain the form in which east microsensor of optical fiber was micromachined, we have planned to break down this process into three main steps: fusion of the ends, getting the preform, and obtaining semidrop. These parameters are described in detail, as follows:

2.3.1 Fusion of the ends

Properly placed the fiber ends, as it was explained previously, the discharge of the 20 W electrical arc is practiced automatically, by a lapse of 13 seconds approximately. The place where the electrical arc is due to apply is just, at the end of the pair of ends, as is shown in the Fig. 2.

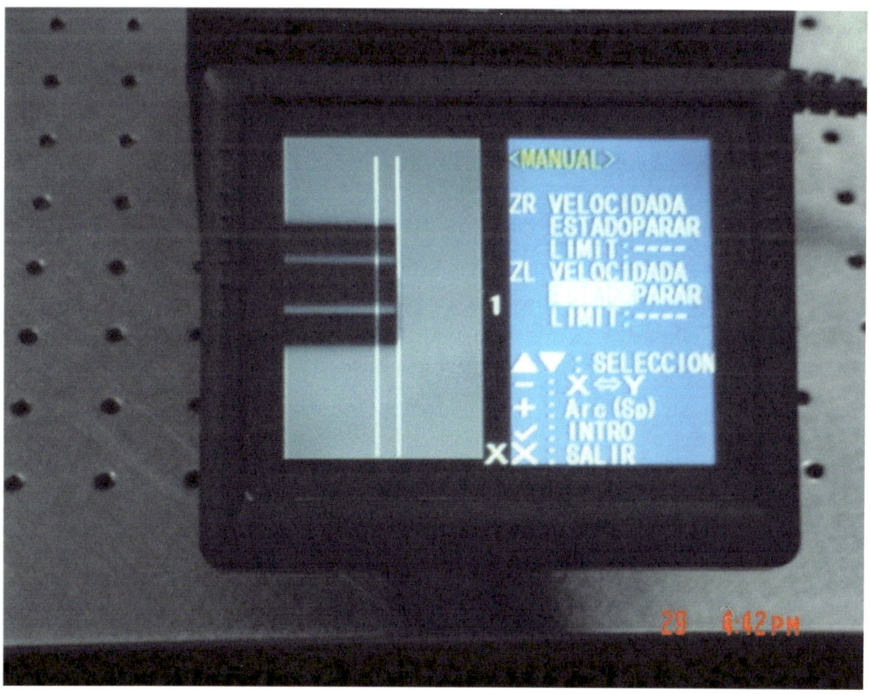

Fig. 2. Application point of the electrical arc (*screen on Spanish*)

This application has like objective realizing a union that fixes both fiber ends. By means of the application of this electrical arc, the pair of fiber ends reaches the temperature of 1900 °C melt merging and reached a preliminary part of the fiber end with the other, as shown in Fig. 3.

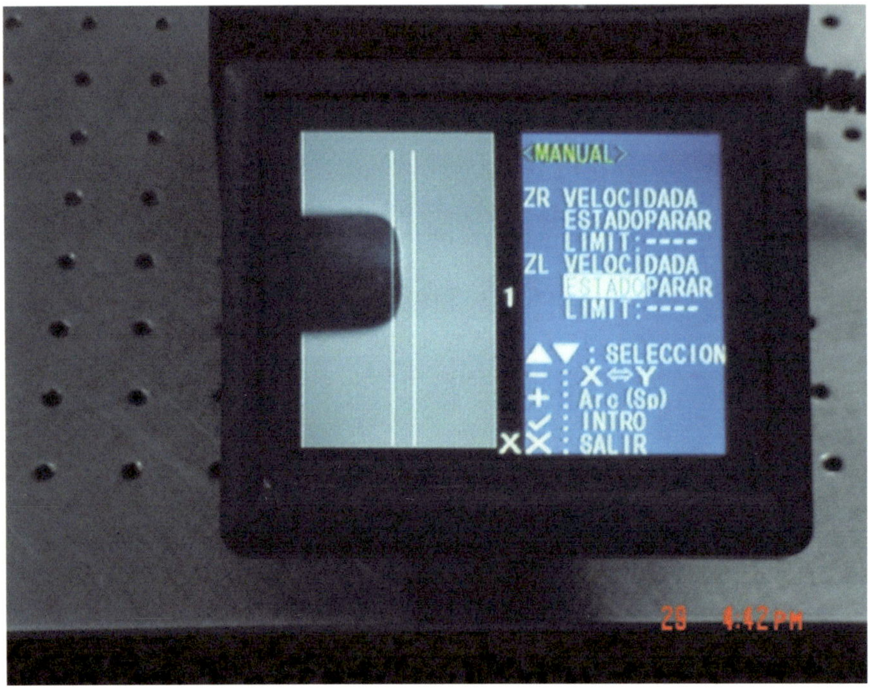

Fig. 3. Fusion of ends

2.3.2 Obtaining of the preform

After the first electrical discharge to have realized, the pair of fiber ends has been fused partially. The following discharging, of the same characteristics, is applied right in the region interface fused fiber-fiber, as it is in the Fig. 4. The objective of this discharging is to approximate the preform obtained towards the physical model of the semidrop.

This is obtained by means of positioning again the pair of fused fibers, just where the discharge will appear, see Fig. 5. 20 W electrical power by 15 seconds is applied after the discharge place has been chosen. Therefore, the following preform of the microsensor has being obtained.

Note physically, the semidrop is already shaped and this stage is called the microsensor preform, see Fig. 6, but is necessary to provide one more a discharging to homogenize, in liquid face, the distribution of refractive indices in the region of the semidrop, to ensure that the microsensor will have a transmittance of not more than 30 dB for the visible or IR region in contact with the air.

2.3.3 Obtaining of the semidrop

The last discharging is applied, just in the curved zone of the preform, as is in the Fig. 7. This discharging also is of the same characteristics, that the previous ones. In this last application,

because the fusion temperature in the region of the semidrop is reached, the splicer machine is due to have in an advisable position to attenuate the effect of the gravity force during the application of the electrical arc, and thus to avoid a deformation in the semidrop geometry.
After the electrical discharging applying, Fig. 8, the final conformation microsensor is obtained with the form shown in the Fig. 9. Final dimensions are width 250 μm, and thickness 125 μm.

3. Physical description of the optical fiber microsensor

The optical fiber microsensor is constituted by a pair of parallel optical fibers, where one of the ends of both fibers are joined by controlled electric arc melting, forming a glass semigota, the remaining two fiber terminals work as input and output ends, see Fig. 1. The fibers used are step index, whose core and cladding diameters are respectively 8,3/125 microns, these fibers are commonly used in optical communications, for a wavelength of 1550 nm. The optical fiber microsensor has a transmittance function or transmitted, this function indicates the optical filter response imposed by this microsensor dependent on wavelength, and refractive index of the liquid tested. The optical fibers when transmitting perfectly in the 1550 vicinity nm, do not present attenuation in the region of the microsensor that corresponds to fibers.
Nevertheless, the microcavity formed in the fused ends, by its dimensions, geometric, wavelength, and the refractive index of external means, *n*; they influence considerably in the behavior of the function of transmitivity of the microsensor. In order to guarantee the same response of the fiber microsensor, the fused ends must describe a glass semidrop.

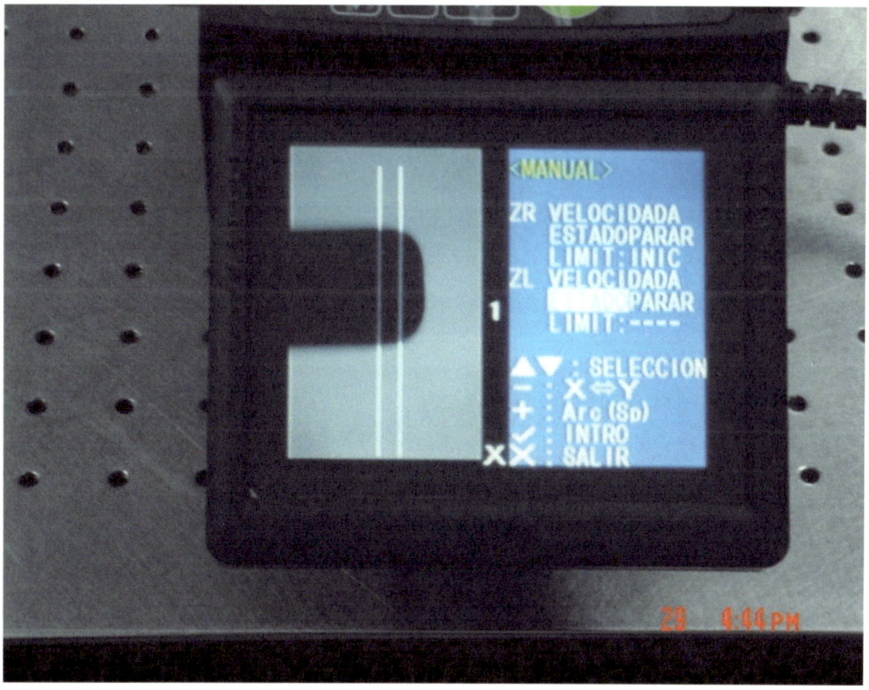

Fig. 4. Place of application of the electrical arc

In the region of interaction of the semidrop, the microcavity or semidrop constitutes to incident medium (glass), and to medium of transmission it constitutes external means. The reflective properties of the semigota- surrounding medium interface, are regulated by the coefficients of Fresnell, which are not evaluated due to the limitations that impose the dimensions of the microcavity, nevertheless for experimental intentions is advisable to determine the transmittance of the fiber microsensor, $T(\lambda, n)$. This means that if we keep fixed the wavelength of radiation, the transmittance will only be based on the refractive index of surrounding medium, $T(n)$.

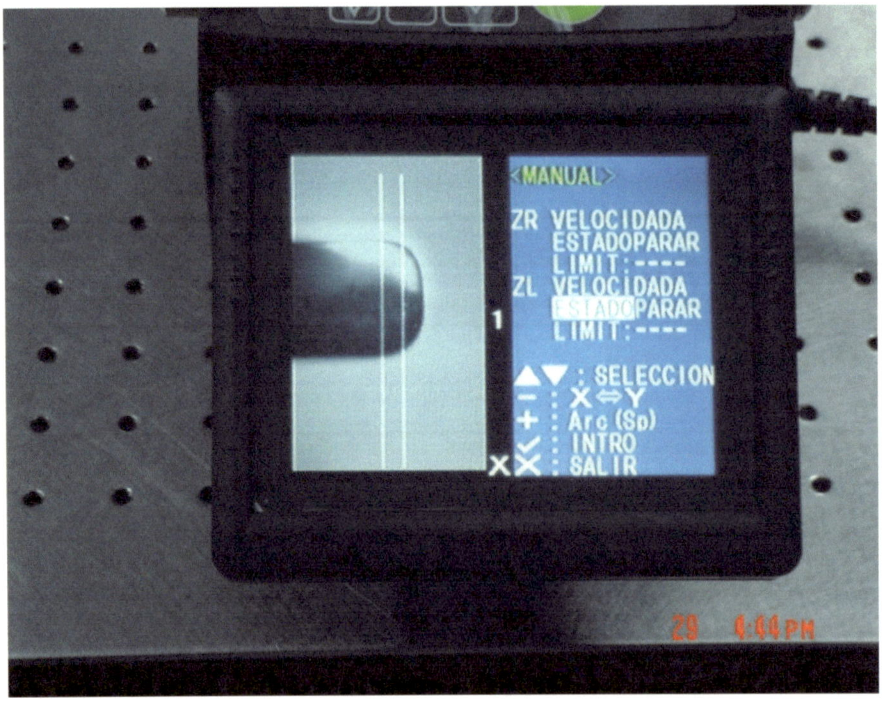

Fig. 5. Application of the electrical arc

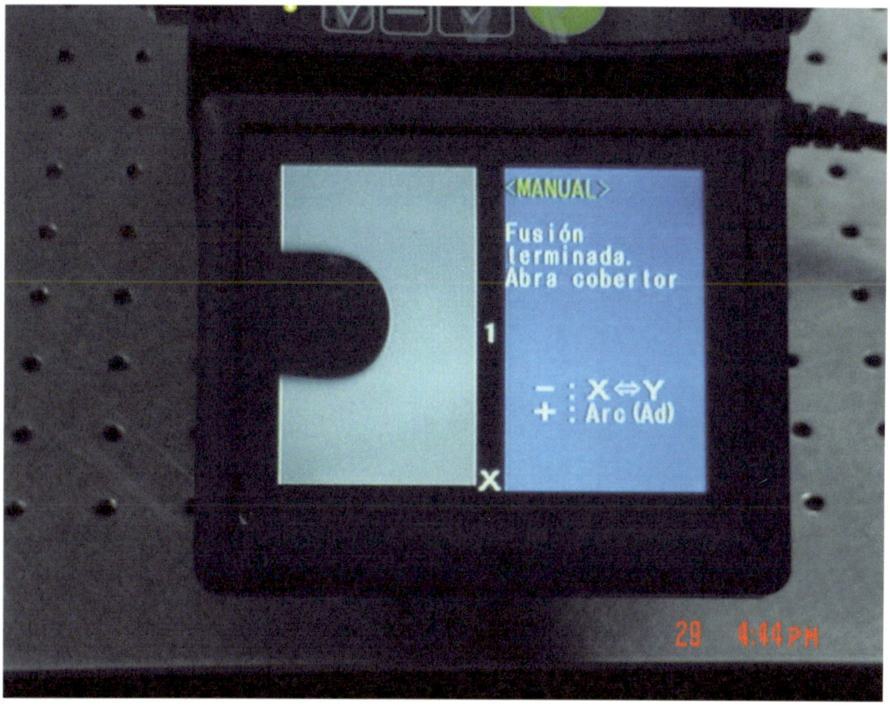

Fig. 6. Obtain of the preform

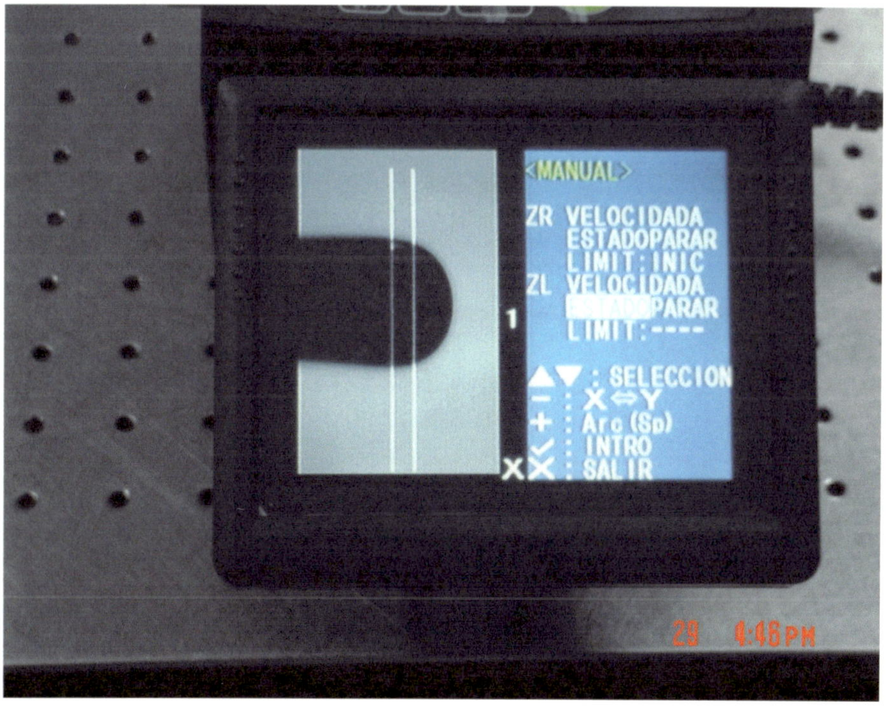

Fig. 7. Place of application of the discharging

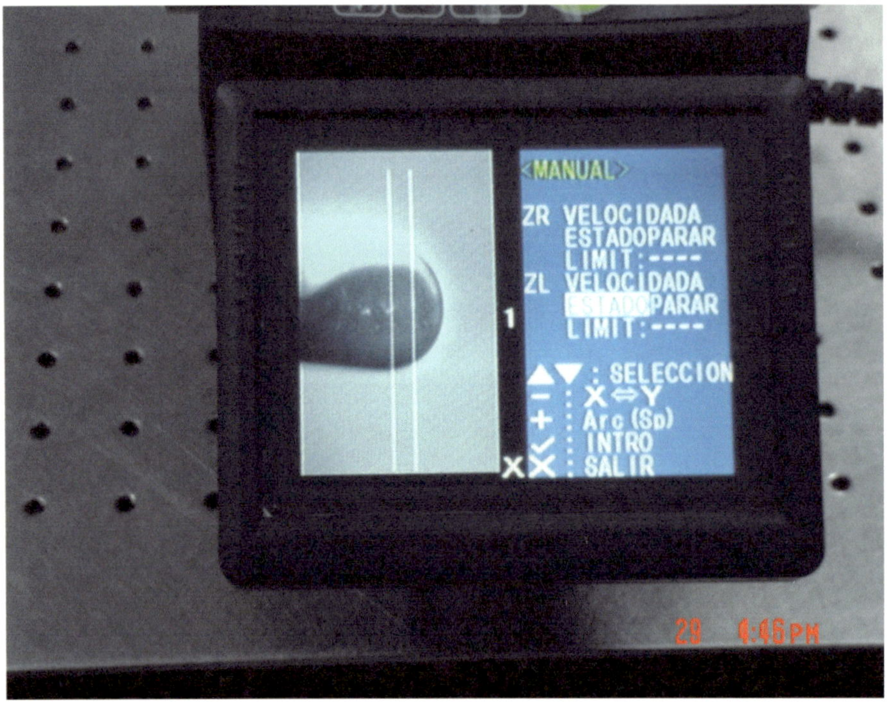

Fig. 8. Application of the electrical arc

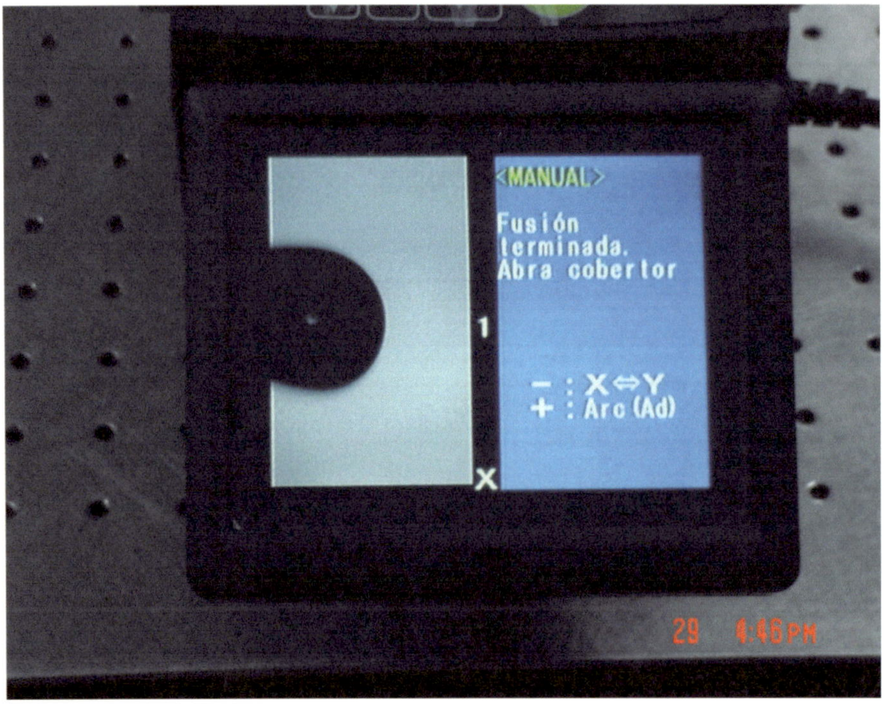

Fig. 9. Obtain of the semidrop

4. Theoretical model of the performance of the optical fiber microsensor

The theory of operation of the fiber microsensor starts off of the fact that each pair of immiscible liquids owns its respective refractive indices: n_1 y n_2. The microsensor of fiber, like optical device, owns its transmittance function that in this case is based on the refractive index of the liquid within which the semidrop of the microsensor is immersed, Eq. 1.

$$T(n) = \frac{P_{out}(n)}{P_{in}} \tag{1}$$

Since the optical power of input, P_{in}, is modified by the reflective properties of the semidrop interface liquid-microsensor in contact, the P_{out} modifies, based on n, of the liquid, and is translated to a V_{pho} of output, by means of a photodetector, that owns a responsivity of voltage, R_v. Eq. (2), illustrates the relation of P_{in} with respect to the measured voltage at the output of the microsensorial system, V_{pho}. A dynamic resistance of the photodetector of 1 Ω, is assumed, by simplicity.

$$V^2_{pho} = R_{volt}P_{out} = R_{volt}T(n)P_{in} \tag{2}$$

Since for each one of the immiscible liquids, including the air, corresponding output voltages, V_{out}. A identification free of errors, offers a high contrast between the associated voltages to each liquid, ΔV , as shown in Eq. 3.

$$\Delta V = V_{phoout2} - V_{phoout1} \tag{3}$$

Considering that the guarantee to operate to the microsensor with two average immiscible liquids, is that the transmittance differences are $\Delta T \neq 0$, implying a nonlinear relation, Eq. 4.

$$0 < \Delta V = \sqrt{R_{volt}\Delta T P_{in}} \tag{4}$$

Sensitivity to both immiscible liquid refractive index, *S(n)*, is described on the basis of which sensitivity in the rapidity of change of output, *T (n)*, with respect to the input *n*, for the case of our microsensor, Eq. 5.

$$S(\Delta n) = \frac{\Delta T}{\Delta n} \tag{5}$$

where,

$$\frac{\Delta T}{\Delta n} = \frac{T(n_{water}) - T(n_{oil})}{n_{water} - n_{oil}} \tag{6}$$

$$S(\Delta n) = \frac{T(n_2) - T(n_1)}{n_2 - n_1} \tag{7}$$

with respect to the rapidity of commutation of the microsensor, this property rests in the bandwidth of the photodetector used in the microsensorial system. Nevertheless in theoretical terms, the rapidity of commutation, Δt_{rise}, is shown through Eq. 8.

$$\Delta t_{rise} = 0.9(t_2 - t_1) \tag{8}$$

where, t_2, is the instant where it happens the maximum value, and t_1, is the instant where the minimum voltage of V_{phoout} happens. Nevertheless in practical terms, and valid for optoelectronic devices, it is possible to be calculated, through Eq. 9.

$$\Delta t_{rise} = \frac{0.35}{BW} \tag{9}$$

where BW is the bandwidth of the photodetector.

5. Experimental results and discussions

Like a pair of immiscible liquids, it has been selected to oil and to the water, by the importance that it has in the economic growth in the world. In this work it was investigated firstly, the effect of the state of polarization of the electric field to the input of the fiber microsensor on the voltage induced in the photodetector, V_{phoout}. The configuration used in this measurement is shown in Fig. 10.

According to Eq. 2, the voltage that appears in the photodetector, Vphoout, does not have to depend on the polarization of the incident electric field to the microsensor input. However in the interface semigota-liquid, reflections of Fresnell take place, where the polarization of the incident beams is polarimetrically influenced by the characteristics of the same. In order to verify experimentally, that the voltage that appears in the photodetector, Vphoout, is independent of the polarization, it was made vary the state of polarization of the field, $\vec{E}in$, by means of the use of a polarization controller of three coils.

When varying the angular position of the coil modifies the angular inclination of state of linear polarization of the field, $\vec{E}in$, to the input of the microsensor. The dependency of the voltage that appears in the photodetector, V_{phoout}, with respect to angle θ, of polarization at the input of the microsensor, as shown in Fig. 11. This experiment was realized at the wavelength of 1550 nm, with interface semidrop-water.

It is appraised that the voltage that appears in the photodetector, V_{phoout}, has small dependency of the angle of polarization. Probably in the region of the semidrop, the dependency of the polarization is remarkable, but this is inhibited by the fiber segments, the input fiber segment, and the output fiber segment, which are standard and non-preservers of the polarization, whose lengths are of not less than 30 cm. In this experiment, the voltage that appears in the photodetector, V_{phoout}, presents fluctuations within the width vicinity ± 0,1 μV, which demonstrates that, V_{phoout}, is independent of the polarization.

In order to research the corresponding transmittances for each interaction medium: n_{air}, n_{water}, y n_{oil}, was used an irradiation source that operates in the vicinity of the 1550 nm, for these reason it implemented the gain spectrum of a 10 m segment of erbium doped fiber (EDFA). This irradiation source is an EDFA arrangement without signal at the input, is unidirectional, reason why it emits at the output the spectrum gain of the EDFA, as shown in Fig. 12. EDFA spectrum output is shown in Fig. 13. These spectrums own intensities different from zero, in the interval from 1500 from 1600 nm. In order to enter this known spectrum at the input of the fiber microsensor, the output of EDFA was coupled with the input end of fiber microsensor

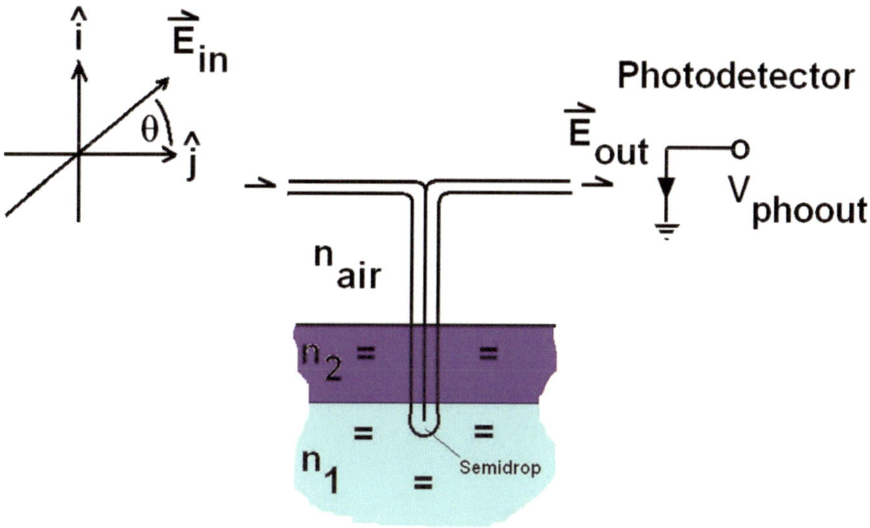

Fig. 10. The semidrop is immersed in two immiscible mediums, whose refractive indices are n_1, n_2, additionally n_{air}

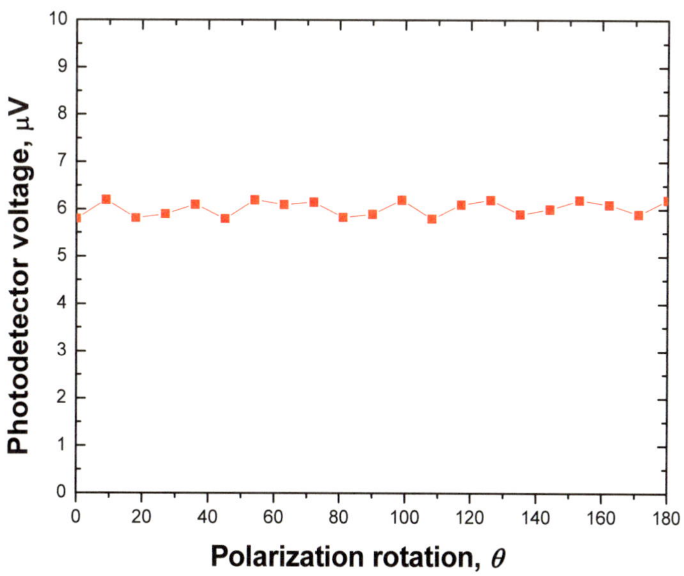

Fig. 11. Voltage variations of the microsensorial system output

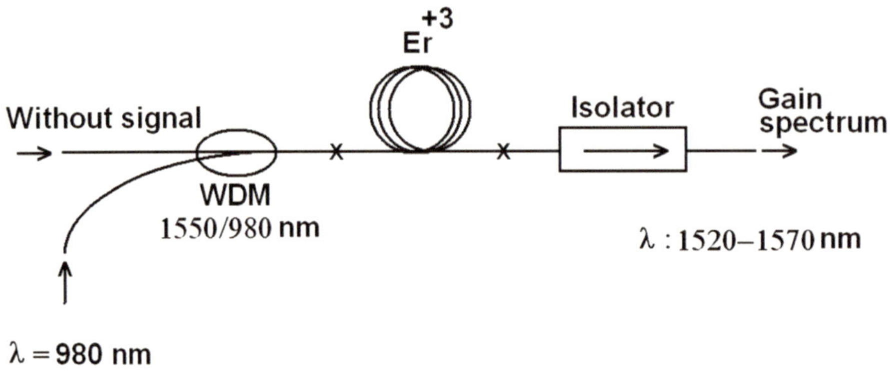

Fig. 12. Setup of EDFA, without input signal

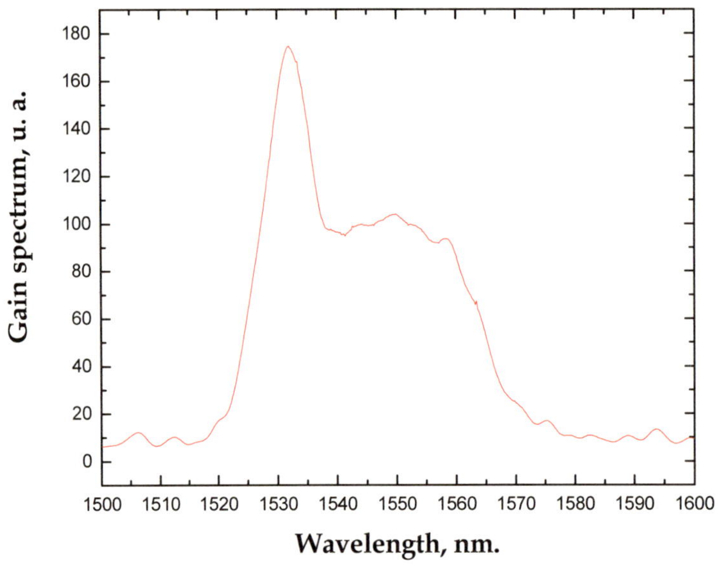

Fig. 13. Irradiation spectrum used to determine the spectral response of optical fiber microsensor

The experimental scheme used to obtain the output spectra is shown in the Fig. 14. The output spectra are useful to determine the functions of transmittance of the microsensor corresponding to each surrounding mediums, according to Eq. 1, for each refractive index n_{air}, n_{water}, and n_{oil}.

The Fig. 15 shows to the value of transmittance functions for each system semigota-water, semidrop-air, and semidrop-oil, in the region of the 1500-1600 nm. These functions are standardized with respect to the transmittance spectrum of the system semidrop-air, which presents the higher peak of transmission in 1527 nm, and it is different from zero from 1508 nm to 1560 nm. The transmittance spectrum of the system semidrop-water presents a peak in 1522 nm, and it is different from zero from 1515 nm to 1560 nm. The transmittance spectrum of the system semidrop-oil is equal to zero from 1500 nm to the 1600 nm. Reason why to distinguish oil of air, is propitious to use wavelengths in the 1508-1560 nm range, and to distinguish oil of water, is propitious to use wavelengths in the 1515-1560 nm range. Nevertheless, to discriminate the water of oil, the spectral region that offers this possibility, we divided it in two: the one of low contrast, and the one of high contrast. The one of low contrast begins from 1500 to 1515 nm, and from 1534 nm to 1562 nm, as it is possible to be appreciated in the Fig. 15.

The spectral region of high contrast begins from 1523 nm to 1534 nm. This procedure ensures the success of operation of the optical fiber microsensor, and it is attainable for any other pair of immiscible liquids of industrial or biomedical interest.

Remarkable contrast exists in the values of *T(n)* for the mediums air, oil and air for the wavelength of 1550 nm, as is shown in Fig. 16. During the experiment and measuring, the decays of power, P_{out}, were measured in percentage of the power compared with the output power. In other words, the P_{out} of the system semigota-air represents the 100% of the microsensor.

When submerging the microsensor towards oil, was observed a reduction to 0,0% of the la P_{out} power of the air. Nevertheless, when the microsensor made contact with the water, the power was restituted to 18,5% with respect to the P_{out} of the air. These results are in Table 1. These data show the capacity of discrimination of the microsensor for the wavelength of 1550 nm.

In order to demonstrate the effectiveness of microsensor of fiber for the detection of the interfaces formed by immiscible liquids, the graph shown in the Fig. 17 was developed. In this graph the changes of transmittance acquire knowledge that detects the microsensor of fiber, when realizing an orthogonal sweeping, from the bottom towards the surface, crossing the interfaces that form in a tank cistern that contains interfaces water-oil, , h_{water}, oil-air, h_{oil}.

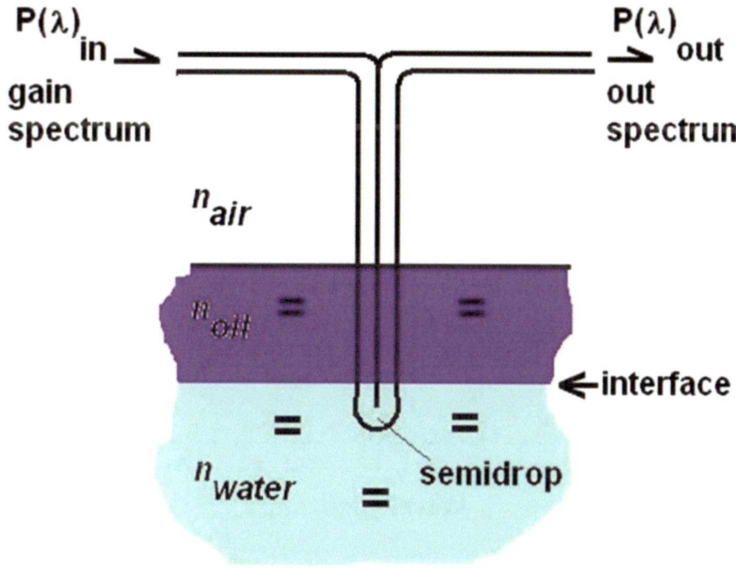

Fig. 14. Experimental setup to obtain output spectrum in the IR region (1550-1600nm)

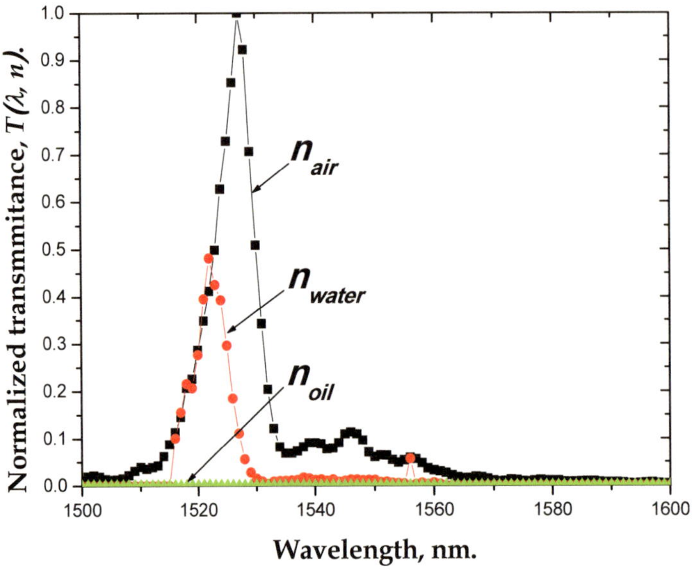

Fig. 15. Transmittance spectra for semigota-water systems, semigota-air, and semigota-oil

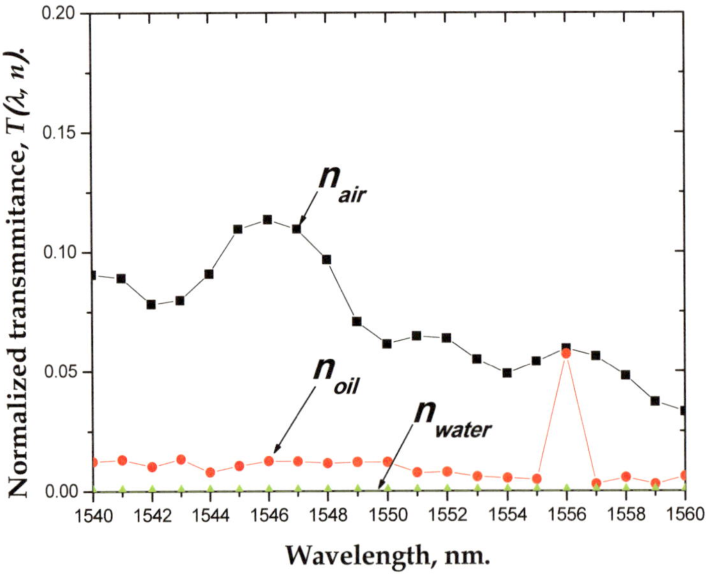

Fig. 16. Spectral region of low contrast to n_{water} and n_{pil} detection

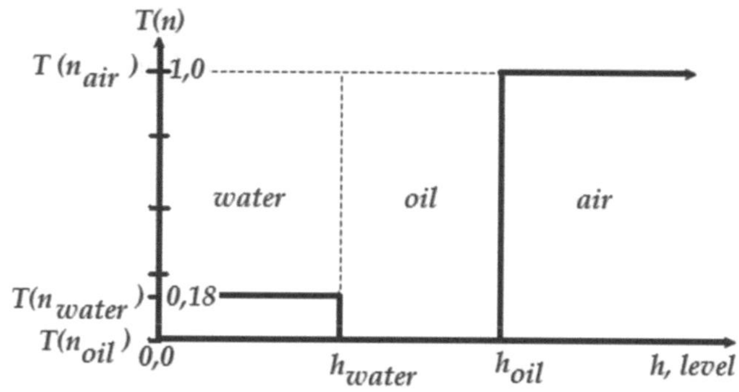

Fig. 17. Transmittance values corresponding to levels of water, oil, and air

It is observed that in these levels, a commutation in the transmittance of the fiber microsensor happens, that is interpreted like the existence of each of these interfaces that appear in the axis of *h*, level. This evaluation was performed for a wavelength of 1550 nm. Note that the maximum transmittance for semidrop-air system, considered as 100% for the semidrop-water system is 18,5%, and semidrop-oil system is zero. For this reason, the fiber microsensor is functional for the detection of interfaces formed between the two immiscible liquids.
To research the sensitivity of the optical fiber microsensor, referred to Eq. 7, which requires the information specified in Table 1, and the indices of refraction of water 1,3, oil 1,48, and air 1,0. Note in Table 2, the fiber microsensor has a high sensitivity to cross the air-oil, because the change of transmittance is from 1,0 to 0,0, for a small variation of refractive indices. In the case of oil-water interface, transmittance values are very close, as their refractive indices, so that the sensitivity of fiber microsensor suffers a significant reduction.

Surrounding medium	*T(n) to 1550 nm, (%)*
air	100
water	18,5
oil	0,0

Table 1. Percentage decay of the microsensor *T(n)*

Interfaces	*S(Δn)*
water-oil	1,0
oil-air	2,0

Table 2. Optical fiber microsensor sensitivity

With a similar procedure and by using a white radiation source (light emitting diode of high brightness) and a silicon photodetector, gives the percentage values of *T(n)* of the system interacting with the same mediums, see Table 3. Note that the contrast in the detection for the air, water and oil has improved. The advantage to radiate with white light source is that, all wavelengths of the interval from 390 to 780 nm have contributed, so photodetector collect more power.

Surrounding medium	*T(n) to 390-780 nm, (%)*
air	100
water	13
oil	20

Table 3. Percentage decay of the microsensor *T(n)* for visible region

Like any proximity sensor device, the optical fiber microsensor when passing the interface of air, oil, and water, indirectly reports the location of the interface, by location or area of switching their output voltages: H (high) or L (low). Knowledge of minimum time, that the microsensor can resolve a change of medium, is of importance to the location of the interfaces of immiscible liquids. The switching speed of the microsensor is strongly restricted by the bandwidth of the photodetector to the optical output of the microsensor (13 kHz). Under this, to know the response time of the microsensor, it applies the criterion of rise time, tr, (risetime) with respect to bandwidth of LTI system, which in this case is the bandwidth, BW, of InGaAs photodetector. The relationship applied in this deduction is for Eq. 9, so that, tr = 26,92 μm. Finally, the finish of fiber optic microsensor is shown in Fig. 18, and specifications of optimized optical fiber microsensor are shown in Table 4.

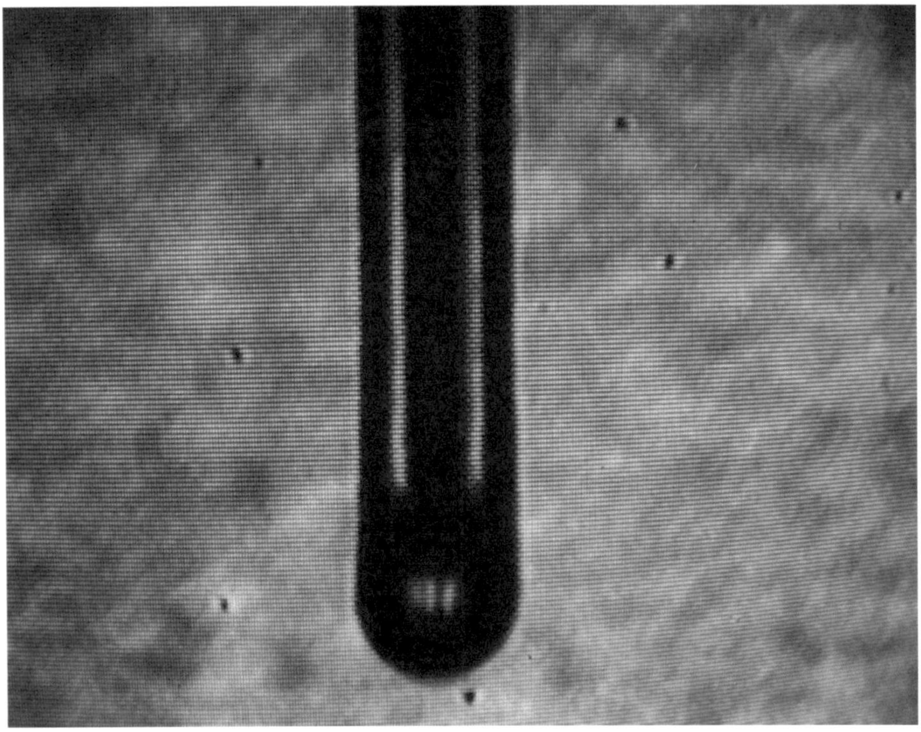

Fig. 18. Physical illustration of the optical fiber microsensor, increased 220 times in a monitor screen

Elements	*Characteristics*
Optical fibers	8,3/125 μm
ILD	1550 ± 0,5 nm
Power range	5 mW - 6,5 mW
Photodetector	Si low power, BW= 1,75 MHz
Max. response time	< 2 μs
Dynamic range	Refractive index < 1,9
Sensitivity	- 5,714 x 10^{-4} μV, water-oil
Temperature	0° a 50 °C

Table 4. Specifications of optimized optical fiber microsensor

6. Conclusions

This research has shown that the fiber optical microsensor is possible to manufacture by electric arc technology. We demonstrated that is possible to improve the response detection of optical fiber microsensor, for certain immiscible liquids, by proper choice of a specific wavelength. This wavelength or radiation spectral region should provide a high discrimination of the two liquids. Speed detection of the microsensor, strongly dependent on the bandwidth of the photodetector. The method used in this research can be applied in the detection of other mediums, including other regions of the optical spectrum. Due to geometrical and physical properties of fiber microsensor, it can be applied to biomedicine, because it is micrometer-scale, flexible like being guided in an artery or a catheter inserted, and it is easy to sterilize. As future research work, we have considered using the microsensor to implement a system of characterization and recognition of organic and inorganic liquids (alcohols, acids, neutrals, blood, urine, water, etc.) by using spectroscopy in the infrared and/or visible region. Optical fiber microsensor represents an alternative in the industrial applications, and mainly in the detection of explosive, corrosive and/or highly harmful liquids, for the human being.

7. Acknowledgment

We are thankful to the Benemérita Universidad Autónoma de Puebla, and the group of full professors of the Facultad de Ciencias de la Electrónica, Optoelectronic department, by to have looked for at the time, the ways and possibilities, to acquire these efficient equipments of measurement, to have been able to make reality this project.

8. References

De Leon , V. & Khotiaintsev, S. (1996). Raytracing in the design of microoptical sensors used for the determination of refractive index of surrounding media. *Proceeding of SPIE, Infrared Spaceborne Remote Sensing IV* ,Vol. 2817, pp. 299-311, ISBN 9780819422057, Denver, CO, USA, 1996

Jones, O. C. & Zimmer, G. A. (1978). Optical probe for local void fraction and interface velocity measurements, *Review of Scientific Instruments,* Vol. 49, No. 8, (August 1978), pp. 1090-1094, ISSN 0034-6748

Molina-Flores, E. et al, (1995). Fiber-optic multipoint high-resolution level-sensor for biomedical applications. *Proceeding of SPIE, Medical and Fiber Optic Sensors and Delivery Systems,* Vol. 2631, pp. 121-126, ISBN 9780819419958, Barcelona, Spain, 1995

Molina-Flores, E., Aguilar-Jiménez, O. & Molina-Flores, D. (2006). Microsensor refractométrico discreto de fibra óptica para detección de interfaces petróleo-agua, operando en la región visible. *Internet Electrón. J. Nanocs. Moletrón,* Vol. 4, No. 3, (December 2006), pp. 815-826, ISSN 0188-6150

Svirid, V.; Khotiaintsev, S. & Swart, P. L. (2002). Novel optical fiber refractometric transducer employing hemispherical detection element, *Optical Engineering,* Vol. 41, No. 4, pp. 779–787, ISSN 0091-3286

Volodymyr, S., & Khotiaintsev, S., (2000). Optoelectronic multipoint liquid level sensor for light petrochemical products, *Proceedings of SPIE, Optoelectronic and Hybrid Optical/Digital Systems for Image and Signal,* Vol. 4148, pp. 262-268, ISBN: 9780819437952, 2000

9

A Glass Capillary-based Microsensor for L-Glutamate in *in vitro* Uses

Masao Sugawara and Atushi Shoji
Department of Chemistry, College of Humanities and Sciences, Nihon University, Tokyo Japan

1. Introduction

Since the pioneering work by R. Adams (1976), who detected a neurotransmitter catecholamine in mammalian brain by implanting a solid carbon electrode directly in animal brains, a lot of miniaturized *in vivo* and *in vitro* sensors for neurotransmitters have been poposed (Hirano&Sugawara, 2006; Sugawara, 2007; Zeyden et al. 2008; O'Neill et al., 1998). The field is slowly, but continually, expanding. Most of researches aim at developing electrochemical sensors with spatial and temporal resolution, which enable us to discern distribution of neurotransmitters within each neuronal subfield and estimate its concentration level and temporal changes in intact brains, acute slices and cultured neurons.
In the central neuronal system of mammalian brain, L-glutamate is released from synaptic terminals and plays a vital role in brain development, synaptic plasticity, neurotoxicity, and neuropathological disorders (Reis et al., 2009; Bliss&Collingdige, 1993; Malenka&Nicoll, 1999). L-Glutamate is also involved in neuropathological disorders such as epilepsy, stroke, Parkinson's disease and Alzheimer's disease (Nishizawa, 2001; Mattoson et al., 2008). L-Glutamate may activate transmitter receptors located extrasynaptically on neurons and glia at greater distance from the place of exocytosis of synaptic vesicles, though the concentration level of such spillover of L-glutamate from synaptic cleft is not clear yet (Volterra&Meldolesi, 2005). The basal and enhanced level of extracellular L-glutamate plays a key role in neuronal functions, because its level will determine whether L-glutamate has actions or negligible actions on most glutamate receptors (Herman&Jahr, 2007).
Up to date, the *in vivo* level of L-glutamate in brain has been reported mostly for corpus striatum, while no *in vivo* data have been reported for the hippocampus. On the other hand, acute slices of hippocampal tissue offer experimental control of the neuronal network environment. In the *in vitro* case as well, a very limited number of microsensors have been applied to acute brain slices, probably because of technical difficulties in handling thin living slices (thickness 200-400 μm) and the lack of suitable miniaturized sensors.
A glass capillary-based enzyme sensor has been developed in which a three-electrode system is built in the capillary (Nakajima et al., 2003), hence outer reference and auxiliary electrodes are not necessary to be set in brain tissue. The sensor with the tip diameter of approximately 10 μm is promising as a microsensor for monitoring the enhanced extracellular level of L-glutamate release in each neuronal region of acute hippocampal slices

under chemical and electric stimulation. In this review, we describe the principle, properties and application of a glass capillary-based sensor for *in vitro* monitoring of L-glutamate in hippocampal slices.

2. Preparation and response principle of a glass capillary sensor for L-glutamate

2.1 Preparation of a glutamate oxidase (GluOx)-coated Au electrode

A working Au electrode, which is to be set in a capillary pipette, is prepared in the conventional manner (Oka et al., 2007). The one end of a gold wire (ø 0.30 mm) is coated with 0.3 μl of a detergent solution supplied as refill kit peroxidase (Os-gel-HRP, Bioanalytical systems, USA). The gold is then coated twice with each 0.3 μl of the Os-gel polymer followed by air-drying overnight. The surface of the Os-gel-HRP is coated with 1 μl of ACSF (Mg^{2+}, Ca^{2+}-free) solution containing 2% bovine serum albumin (BSA), 0.2% glutaraldehyde and 65 U/ml GluOx. The electrode is necessary to be stored at 4°C until use. In this protocol, Nafion coating and ascorbate oxidase immobilization, which are commonly used for eliminating interference from L-ascorbate, are not employed, because the inner solution of a glass capillary sensor contains L-ascorbate oxidase (*vide infra*).

2.2 Preparation of a capillary sensor

The structure and photo of a glass capillary microsensor is shown in Fig. 1. The capillary sensor is composed of a Borosilicate glass capillary (outer diameter 1.5 mm and inner diameter 0.86 mm) having a tip diameter of approximately 10 μm, prepared by using a three-pull technique with a micropipette puller. The tip diameter can be measured under a microscope. Before use, the glass capillary is filled with a Mg^{2+}, Ca^{2+}-free ACSF (approximately 3 μl) containing $1x10^3$ U/ml ascorbate oxidase (AsOx). A Teflon-coated Pt wire (ø 0.127 mm) with ~2 mm of exposed Pt and a Teflon-coated Ag/AgCl wire (ø 0.127 mm) served as counter and reference electrodes, respectively. The working, counter and reference electrodes are inserted into the capillary pipette. The distance between the tip of the glass pipette and the working electrode is usually ~1.5 mm or shorter, as observed under a microscope. A larger distance leads to a delay in the response to L-glutamate.

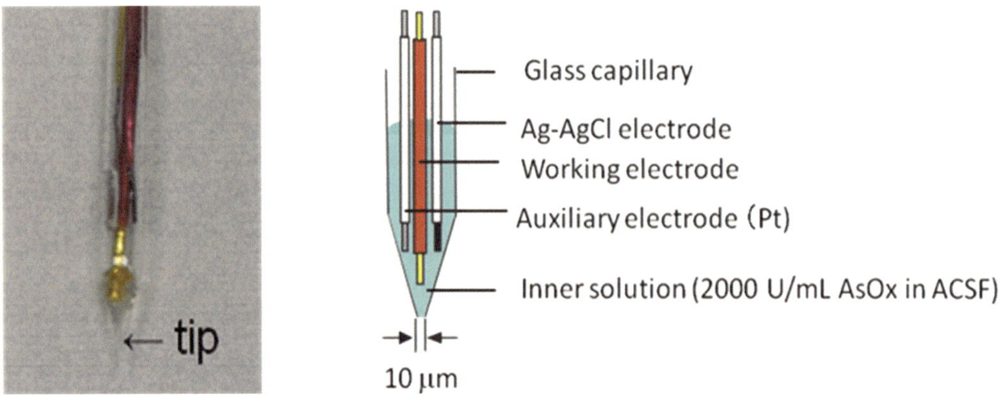

Fig. 1. A photo of a glass capillary sensor and its structure.

2.3 Electrochemical reaction at a glass capillary sensor

L-Glutamate released in brain slices diffuses into the inner solution of a capillary sensor to reach the top layer (GluOx-BSA layer) of the underlying working electrode. L-Glutamate oxidase (GluOx) catalyzes the oxidation of L-glutamate into α–ketoglutarate, producing electro-active H_2O_2 (Kusakabe et al., 1983).

$$\text{L-glutamate} + H_2O + O_2 \rightarrow \alpha\ \text{ketoglutarate} + NH_4^+ + H_2O_2 \qquad (1)$$

The Os-gel-HRP on the working electrode mediates the reduction of H_2O_2 in the following way (Vreeke et al., 1992).

$$2\text{Os(II)} + H_2O_2 + 2H^+ \rightarrow 2\text{Os(III)} + 2\ H_2O \qquad (2)$$

$$\text{Os(III)} + e^- = \text{Os(II)} \qquad (3)$$

The Os(III) produced is reduced at the underlying electrode, giving a reduction current, which is used as a response of the present sensor. The operation potential is 0 V vs. Ag-AgCl.

The capillary electrode has the advantage that the inner solution can contain various enzymes. The interference from ascorbic acid, one of the major components in the brain, is removed by adding ascorbate oxidase to the inner solution. The enzyme catalyzes the oxidation of L-ascorbate to 2-dehydroascorbate (Tokuyama et al., 1965; Nakamura et al., 1968) according to

$$\text{L-ascorbate} + 1/2\ O_2 \rightarrow \text{dehydroascorbate} + H_2O \quad (4)$$

The pH range (pH 5.6-7.0) for the catalytic action of ascorbate oxidase is very close to that (pH 5.5-10.5) of GluOx (Kusakabe et al., 1983) , and hence both enzymes are active at pH 7.0.

2.4 Monitoring L-glutamate with a capillary sensor

Our protocol for monitoring L-glutamate in brain slices is as follows. Prior to its implantation into a hippocampal slice, a glass capillary sensor is operated in air at 0 V vs. Ag-AgCl until a steady current is obtained. Then, the sensor is positioned above the surface of a target neuronal region of a hippocampal slice in interface preparation, followed by lowering into the target region of the slice at a depth of ~100 μm with a manipulator. Appearance of a sharp electric noise indicates the implantation of the sensor into the slice. The sensor is kept in the slice until a steady current is attained. After attainment of a steady current, recording an L-glutamate current is started. Since the volume of the sensor inner solution is maintained, continuous and long-time monitoring of L-glutamate in a brain slice is feasible.

3. Response principles of a capillary sensor in bulk solutions and brain slices

The response profiles of a capillary sensor are categorized into two cases, depending on whether it is used in an aqueous solution just above the target region of a brain slice or it is implanted in the target region of a brain slice (Sugawara, 2007). When a capillary sensor is positioned in a bulk aqueous solution just above a brain slice, capillary action is essentially important for its operation. In an aqueous solution, a small volume of a sample solution containing L-glutamate is spontaneously sampled into an inner solution by capillary action. On the other hand, such capillary action does not work in brain slices because of viscous nature of extracellular fluid. It is noted that a small fraction of GluOx is leached from the top surface of a GluOx-immobilized Au electrode into a capillary inner solution (Oka et al., 2007) and hence, leached GluOx catalyzes the oxidation of L-glutamate, producing hydrogen peroxide, which is detected at the working electrode.

3.1 Capillary action of a pulled glass capillary in an aqueous solution

Figure 2 shows the photos that demonstrate the capillary action of a pulled glass capillary in an aqueous solution. The capillary inner solution contained a visible dye, i.e., methylene blue(MB). One can see that an aqueous solution comes into the capillary with time. When an aqueous solution contained 5% dextrane, the sampling rate was deteriorated significantly due to an increase in viscosity. In another set of experiments, we quantified the capillary action by measuring the weight of a capillary dipped in an aqueous solution (Nakajima et al., 2003). With a tip diameter of 2.5 μm or less, a rise of the solution by capillarity is not observed. On the other hand, in the case that a capillary with a tip diameter of 10 μm is dipped in an aqueous solution, the weight of the solution in the capillary increases linearly with dipping time up to 20 min. The slow rise of the solution is due to the conical tip of the capillary, which decelerates the movement of the solution. Thus, pulled glass capillaries exhibit different magnitudes of capillarity, which depend on tip size and dipping time. The quantitative response of a glass capillary sensor in an aqueous solution is relied on such capillary action.

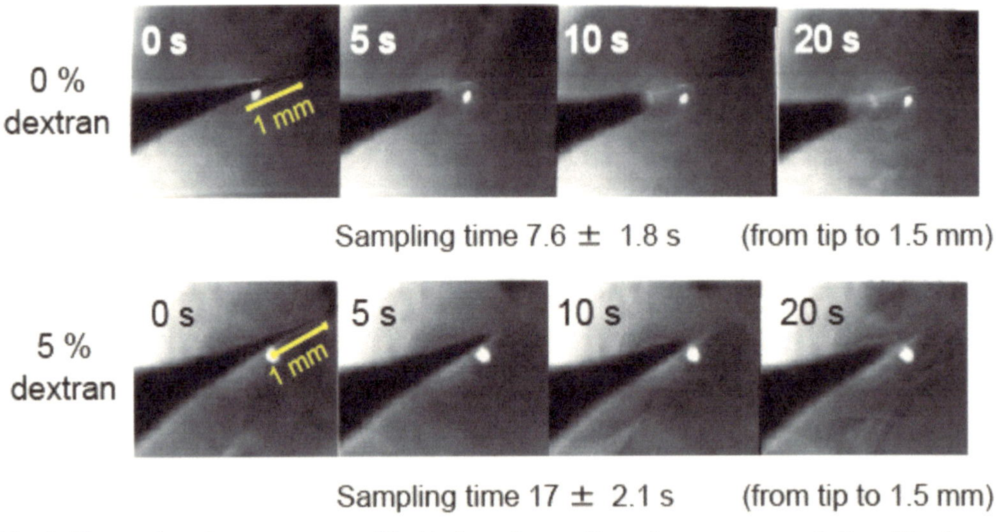

Fig. 2. Photos that demonstrate capillarity-based sampling of an aqueous solution. The inner solution of a glass capillary contained methylene blue.

3.2 Diffusion of L-glutamate into a capillary inner solution

In contrast to its use in an aqueous solution, the capillary action of a glass capillary does not work in a brain slice because of the viscous nature of extracellular fluid. The volume of an inner solution of a glass capillary is maintained, as shown in Fig. 3, even after its implantation into a brain slice. Under such circumstance, the response of a capillary-based sensor is based on diffusional entry of L-glutamate into its inner solution.

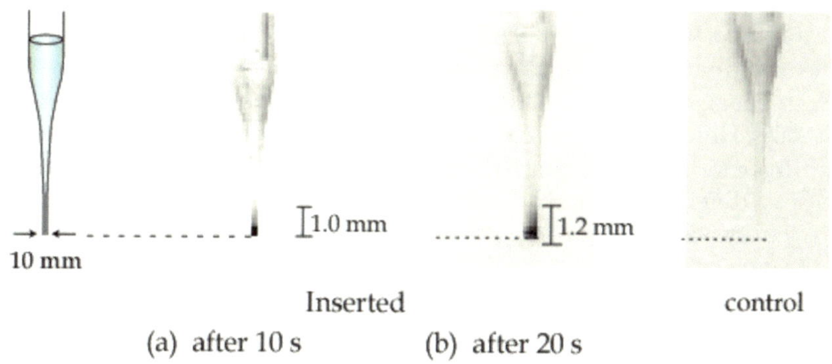

Fig. 3. Fluorometric images of a glass capillary. A pulled glass capillary containing ACSF (Mg^{2+} and Ca^{2+}-free) was inserted into a brain slice, loaded in advance with a fluorescence dye BCECF (Oka et al., 2007).

3.3 *In situ* calibration

One of the essential tasks to be considered is how to correlate the sensor response to final L-glutamate concentration. There are two ways for maintaining brain slices alive (Sugawara, 2007), *i.e.*, a brain slice is fully submerged in a bath solution (Fig. 4a) and a slice is kept alive by passing an ACSF underneath the slice (Fig. 4b). Calibrating the implanted glass capillary sensor is also dependent on how slices were maintained. In the submerged case, calibrating sensor responses and stimulation of the slice can be performed by changing the concentration of L-glutamate or a stimulant in the bath solution. For brain slices in interface preparation, and also in submerge preparation, post-*in vitro* calibration is common for calibrating the responses of an implanted sensor, because the adsorption of extracellular components on the top surface of a sensor deteriorates the sensitivity of the response. In this protocol, an implanted sensor is transferred into an aqueous solution and calibrated with a standard L-glutamate solution. However, the post-in vitro calibration approach is based on the assumption that the sensor exhibits the same degree of deterioration both in a brain slice and an aqueous solution. To improve the uncertainty of the post-*in vitro* calibration, we suggested a method for calibrating an implanted sensor by injecting a small volume of (5 μl) of a standard L-glutamate solution into the close vicinity of the glutamate sensor through a glass capillary (Oka et al., 2007; Chiba et al., 2010). The sensor exhibits a transient current-time profile rather than a steady one (Fig. 5), due to the active reuptake process and diffusional wash out of L-glutamate. Consequently, an instantaneous current is used for calibration. The calibration has to be done at each neuronal region, because the activity of reuptake process is neuronal region-dependent.

It is noted that L-glutamate levels measured with a capillary sensor are dependent on the type of slice preparation (Sugawara, 2007). In the submerged case where a sensor is positioned above the surface of a target neuronal region, an L-glutamate level obtained is the one that diffused out of the slice. Such alignment of a sensor is common for not only capillary sensors but also patch sensors using natural receptors. However, thus-obtained L-glutamate levels are obviously lower than those in the brain slice (Oka et al., 2009). Consequently, a relative change in the response rather than the very magnitude of the response is a matter of concern for monitoring neuronal events. On the other hand, the implantation of a sensor into a brain slice can measure L-glutamate in the vicinity of neurons, but calibrating the sensor response needs a hard task.

The lower detection limit for L-glutamate of GluOx-based sensors has been reported to be sub-μM or better. The detection limit of nM range has also been reported (Tang, et al., 2007; Braeken et al., 2009). However, these values are based on the measurements in an electrolyte solution rather than in brain or brain tissues. The properties of tissue environment, for example viscosity, differ significantly from an electrolyte solution. The diffusion of L-glutamate affected by viscosity may alter the sensitivity of the sensor. Therefore, *in situ* calibration of the sensor response in brain tissue is important for knowing the detection limit of an implanted sensor.

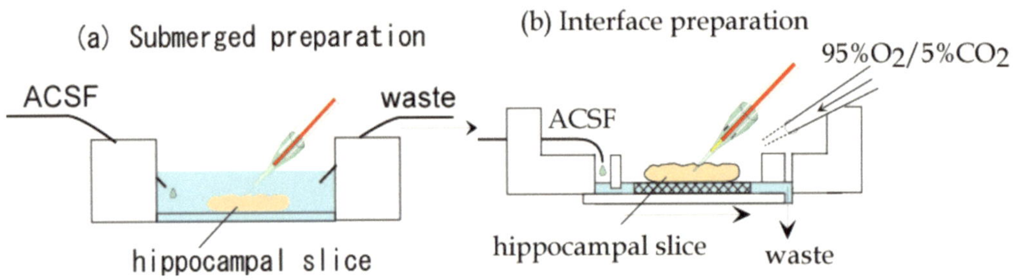

Fig. 4. Two types of slice preparation. (a) A brain slice is submerged in a bath solution and (b) a brain slice is placed on a lens paper through which an ACSF flows (Sugawara, 2007).

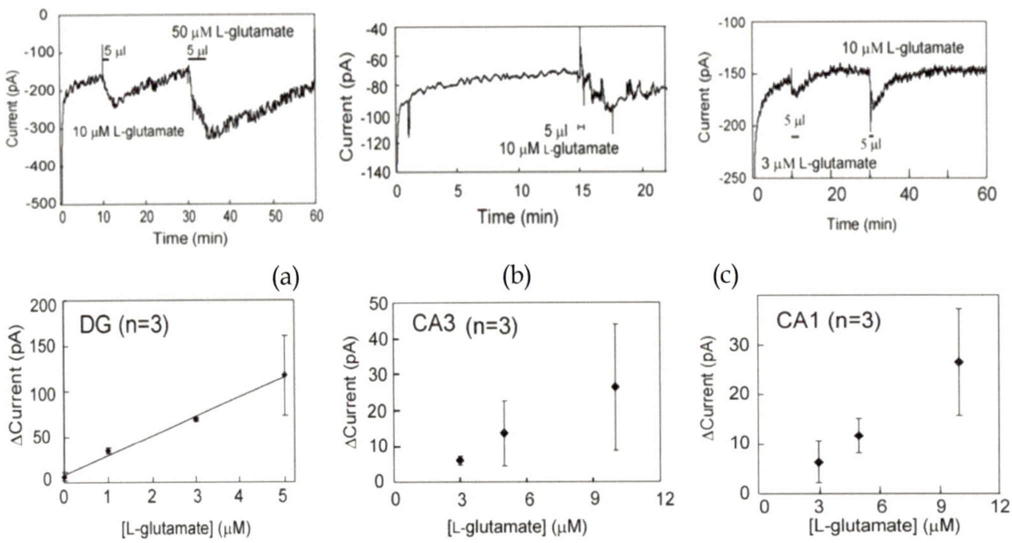

Fig. 5. Current-Time profiles for *in situ* calibration at (a) DG , (b) CA3 and (c) CA1 and corresponding calibration graphs for L-glutamate (Oka et al., 2007, 2009; Chiba et al., 2010).

3.4 Selectivity

A number of interfering species coexist with glutamate in brains. The selectivity issues of GluOx-based sensors for eliminating interference from coexisting ions and molecules have been addressed by several authors. Interference from ascorbate that are present in the brain at concentration much larger than that of L-glutamate has been eliminated by coating Nafion that excludes ascorbate anions electrostatically from the electrode surface (Day et al., 2006; Oldenziel et al., 2006a; Rutherford et al., 2007; Burmeister &Gerhardt, 2001] or by using ascorbate oxidase that mediate the decomposition of ascorbate before it approaches to the surface of an underlying electrode (Oldenziel et al. 2006b; Kulgina et al, 1999; Oka et al. 2007) . Conducting polymer-modified electrodes are also effective for eliminating the interferent (Rahman et al., 2005). Another potential interfering compound is L-glutamine. The interference arises probably from the presence of glutaminase as contaminant in isolated GluOx (Yamauchi et al., 1984). The isolated GluOx-based sensor shows a response to glutamine at a few μM level. However, at large glutamine concentration above 300 μM, which corresponds to the concentration of glutamine in brain (Kanamori&Ross, 2004; Lerma et al., 1986), the response to glutamine is saturated. In addition, such interference disappears in the presence of a small amount of L-glutamate (Oldenziel et al., 2006).

The response to glutamine is significantly modified by using recombinant GluOx (Hozumi et al., 2011). The recombinant GluOx-based sensor suffers interference from glutamine only at high concentration above 300 μM, and the response to glutamine is very weak (1.83 pA/μM an average between 300 and 500 μM) in comparison with the response to L-glutamate (472 pA/μM). The glutamate sensor based on recombinant GluOx does not exhibit responses to glycine, GABA, serotonin, (each 1.0 mM), and L-aspartic acid (200 μM). Since the typical concentration of L-ascorbic acid in brain is 100-500 μM (Walker et al., 1995; Nedergaard et al., 2002) and the estimated basal concentration of glutamine in brain ranges from 200 to 400 μM (Lerma et al., 1986; Kanomori and Ross, 2004) and that of L-aspartic acid is 0.25 μM or less (Robert et al., 1998), the effect of these compounds on L-glutamate currents appears to be of the minor importance.

4. Monitoring of L-glutamate release in hippocampal slices

Since L-glutamate released from nerve terminals into the synaptic cleft is subject to diffusion and dilution into extracellular space and uptake into neurons and glia by excitatory amino acid transporters, the extracellular level of L-glutamate is essentially important for elucidating neuronal signal transmission processes. Although the *in vivo* level of L-glutamate in the neuronal subfield of the hippocampus is important (Table 1), no reports have been published yet on hippocampal levels of L-glutamate. A small number of data have been gathered from *in vitro* study using acute hippocampal slices that consist of freshly isolated brain tissue maintained in a chamber. The data described in this section are mostly taken from our results. The extra-cellular and extra-slice L-glutamate levels in acute mouse hippocampal slices obtained with a capillary sensor are given in Table 2, together with the reported basal levels for hippocampal slices (Oldenziel et al, 2007; MaLamore et al., 2010, Hermann&Jahr, 2007). The reported basal level varies in a wide range from several tens nM to a few μM (Table 1). Therefore, the basal L-glutamate level in brain and brain slices is still a matter of debate.

4.1 Chemical stimulation

The monitoring of enhanced L-glutamate level evoked by physiologically relevant stimuli enables us to discern the role and action of each stimulant as well as the regional distribution of L-glutamate in hippocampal slices. We performed monitoring of L-glutamate release in various neuronal regions of mouse hippocampal slices under stimulation with KCl, tetraethylammonium (TEA) chloride and ischemia.

4.1.1 KCl stimulation

The depolarization evoked by KCl (0.10 M) stimulation enhances extracellular L-glutamate level in hippocampal slices, but the enhanced concentration level at dentate gyrus (DG) (Oka et al., 2007) is much larger than those at cornu ammonis 1 (CA1) (Oka et al., 2009) and cornu ammonis 1 (CA3) (Chiba et al., 2010). The K^+-evoked L-glutamate levels in the CA1 and CA3 regions are very low, i.e., approximately 4 μM, owing to reuptake processes. Sodium-dependent excitatory amino acid transporters (EAATs) (Taxt et al., 1984; Rothstein et al., 1994; Furuta et al., 1997) are present differentially within neurons and astroglia (Taxt et al., 1984): EAAC1(EAAT3) is highly enriched in hippocampus with distribution of CA1, CA3 > DG, while EAAT4 is present in trace amount in hippocampus (Furuta et al., 199). The regional transporter distribution suggests that L-glutamate in the CA1 and CA3 regions is more strongly removed from extracellular space than in the DG region and hence, the L-glutamate level is maintained very low.

4.1.2 TEA stimulation

The stimulation by a K^+ ion channel blocker tetraethylammonium (TEA) chloride is known to elicit chemically induced synaptic potentiation (cLTP) in CA1 of hippocampal slices (Aniksztejn&Ben-Ari, 1991; Hosokawa et al., 1995). The TEA stimulation activates both NMDA receptor channels (Hanse&Gustafsson, 1994) and voltage-dependent calcium channels (Huang et al., 1993). The activation of the NMDA receptor channels induces a calcium influx, often inducing LTP, which is similar to that evoked by a brief afferent tetanus (electrical stimulation). The regional distribution of extracellular L-glutamate in hippocampal slices under TEA stimulation (Oka et al., 2007; Oka et al., 2009, Chiba et al., 2010) is similar to that obtained by K^+ stimulation, though L-glutamate levels at CA1 and CA3 are slightly larger than those obtained by K^+ stimulation.

4.1.3 Ischemia

One of the most enigmatic aspects of ischemic injury is the selective vulnerability of the hippocampal CA1 neurons, whereas the neurons in the CA3 and DG regions are relatively spared. The monitoring of L-glutamate with a glass capillary sensor showed that the extra-slice level of L-glutamate is in the order of CA1 ≈ CA3 > DG (Nakamura, et al., 2005), which is in accordance with the imaging study in terms of an L-glutamate flux (Hirano et al., 2003; Okumura et al., 2009). The concentration level of L-glutamate is much larger than those observed by other chemical stimulation (Table 2), indicating that the high level of L-glutamate is released into the extracellular space and diffuses out of the slice into the bath. The time course of the L-glutamate flux at region CA1 is biphasic and that at region DG is modestly biphasic. Similar biphasic time course of L-glutamate release has been reported for rat striatum (vulnerable to ischemic injury) by using a dialysis electrode (Asai et al., 1996; Kohno et al., 1998).

Region	Concentration (μM)	Method	Size (μm)	Reference
Striatum	29.0 ± 9.0	carbon fiber E	10	Kulagina et al 1999
Brain	23.6 ± 5.3	hydrogel E	10	Oldenziel et al. 2006a
Striatum	18.2 ± 9.3	hydrogel E	10	Oldenziel et al. 2006b
Striatum	7.3 ±0.9	self-referencing E	15	Rutherford et al. 2007
dosal striatum	2.0 ± 0.5	conducting polymer E	25	Rahman et al, 2005
Frontal cortex	1.6 ± 0.3	self-referencing E	15	Day et al. 2007
left striatum	1.5 ± 0.3	microsensor	3.0 (mm)	Zhang et al. 2004
striatum	1-3	ceramic-based multi E	15	Pomerleau et al. 2003
right striatum	3.0 ± 0.6	microdialysis		Miele et al. 1996
striatum	2.7 ± 0.9	microdialysis		Robert et al. 1998
striatum	1.15 ± 0.7	microdialysis		Lada et al. 1997
striatum	1.97 ± 0.7	push-pull perfusion		Kottergoda et al. 2002

Table 1. Examples of basal extracellular L-glutamate levels in the anesthetized rat brain (*in vivo*).

Stimulation	Subfield (concentration, μM)	Method (size , μm)	Reference
Extracellular(acute)			
KCl (0.1M)	DG (22) > CA3 (4.1) ~ CA1 (4.0)	capillary E (10)	Nakajima et al .,2003
			Chiba et al. ,2010
TEA (25 mM)	DG (20) > CA3 (7.5)~ CA1 (6.0)	capillary E (10)	Oka et al., 2009
			Chiba et al., 2010
Electric (2 Hz)	CA1 (1.4)	capillary E (10)	Hozumi et al., 2011
Electric (0.052Hz)	CA1 (32 nM)		
Basal	CA1 (1.7)	hydrogel E (10)	Oldenziel et al., 2007
Basal	CA1 (~25 nM)	patch sensor (whole cell)	Hermann and Jahr, 2007
Extracellular(cultured)			
KCl (0.1M)	hippocampal slice (0.12)	microdialysis	Robert et al. 1998
Extra-slice (acute)			
Basal	P19 cell (5.8)	self-referencing E (2-5)	Maclamore et al. 2010
GABA	CA1(1.5) > CA3	patch sensor (2)	Shimane et al. 2006
Ischemia	CA3 (60) ~CA1 (47) > DG(20)	capillary E (10)	Nakamura et al. 2005

Table 2. Regional distribution of extacellular and extra-slice levels of L-glutamate in mouse hippocampal slices.

4.2 Electric stimulation

Recording field excitatory postsynaptic potentials (fEPSPs) have been a well established method for knowing neuronal activities in electrophysiological studies (Bliss et al., 2007). The potentials are produced by a group of cells and reflect in an indirect way the changes in the synaptic and action potentials. Analysis of fEPSPs provides information on the average activity of the neurons in group, including the induction and expression of long-term potentiation (LTP) and long-term depression (LTD). Simultaneous monitoring of L-glutamate release and fEPSPs evoked by physiologically relevant electric stimulation will provide explicit information on the actual amount of L-glutamate released into the synaptic cleft and extracellular space in the vicinity of the stimulation site. However, such a measurement is still a challenging task, because of the requirement of placing multiple electrodes in a confined neuronal region and fairly low level of glutamate will be released by physiologically relevant stimulation.

A glass capillary sensor has the advantage that all electrodes required for amperometric current measurements are built in the capillary interior and hence, the tip size (approximately 10 μm) is small enough to implant it between stimulation and recording electrodes for fEPSP measurements. Fig. 6 shows the setup of simultaneous recordings of a glutamate current with a glass capillary sensor and fEPSPs with stimulation and recording electrodes. The tip of the capillary sensor can be positioned in the middle between stimulation and recording electrodes for fEPSP measurements. The example of simultaneous measurements at CA1 of a hippocampal slice demonstrates that although no significant changes in a glutamate current is detected by application of 0.052 Hz (test stimuli), a transient change in the current is observed by application of 2 Hz stimulation, indicating enhanced release of L-glutamate in CA1 region (Hozumi et al., 2011). The L-glutamate level in region CA1 at 2 Hz stimulation obtained by *in situ* calibration ranged from 0.8 to 2.2 μM (1.4 μM as an average) from 5 independent measurements. The concentration is slightly smaller than those obtained by KCl stimulation. Although direct monitoring of L-glutamate level at test stimuli (0.052 Hz) is not achieved, we can record an L-glutamate current by changing the intensity of electric stimulation from 1 Hz to 3 Hz. The estimated concentration level at 0.052 Hz is 32 ± 7 nM (n=3), which is very close to the reported one using the whole cell recordings (Herman&Jahr, 2007).

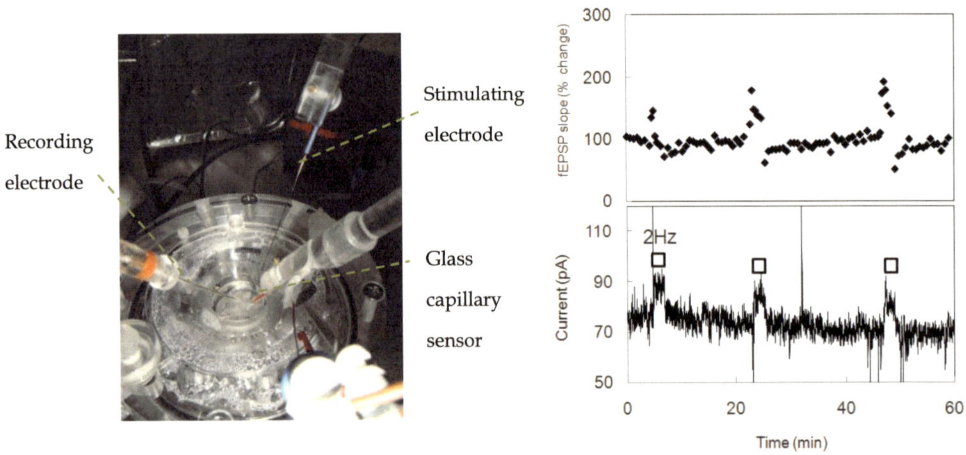

Fig. 6. A photo that shows the setup of simultaneous monitoring of fEPSPs and a glutamate current. The traces of fEPSP and a glutamate current were simultaneously monitored.

5. Conclusions and prospects

The present review demonstrates that a glass capillary-based microsensor is useful for knowing the distribution and level of extracellular L-glutamate in acute hippocampal slices. The L-glutamate levels are markedly dependent on the neuronal regions and types of stimulation. Although discerning the concentration level of L-glutamate in acute brain slices with microsensors is still on the stage of accumulating the local concentration level of L-glutamate, there are increasing efforts for clarifying the sources and places of extracellular L-glutamate release. Since the experimental condition can be controlled easily, microsensors will be promising as a tool for monitoring and estimating the averaged extracellular level of L-glutamate in acute brain slices. On the other hand, recent advances in developing new fluorescent probes have enabled to visualize L-glutamate with spine-sized resolution using a cultured neuron. Such a single synapse study will significantly help the understanding of the molecular events that occur at a synapse. Combining the single synapse data with the extracellular data will significantly advance the understanding of the sources and places of extracellular L-glutamate release. In addition, the sophisticated combination of microsensor studies with an electrophysiological study to correlate L-glutamate level to the neuronal activity will be a promising way for solving debates as to whether LTP is due to presynaptic or postsynaptic changes.

6. References

Adams, R.N. (1976). Probing brain chemistry with electroanalytical techniques. *Anal. Chem.*, 48, 1126A-1138A.

Aniksztejn L. & Ben-Ari, Y. (1991). Novel form of long-term potentiation produced by a K^+ channel blocker in the hippocampus. *Nature, 349*, 67-69.

Asai, S.; Iribe, Y.; Kohno, T. & Ishikawa, K. (1996). Real time monitoring of biphasic glutamate release using dialysis electrode in rat acute brain ischemia. *NeuroReport*, 7, 1092-1096.

Bliss, T.V.P. & Collingdidge, G.L. (1993). A synaptic model of memory: long-term potentiation in the hippocampus. *Nature,* 361, 31-39.

Bliss, T.; Collingridge, G. & Morris, R. (2007). *Sypnaptic plasticity in the hippocampus.* In The Hippocampus book, Andersen A.; Morris, D.; Amaral, D.; Bliss, T. &O'keefe, J. eds., New York, Oxford university Press,, pp343-474.

Braeken, D.; Rand, D.R.; Andrei, A.; Huys, R.; Spira, M.E.; Yitzchaik, S.; Shappir, J.; Borghs, G.; Callewaert, G. & Bartic, C. (2009). Glutamate sensing with enzyme modified floating gate field effect transistors. Biosensors and Bioelectronics, 24, 2384-2389.

Burmeister, J.J. & Gerhardt, G.A. (2001). Self-referencing ceramic-based multisite microelectrodes for the detection and elimination of interferences from the measurements of L-glutamate and other analytes. *Anal. Chem.*, 73, 1037-1042.

Chiba.H.; Deguti,Y.; Kanazawa, E.; Kawai, J.; Nozawa, K.; Shoji, A. & Sugawara, M. (2010). *In vitro* measurements of extracellular L-glutamate level in region CA3 of mouse hippocampal slices under chemical stimulation. *Anal Sci.*, 26, 1103-1106.

Day, B.K.; Pomerleau, F.; Burmeister, J.J.; Huettl, P. & Gerhardt, G.A. (2006) Microelectrode array studies of basal and potassium-evoked release of L-glutamate in the anesthetized rat brain. *J. Neurochem.*, 96, 1626-1635.

Furuta, A.; Martin, L.J.; Lin, C.-I.; Dykes-Hoberg, M. & Rothstein, J.D. (1997). Cellular and synaptic localization of the neuronal glutamate transporters excitatory amino acid transporter 3 and 4. *Neuroscience*, 81, 1031-1042.

Hanse, E. & Gustafsson, B. (1994). TEA elicits two distinct potentiation of synaptic transmission in the CA1 region of the hippocampal slice. *J. Neurosci.*, 14, 5028-5034.

Herman, M.A. & Jahr, C.E. (2007). Extracellular glutamate concentration in hippocampal slice. *J. Neurosci.*, 27, 9736-9741.

Hirano, A. & Sugawara M. (2006). Receptors and enzymes for medical sensing of L-glutamate. *Mini-review in Medicinal Chemistry*, 6, 1091-1100.

Hirano, A.; Moridera, N.; Akashi, M.; Saito, M.& Sugawara, M. (2003). Imaging of L-glutamate fluxes in mouse brain slices based on an enzyme-based membrane combined with a differential-image analysis. *Anal. Chem.*, 75, 3775-3783.

Hosokawa, T.; Rusakov, D.A.; T. V. P. Bliss, T.V. P. & Fine, A. (1995). Repeated confocal imaging of individual dendritic spines in the living hippocampal slice: evidence for changes in length and orientation associated with chemically induced LTP. *J. Neurosci.*, 15, 5560-5573.

Hozumi, S.; Kana Ikezawa, K.; Atushi Shoji,A.; Hirano-Iwata,A.; Bliss,T.& Sugawara, M. (2011). Simultaneous monitoring of excitatory postsynaptic potentials and extracellular L-glutamate in mouse hippocampal slices. Biosensors&Bioelectronics, 26, 2975-2980.

Huang, Y.-Y. & Malenka, R. C. (1993). Examination of TEA-induced synaptic enhancement in area CA1 of the hippocampus : the role of voltage-dependent Ca^{2+} channels in the induction of LTP. *Neurosci.*, 13, 568-576.

Kanamori, K. & Ross, B. D. (2004). Quantitative determination of extracellular glutamine concentration in rat brain and its elevation in vivo by system A transport inhibitor, alpha-(methylamino)isobutyrate. J. Neurochem., 90, 203-210.

Kohno, T.; Asai, S.; Iribe, Y,; Hosoi, I.; Shibata, K. & Ishikawa, K. (1998). An improved method for the detection of changes in brain extracellular glutamate levels. *J. Neurosci. Methods*, 81, 199-205.

Kottegoda, S.; Shaik, I. & Shippy, S. A. (2002). Demonstration of low flow push-pull perfusion. *J. Neurosci. Methods*, 121, 93-101.

Kulagina, N.V.; Shankar, L. &Michael, A.C. (1999) Monitoring glutamate and ascrobate in the extracellular space of brain tissue with electrochemical microsensors. *Anal. Chem.*, 71, 5093-5100.

Kusakabe, H.; Midorikawa, Y.; Kuninaka, A.; Fujisima, T. & Yoshino, H.(1983). Purification and properties of a new enzyme, L-glutamate oxidase, from Streptomyces sp. X-119-6 grown on wheat bran. *Agric. Biol. Chem.*, 47, 1323-1328.

Lada, M.W.; Vickroy, T. W. & Kennedy, R.T. (1997). High temporal resolution monitoring of glutamate and aspartate in vivo using microdialysis on-line with capillary electrophoresis with laser-induced fluorescence detection. *Anal. Chem.*, 69, 4560-4565.

Lerma, J.; Herranz, A. S.; Herreras, O.; Abraira, V. & Martin del Rio, R. (1986). In vivo determination of extracellular concentration of amino acids in the rat hippocampus. A method based on brain dialysis and computerized analysis. *Brain research*, 384, 145-155.

Malenka, R. C. & Nicoll, R. A. (1999). Long-term potentiation-a decade of progress?. *Science*, 285, 1870-1874.

McLamore, E. S.; Mohanty, S.; Shi, J.; Claussen, J.; Jedlicka, S.S.; Rickus, J. L. & Porterfield, D. M. (2010). A self-referencing glutamate biosensor for measuring real time neuronal glutamate flux. J. Neurosci. Method, 189, 14-22.

Mattson M. P. (2008). Glutamate and neurotrophic factors in neuronal plasticity and disease. *Ann. N. Y. Acad. Sci.*, 1144, 97-112.

Miele, M.; Boutelle, M.G. & Filenz, M. (1996). The source of physiologically stimulated glutamate efllux from the striatum of conscious rats. *J. Physiol.*, 497, 745-751.

Nakajima, K.; Yamagiwa, T.; Hirano, A. & Sugawara, M. (2003). A glass capillary microelectrode based on capillarity and its application to the detection of L-glutamate release from mouse brain slices. *Anal. Sci.*, 19, 55-60.

Nakamura, N.; Negishi, K.; Hirano, A. & Sugawara, M. (2005). Real time monitoring of L-glutamate release from mouse brain slices under ischemia with a glass capillary based enzyme electrode. *Anal. Bioanal. Chem.*, 383, 660-667.

Nakamura, T.; Makino, N. & Ogura, Y (1968). Purification and properties of ascorbate oxidase from cucumber. J. Biochem. 64, 189-195.

Nedergaard, M.; Takano, T. & Hansen, J. (2002). Beyond the role of glutamate as a neurotransmitter. *Nat. Rev. Neurosci.*, 3, 748-755

Nishizawa, Y.(2001). Glutamate release and neuronal damage in ischemia. *Life Sci.*, 69, 369-381.

Oldenziel, W.H.; Dijkstra, G.; Cremers, T.I.F.H. & Westrink, B.H.C. (2006). Evaluation of hydrogel-coated glutamate microsensors. *Anal. Chem.*, 78, 3366-3378.

Oldenziel, W.H.; Zeyden M. Dijkstra, G., Chijsen, W.E.J.M.; Karst, H.; Cremers, T.I.F.H. & Westerink, B.H.C. (2007). Monitoring extracellular glutamate in hippocampal slices with a microsensor. *J. Neurosci. Methods.*, 160, 37-44.

Oldenziel, W.H.; Dijkstra, G.; Cremers, T.I.F.H. & Westrink, B.H.C. (2006). In vivo monitoring of extracellular glutamate in the brain with a microsensor. *Brain Res.*, 1118, 34-42.

O'Neill, R. D.; Lowry, J. P. & Mas, M. (1998). Monitoring brain chemistry in vivo: voltammetric techniques, sensors, and behavioral applications. *Crit. Rev. Neurobiol.*, 12, 69-127.

Oka, T.; Tasaki, C.; H. Sezaki, H. & Sugawara, M. (2007). Implantation of a glass capillary electrode in mouse brain slices for monitoring of L-glutamate release. *Anal. Bioanal. Chem.*, 388, 1673-1679.

Oka, T.; Tominaga, Y.; Wakabayashi, Y. Shoji, A. & Sugawara, M. (2009). Comparison of the L-glutamate level in mouse hippocampal slices under tetraethylammonium chloride stimulation as measured with a glass capillary sensor and a patch sensor. *Anal. Sci.*, 25, 353-358.

Okumura, W,; Moridera, N.; Kanazawa, E.; Shoji, A.; Hirano-Iwata, A. & Sugawara, M. (2009). Visualizing L-glutamate fluxes in acute hippocampal slices with glutamate oxidase-immobilized cover slips. *Anal. Biochem.*, 385, 326-333.

Pomerleau, F.; Day, B., K.; Huettl 、P.; Burmeister, J. J. & Gerhardt, G. A. (2003). Real time in vivo measures of L-glutamate in the rat central nervous system using ceramic-based multisite microelectrode arrays. *Ann. N. Y. Acad. Sci.*, 1003, 454-457.

Rahman, M.A.; Kwon, N.-H.; Won, M.-S.; Choe, E.S. & Shim, Y.-B. (2005). Functionalized conducting polymer as an enzyme immobilizing substrate: an amperometric glutamate microbiosensor for in vivo measurements. *Anal. Chem.*, 77, 4854-4860.

Reis, H.J.; Guatimosim, C.; Paquet, M.; Santos, M.; Ribeiro, F.M.; Kummer. A.; Schenatto, G.; Vsalgado, J.V.; Vieira, L.B.; Teixeira, A. L. & Palotás, A. (2009). Neuro-transmitters in the central nervous system & their implication in learning and memory processes. *Curr. Med. Chem.*, 16, 796-840.

Robert, F.; Bert, L.; Parrot, S.; Denoroy, L.; Stoppini, L. & Renaud, B. (1998). Coupling on-line brain microdialysis, precolumn derivatization and capillary electrophoresis for routine minute sampling of O-phosphoethanolamine and excitatory amino acids. *J. Chormatog. A*, 817, 195-203.

Rutherford, E.C.; Pomerleau, F.; Huettl, P.; Strömberg, I. & Gerhardt, G.A. (2007). Chronic second-by-second measures of L-glutamate in the central nervous system of freely moving rats. *J. Neurochem.*, 102, 712-722.

Shimane, M.; Miyagawa, K. & Sugawara, M. (2006). Detection of γ-aminobutyric aid-induced glutamate release in acute mouse hippocampal slices with a patch sensor. *Anal. Biochem.*, 353, 83-92.

Sugawara, M. (2007). Methodological Aspects of *in vitro* sensing of L-glutamate in acute brain slices. *The Chemical Records*, 7, 317-325.

Tang, L.; Zhu, Y.; Yang, X. & Li, C. (2007). An enhanced biosensor for glutamate based on self-assembled carbon nanotubes and dendrimer-encapsulated platinum nanobiocomposite-doped polypyrrole film. *Anal. chim. Acta*, 596, 145-150.

Taxt, T. & Storm-Mathisen, J. (1984). Uptake of D-aspartate and L-glutamate in excitatory axon terminals in hippocampus: autoradiographic and biochemical comparison with γ–aminobutyrate and other amino acids in normal rats and in rats with lesions. *Neuroscience*, 11, 79-100.

Tokuyama, K.; Clark, E.E.; Dawson, C.R. (1965) Ascorbate oxidase: a new method of purification. Characterization of the purified enzyme. *Biochemistry*, 4, 1362-1370.

Volterra, A. & Meldolesi, J. (2005). Astrocytes from brain glue to communication elements: the revolution continues. *Nature review Neurosci.*, 6, 626-640.

Vreeke, M.; Maiden, R. & Heller A. (1992). Hydrogen peroxide and β–nicotinamide adenine dinucleotide sensing amperometric electrodes based on electrical connection of horse radish peroxidase redox centers to electrodes through a three-dimensional electron relaying polymer network. *Anal. Chem.*, 64, 3084-3090.

Walker, M.C.; Galley, P.T.; Errington, M.L.; Shorvon, S.D. & Jefferys, J. G. R.(1995). Ascorbate and glutamate release in the rat hippocampus after perforant path stimulation: a "dialysis electrode" study. *J. Neurochem.*, 65, 725-731.

Yamauchi ､ H.; Kusakabe, Y.; Midorikawa, T.; Fujisima, T. & Kuninaka, A. (1984). Enzyme electrode for specific determination of L-glutamate. *Eur. Congr. Biotechnol.*, 1, 705.

Zhang, F.-F.; Wan, Q.; Li, C.-X.; Wang, X.-L.; Zhu, Z.-Q.; Xian, Jin, L.-T. & Yamamoto, K. (2004). Simultaneous assay of glucose, lactate, L-glutamate and hypoxanthine levels in a rat striatum using enzyme electrodes based on neutral red-doped silica nanoparticles. *Anal. Bioanal. Chem.*, 380, 637-642.

Zeyden, M.; Oldenziel, W.H.; Rea, K.; Cremers, T.I. & Westerink, B.H. (2008). Microdialysis of GABA and glutamate: Analysis, interpretation and comparison with microsensors. *Pharmacology, Biochem and Behavior*, 90, 135-147.

10

Diffractive Optics Microsensors

Igor V. Minin and Oleg V. Minin
Novosibirsk State Technical University,
Russia

1. Introduction

The term Microsensor is typically used to mean a sensing device that is fabricated using microelectronic technology. The field of Microsensors has a fifty-year history starting with several key developments in the 1950's:

- The invention of integrated circuits.
- The discovery of piezoresistance in silicon.
- The discovery of selective etching of single-crystal silicon.
- The development of thin-film read heads for magnetic recording.

During the last twenty years, emphasis in this field has gradually shifted from basic research on materials and process technologies toward product development. Each year new products appear, and with these, an expansion of potential future opportunities. Microsensors have a bright future, both in the commodity arena and in the MEMS-enabled arena. The physical sensors, pressure, acceleration, rotation, and acoustic (microphones) continue to find new commodity-level markets. Sensors, whether commodity sensors such as the cell-phone microphone, or system sensors, are becoming smarter, more capable, and are finding new markets every day. In the present review we describe a wide class of microsensors based on diffractive optics element (DOE).

Diffractive optics is very versatile since any type of wave can be considered for computation within the computer. Digital holograms created with such technology are more commonly called DOE. Another name for DOE is *computer-generated hologram* (CGH). Because of these equivalent terminologies, the word DOE will be used in this chapter.

For their operation the diffractive elements depend on diffraction effects: DOEs are based on the effect of radiation diffraction on a periodic or quasiperiodic structure rather than on refraction as it is in the classical optics. The optical depth of a focusing element ranges within a radiation wavelength. In this respect, the zone plates (FZP) like lens may be referred to diffraction elements, i.e., to a class of quasioptical focusing systems, since according to the definition of quasioptics they are calculated, as a rule, by laws of geometrical optics, and the principle of their operation is based on diffraction effects.

There basic types of DQE are distinguished (according to the principle of their location relative to the direction of electromagnetic wave propagation) [1], they are: a transverse element (implemented primarily on a plane surface), a longitudinal-transverse element (on an arbitrary curved surface), and a longitudinal element (representing a set of screens situated along the direction of electromagnetic wave propagation). DQE can operate by the principle of "transmission" or "reflection".

The main advantages of using diffractive microlenses included:
1. The focal length is precisely defined by photolithography.
2. A wide range of numerical aperture value from F/0.3 to F/5 can be made.
3. Lens diameter can be in the order of few ten micrometer
4. Thickness is on the order of an optical wavelength 100% fill factor can be achieve
5. A DOE can perform more than one function, for example, have multiple focal points, corresponding to multiple lenses on a single element. It can also be designed for use with multiple wavelengths.
6. DOEs are generally much lighter and occupy less volume than refractive and reflective optical elements.
7. With mass manufacturing, they can be manufactured less expensively for a given task.

The simplest binary diffraction lens only required one mask to fabricate. However, the maximun focusing efficiency for a binary lens is less than 10%. Multi phase levels diffraction lens is required for higher efficiency. Theoretical efficiency of a multi phase levels diffractive lens is given by the equation.

$$\eta = \left[\frac{Sin(\pi / M)}{\pi / M} \right]^2$$

Where $M=2^m$ phase levels.
It is clear that by increasing the number of levels, the focusing efficiency also increases. An efficiency of 41% can be reach with two levels. By increasing the number of levels to 4, an efficiency of 81% is reach. 8 phase levels is a practical number of levels that allow high efficiency while keeping the number of mask alignment low. The efficiency for this value is 95%, while the number of masks required is 3.
To produce the desire focus length, one needs to calculate the required ring radius on each mask.

$$r(p,m) = \left[\left(\frac{p\lambda}{n2^m} \right)^2 + 2F\left(\frac{p\lambda}{n2^m} \right) \right]^{1/2}$$

Where F is the desired focus length, p=0,1,2,3..., n is the refractive index of the substrate, r(p,m) is the edge of opaque and transparent rings. For a positive lens with positive photoresist, the odd numbered zone gives the outside radii of the opaque rings, while the even numbered zone gives inner radii. During the fabrication, it is required to quantized the phase step height and hence the phase shift by $d(m) = \frac{\lambda}{(n-1)2^m}$.

A few care is required for fabrication of a diffractive lens:
1. Due to the fact that the focusing efficiency of a diffractive lens is affected by the optical wavelength, a custom design for a particular wavelength is required in order to achieve desire performance.
2. To achieve maximum efficiency, the resolution of the fabrication tools is required to be smaller than the critical dimension of the lens.
3. All masks should align properly.

Let us briefly consider the main frequency properties of DOEs.

2. Zone plate as a frequency-selective optical element

The structure of the diffraction field in the focal region for the axial position of a pointlike radiation source was experimentally studied at wavelengths which are denoted below by λ_j, j = 0, 1, 2, 3..

The longitudinal distribution of field intensity for the actual position of the source reveals the maximum whose position on the OZ axis depends on λ. The dependence of the rear interval on wavelength can be found from expression [1]:

$$B_\lambda(n) = \frac{A^2 + (A + n\lambda/2)^2 - 2(A^2 + r_n^2)^{1/2}(A + n\lambda/2)}{2(A^2 + r_n^2)^{1/2} - 2A - n\lambda}.$$

where A and B are the front and back sections or the design distances from FZP to a source and the positions, respectively, λ_0 is the design wavelength which in the general case differs from the wavelength λ diffracting on FZP, n - number of Fresnel zone. If B is averaged between its extreme values:

$$B = 0.5(B_\lambda(1) + B_\lambda(N)), \tag{1}$$

then expression (1) coincides, within the accuracy of calculations and measurements, with calculations by the Kirchhoff scalar theory as $\Delta\lambda/\lambda_0$ varies within ±20% and with experimental values for all λ_j. These results are plotted in fig. 1.

A similar dependence for a zone plate designed for focusing a wave with planar wavefront is:

$$B_\lambda(p)=r^2p/2N\lambda - N\lambda/2p.$$

Here p is the number of phase quantization levels and N is the total number of Fresnel zones.

From the above is it followed that the FZP may be considered in the first approximation as a circular grating characterized be a constant L=F*λ, where F is the focal length of FZP. Thus the focal length of FZP is inversely proportional to the wavelength and is determinate by F=L/λ. The FZP resolves the incident radiation into monochromatic components. Since each component has its own focal point, it will be projected onto one element of the sensing array along optical axis. This means that the N-element linear array can detect N different wavelength components simultaneously.

It could be noted that a problem of synthesizing an amplitude-phase profile even of the simplest diffraction element such as FZP has no unique solution [1]. Usually, radii of Fresnel zones are determined from the condition of multiplicity to a half-wavelength of the difference of eikonais of a diffracted direct wave and a reference wave. This solution, however, is merely a particular case of a more general one. The essence of constructing such a general solution is that the notion of a *reference radius* R_0 on the aperture of a focusing element is introduced [1].

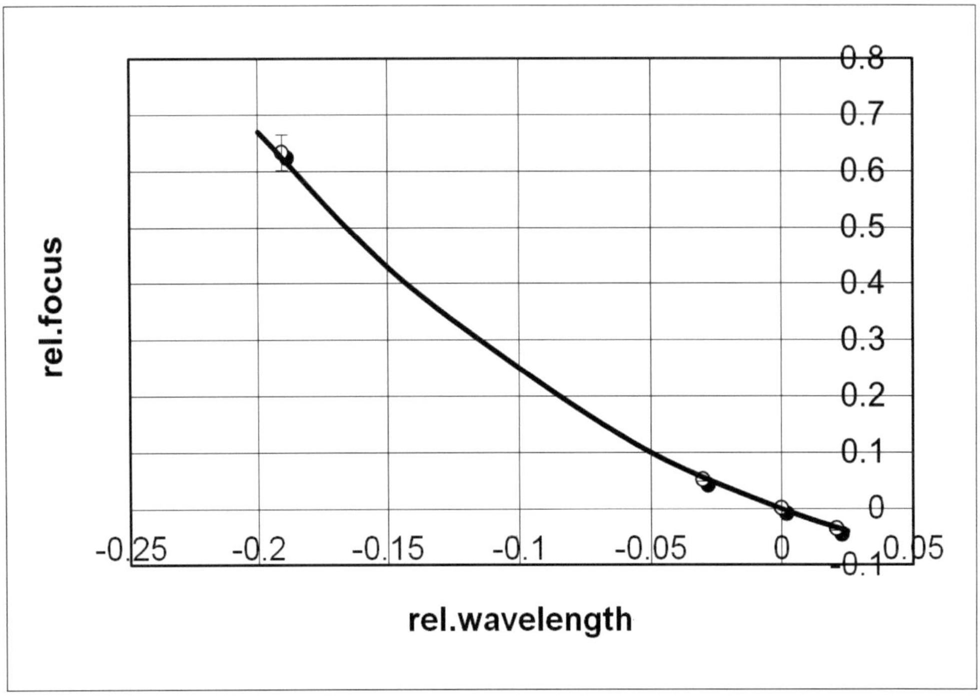

Fig. 1. The position of the focusing area as a function of wavelength of the incident radiation

3. Multiorder diffractive optics properties

Although it is often useful to think of a diffractive lens as a modulo 2π lens at the design wavelength, the spectral properties or wavelength dependence of a diffractive lens are drastically different from those of a refractive lens. The dispersion of a diffractive lens is roughly 7 times larger (in optical waveband) than the strongest flint glass currently available and is opposite in sign. The quantization of the phase function of a diffractive element results in new properties of the element. One of these manifestations of discreteness of the phase function of the elements that we discuss is the possibility of selecting harmonics of coherent radiation. Analytical expressions for a four-level diffractive element were obtained in [1] for the gain of the axisymmetric element G at the main focal point corresponding to the reference wavelength:

$$G(N) = \left[\sum_{n=1}^{N} A(n)\left\{\cos(2\pi M\frac{\lambda_0}{\lambda}(MGiv(n)+1)) - \cos(2\pi M\frac{\lambda_0}{\lambda}(Giv(n))\right\}\right]^2 +$$

$$+\left[\sum_{n=1}^{N} A(n)\left\{\sin(2\pi M\frac{\lambda_0}{\lambda}(MGiv(n)+1)) - \sin(2\pi M\frac{\lambda_0}{\lambda}(Giv(n))\right\}\right]^2$$

$$A(n) = \left[\frac{4MF}{\lambda_0} + 2n - 1\right] / \left[\frac{4MF}{\lambda_0} + M(2n-1)\right]$$

where N is the number of Fresnel zones within the aperture of the element; λ_0 and λ are the reference and the actual radiation wavelength; F is the focal length and Giv(n) is the function equal to the maximum integral value of its argument. So diffractive lens offers the potential for several wavelength components of the incident spectrum to come to a common focus.

An analysis of this relation shows that the maximum gain (for constant signal-to-noise ratio) is achieved for the harmonic with the number $\lambda_0/\lambda = M/2$, while there is no focusing of radiation at the wavelength satisfying $\lambda_0/\lambda = M$, that is, this radiation is selected (fig. 2-3).

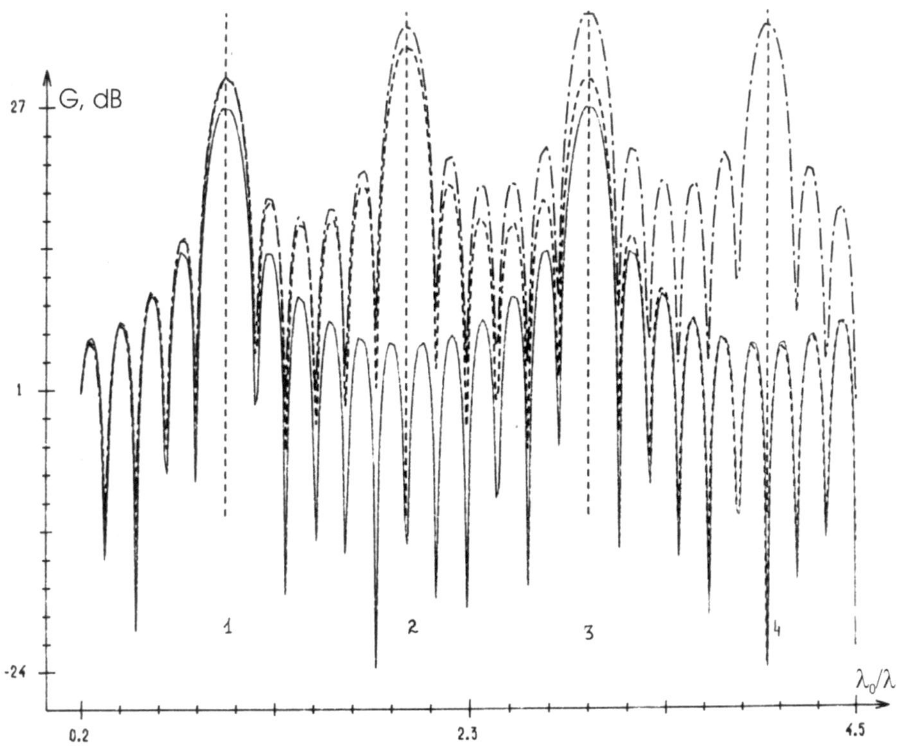

Fig. 2. Gain of a diffractive optical element for three numbers of phase quantization levels [20]:

__________ two levels,

_ _ _ _ four levels,

_._._._._ six levels

It is important to note a well-known property of operation when higher diffracted orders are used. The wavelength bandwidth of the diffraction efficiency around a given diffracted order narrows with increasing values of *n*.

Spectrally selective properties of diffractive elements with a discrete phase function make it possible to use them for mixing a discrete set of wavelengths into the same focusing zone [1]. For instance, the following *theorem* is valid:

For radiation with harmonics $\Lambda_i = \lambda_0/i$, i = 1, ..., N in the interval $Q = [\lambda_1 ... \lambda_2]$ to fall within this interval, that is, $\Lambda_i \in Q$, it is necessary and sufficient for the maximum number of the harmonic to be $N = i_{max} = \text{entier}(\lambda_2/\lambda_1)$, where entier(x) is the integral part of the real x.

The *proof* of this theorem is self-evident. The property of diffractive elements formulated above is also important for practical applications. Thus it may be possible to use this principle and design elements for the x-ray range of wavelengths, to design novel optical elements for optical polychromatic computers, etc.
As for the frequency and focusing properties of diffractive elements using radiation harmonics, their numeric and experimental analysis [1] revealed the following behaviour. When an element operates on a harmonic, the frequency properties expressed in arbitrary units are the same as the frequency properties of the diffractive element in the range of the main wavelength (reference wavelength)– see fig. 3 (in this figure n is the number of the frequency harmonic). This is also true for the transversal and longitudinal resolving powers of a diffractive element if we appropriately replace the current (working) wavelength in the expression.

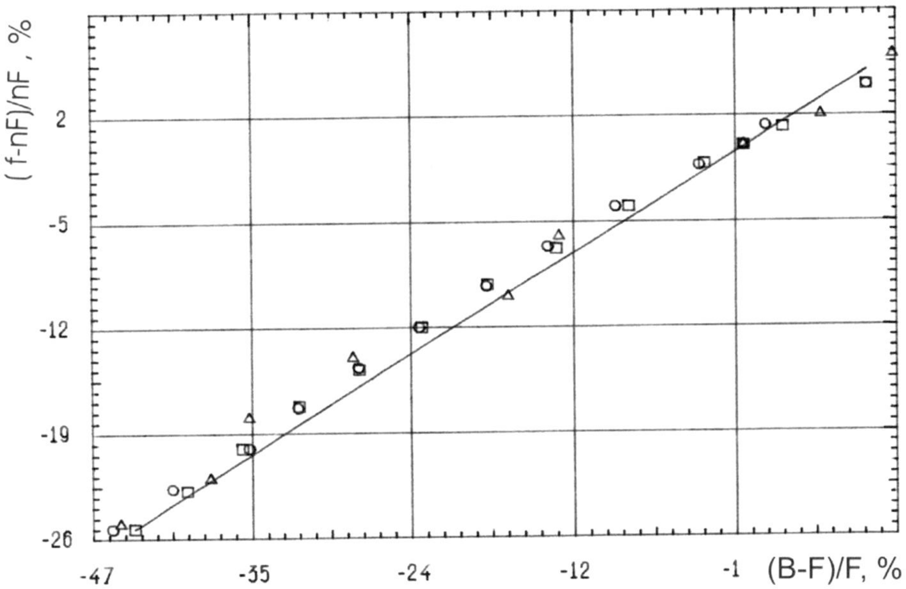

Fig. 3. Frequency properties of a zone plate for the first three harmonics:
$dnF=(f-nf_0)/nf_0$
opensquaresn=1
open triangles n=2
open circles n=3

3.1 Chromatic confocal sensor

The frequency properties of a diffractive optics allow to modify, for example, a confocal (multifocal) sensor [2]. Working principle of frequency-scanning confocal sensor (microscopy) is as follows: the mechanical z-scan (Figure 4a) in conventional confocal systems is replaced by a simultaneously generated series of foci at different wavelength (Figure 4b).
This principle of frequency-scanning confocal microscopy may be applied at any frequency waveband.

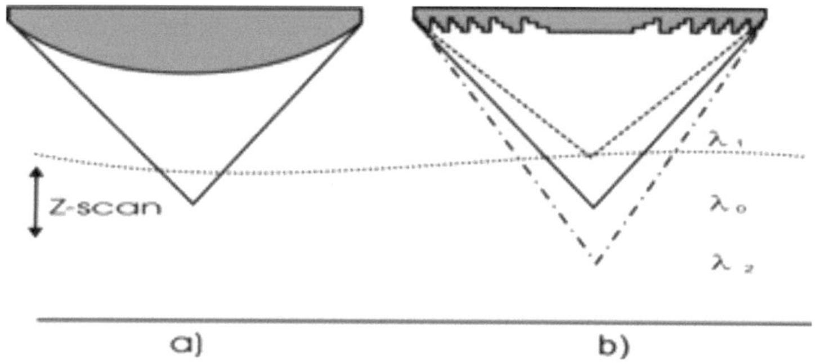

Fig. 4. Physical principle of frequency-scanning confocal sensor.

3.2 Wavefront sensor

Phenomena such as focusing or defocusing of a collimated beam by a lens and the free propagation of a Gaussian beam can be well described as changes of the curvature of a wavefront. The wavefront approach is also the most natural way of describing the operation of optical elements made of a material with a spatially varying index of refraction. Wavefront analysis is the enabling technology in the rapidly developing field of adaptive optics, etc. So the concept of a wavefront is important in optics and in the physics of waves in general.

A Shack-Hartmann Wavefront Sensor is a device that uses the fact that light travels in a straight line to measure the wavefront of light. The device consists of a lenslet array that breaks an incoming beam into multiple focal spots falling on a optical detector. By sensing the position of the focal spots the propagation vector of the sampled light can be calculated for each lenslet. The wavefront can be reconstructed from these vectors. Shack-Hartmann sensors have a finite dynamic range determined by the need to associate a specific focal spot to the lenslet it represents.

One of the key problems with the development of a Shack-Hartmann wavefront sensor is the fabrication of the lenslet array needed. The concept of using binary optics to fabricate these arrays are perspective. The binary optics technique has proven to be successful and is capable of making a large number of different devices with high fidelity.

The similar sensor may be used, for example, for flow measurements. One-dimensional linear sensor arrays that are capable of measuring the phase of optical radiation after it has propagated through the flow field are arranged on the opposite side. These wavefront sensors measure the optical path errors induced on the laser beams by the density variations in the flow caused by the mixture of heated air and entrained cooler room air in the jet. Although each sensor array detects the path-integrated phase in a particular direction through the flow, the set of sensor arrays collects information along many different directions through the flow simultaneously, enabling the inversion ofthe data set and yielding a detailed spatially resolved picture of the plane of the flow through which all the lasers propagated. When the set of sensor arrays in this optical tomography system is operated at speeds of several kilohertz, a tomographic movie of the flow structure in a 2D plane of the flow can be obtained frame by frame [3].

3.3 Fresnel zone plate spectrometer.

DOE can be made to simultaneously disperse the wavelengths of the incoming light - like a grating- at the same time they focus the energy. Figure 5 shows the set-up of Fresnel micro-spectrometer with the opaque-center zone plate. A binary zone plate was typically designed with a concentric series of transparent and opaque rings. It is very interesting to note that when the radius of the zone plate is miniaturized by ***M*** times with a linear miniaturization factor ***M*** such that in paraxial approximation $r_n = \frac{K}{M}\sqrt{n}$, the focal length becomes $\frac{K^2}{M^2\lambda}$, which is shortened by M^2 times. Therefore when the zone plate is miniaturized by five times, the overall thickness due to the focal length is shortened by 25 times, thus approaching a thin-film structure. The focusing properties of the "scaled" FZP is considered in [1]. This miniaturization relation is valid in the Fresnel regime in paraxial approximation. When the optical length Z is changed with a linear scanning actuator, the photon intensity at the detector behind the round aperture slit is recorded with Z which corresponds to $K^2/$ [4].

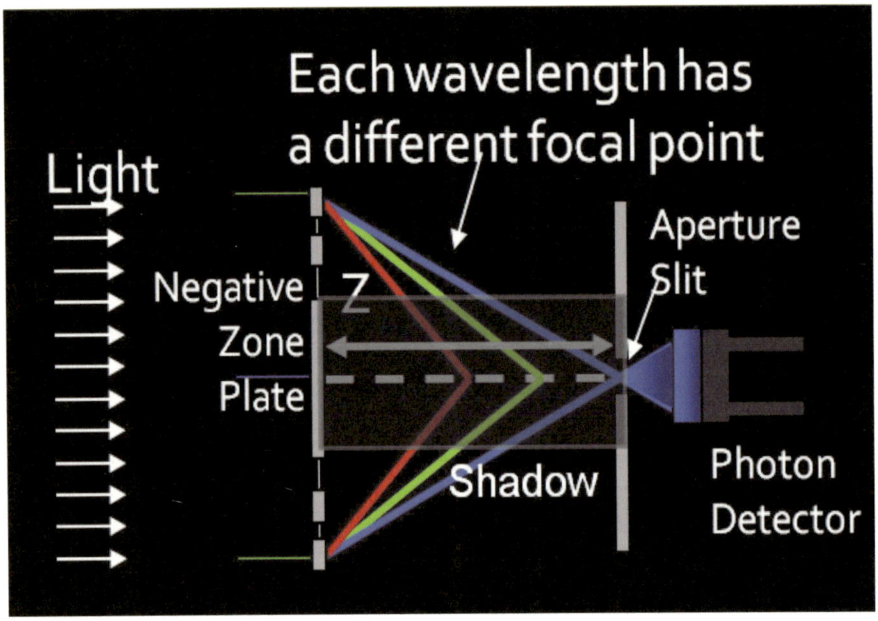

Fig. 5. Optical set-up diagram of a Fresnel micro-spectrometer with negative zone plate: cross-sectional view [4].

The authors [4] have demonstrated that the spectral resolution and resolving power of an ideal Fresnel micro-spectrometer do not depend on the miniaturized size but only on the total number of rings in the Fresnel regime. This is analogous to the spectral resolution relationship of a linear grating with Fraunhofer diffraction. The authors conclude that further miniaturized and improved spectrometers based on Fresnel diffraction can be built if the total number of rings is kept constant or increased.

Low-cost spectrometric sensors utilizes both the dispersing and the focusing abilities of Fresnel zone-like diffractive structures [5]. The standard grating and collimating optics of a scanning spectrometer are replaced with a dispersive and focusing element. At the detector scan area the selected wavelength bands pass over the detector in a predetermined order giving rise to peaks in the detector signal. A DOE may be designed to focus two or more wavelength bands simultaneously onto the same detector position, and the power in the bands may be weighted relative to each other.

Reconfigurable components are well adapted to NIR spectroscopic sensors. In a feasibility study a diffraction filter was fabricated in Si [6] that is both re-configurable and focuses the light. It consists of an array of moveable segments of a Fresnel zone plate; each segment having a few grating grooves (Fig. 6). Each segment is actuated vertically by electrostatic parallel-plate actuators situated outside the optically active part.

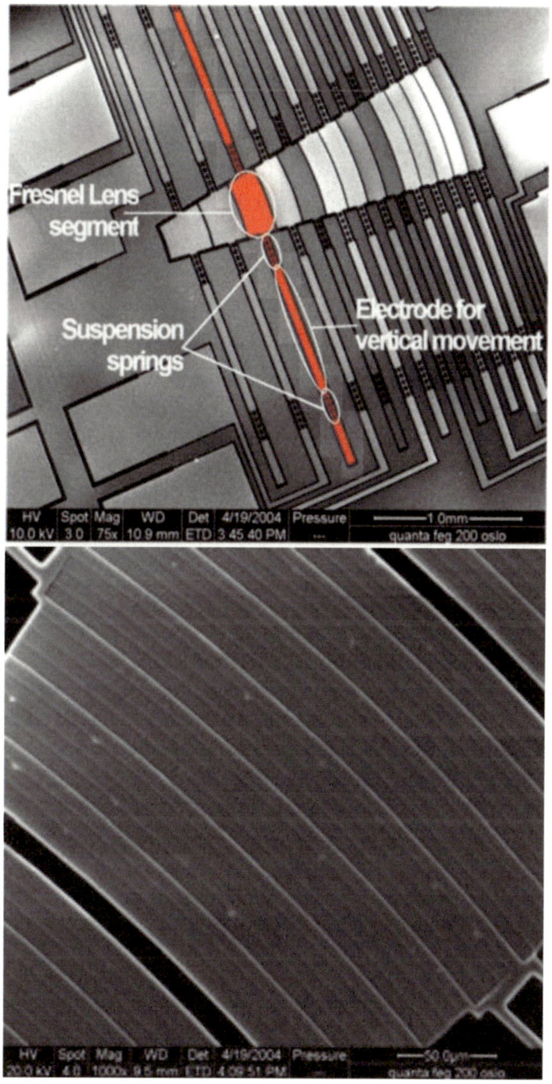

Fig. 6. A micro-mechanical Fresnel lens with 14-segment (left) and a closer view at one segment consisting of 7 grating grooves (right),as seen with a scanning electron microscope [5-6].

In the idle mode the zone plate focuses a target wavelength at the detector. When every other segment of the structure was pulled down, a destructive interference occurs at the centre wavelength giving rise to a sideband on each side of the target wavelength. The target wavelength represents e.g. the absorption line of a gas and will be measured in the idle mode of the actuator, whereas in the activated mode the background on both sides of the target line is measured.

3.4 Diffractive optical microphone

Optical microphones are able to solve major problems that capacitive sensors are suffering from. Optical sensors have no problems with high voltage biasing or need of electrical isolation. They are able to achieve equal or better sensitivity than capacitive displacement sensors with less demanding electronics. The authors describe a new approach where they integrate a diffractive lens into the microphone transducer [7]. An optical microphone consists of a membrane (based on diffractive optical element or Fresnel zone plate) that is deflected by sound pressure or an acoustic signal, where the movement of the membrane is measured by the change in transmitted light intensity due to diffraction. For a given displacement Δd *of diffractive element surface* we can maximize the change in measured optical power, given by the change in diffraction efficiency. The lens consists of a metallic pattern on the substrate, an air gap and a reflecting membrane as shown in Fig 7. The diffractive lens allows to use a simple optical readout system with no additional collimating or focusing optics.

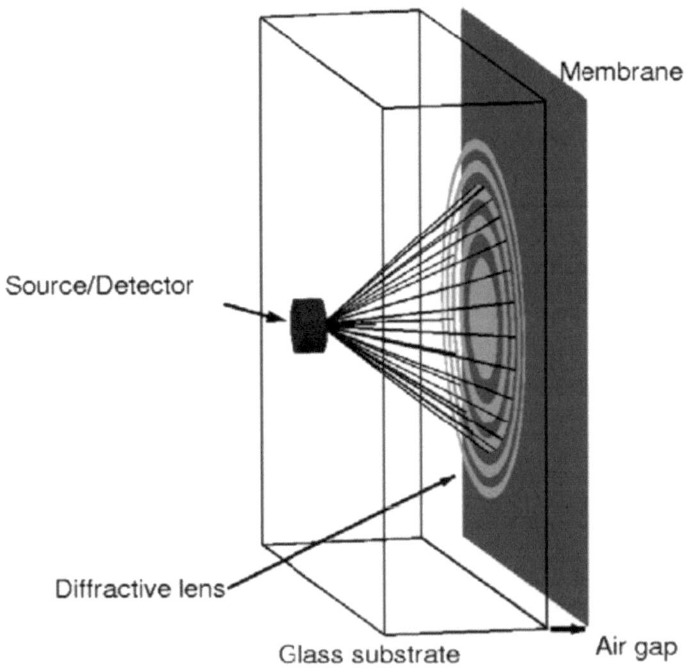

Fig. 7. Simplified geometry of the diffractive microphone. The optical fiber may be replaced by an LED source and a photodiode detector attached to the substrate [7].

By placing a source and detector in the focal plane of the diffractive lens, the measured intensity will be highly sensitive to the position of the membrane.
The diffractive lens was a glass substrate with a 2mm□2mm chrome pattern in the form of a diffractive lens. The authors believed that the microphone can be cheaper, smaller, and are

well suited for integrated design and mass production with photolithography. The fiberbased diffractive lens microphone is suited for demanding applications such as nuclear radiation, medicine and areas with danger of explosion.

4. Diffractive optic fluid shear stress sensor [8]

Accurate measurement of the wall shear stress is needed for a number of aerodynamic studies. Light scattering off particles flowing through a two-slit interference pattern can be used to measure the shear stress of the fluid. The authors [8] designed and fabricated a miniature diffractive optic sensor based on this principle. The developed sensor provided an integrated diffractive optical element (DOE) sensor that integrated the transmitter and receiver optics on the same substrate, as shown in Figure 8. The surface mounted sensor generates a pattern of diverging interference fringes, originated at the surface, and extended into the boundary layer region.

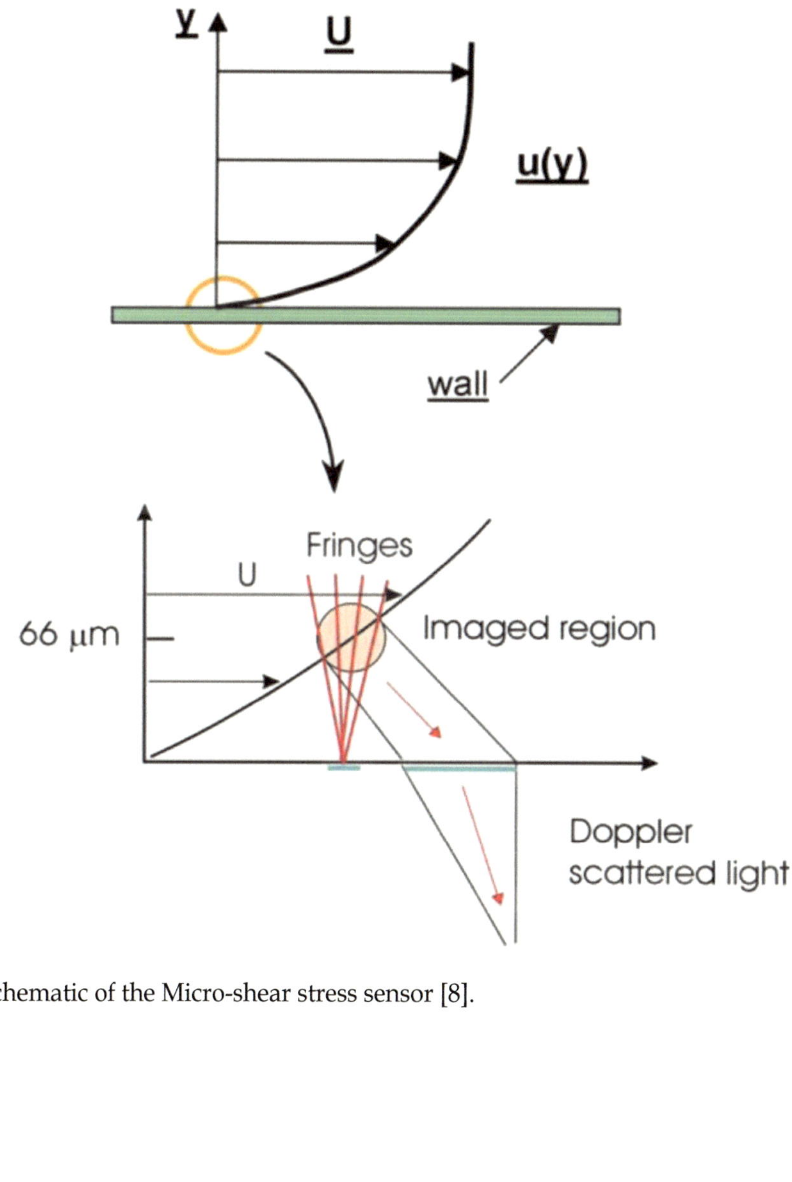

Fig. 8. Schematic of the Micro-shear stress sensor [8].

The intersection region of the transmission fringe pattern and the receiver field of view defined by the backscatter collection optics defined the overall measurement volume location and dimension.

A conceptual drawing of the micro-shear stress sensor is shown in Figure 9. The diverging light from a diode laser is focused by a DOE to two parallel line foci. These foci are coincident with two slits in a metal mask on the opposite side of a quartz substrate. The light diffracts from the slits and interferes to form linearly diverging fringes to a good approximation. The light scattered by particles traveling through the fringe pattern is collected through a window in the metal mask. Another DOE on the backside focuses the light to an optical fiber connected to a detector.

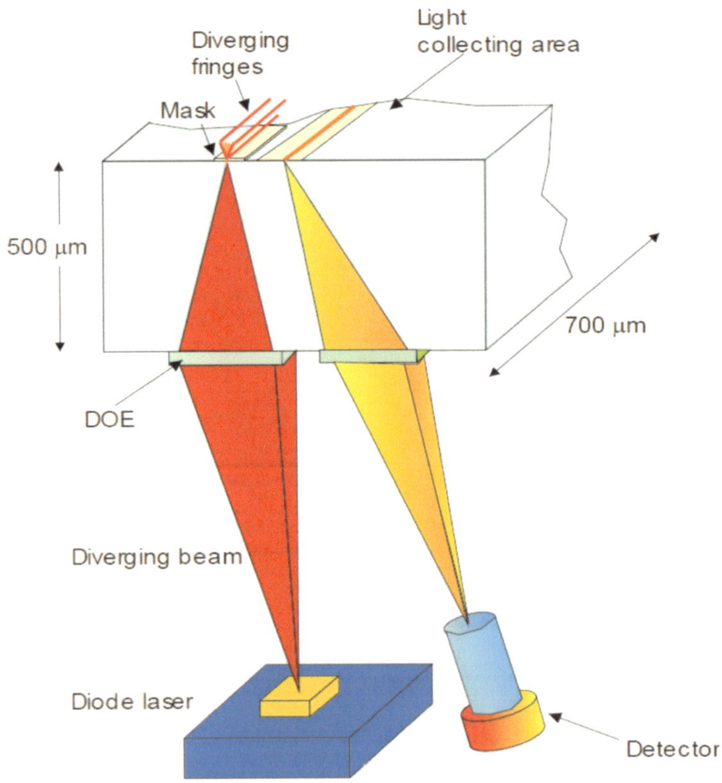

Fig. 9. Schematic of the shear stress sensor assembly [8].

The main sensor element was fabricated by two-sided lithography on a 500 µm thick quartz substrate. The slits and collecting window on the front were fabricated by direct-write electron-beam lithography followed by wet etching of evaporated chrome. The polymethyl methacrylate (PMMA) diffractive optical elements on the back were fabricated by analog direct-write electron-beam lithography followed by acetone development [8].

The shear stress sensor's elements were assembled into a package (Fig. 10 left) with a diode laser (660 nm) and a port for the collection fiber. The overall size of this prototype is 15 mm in diameter and 20 mm in length. The fringes were imaged with a CCD camera using a microscope objective and are shown in Fig. 10. The fringe divergence was measured to be linear with a slope in close agreement with theory. The contrast is very satisfactory and preliminary tests using a moving surface through the fringe pattern yield a clear signal. Testing of the receiver side of the sensor element is underway.

A photograph of the shear stress sensor is shown in Figure 10 (bottom). The sensor substrate, shown in, is mounted into the sensor element location shown on the front face of the assembly. The diverging fringe pattern is visible away from the surface. In this configuration, the light source and the detection electronics are located remotely and are connected to the sensor through optical fibers.

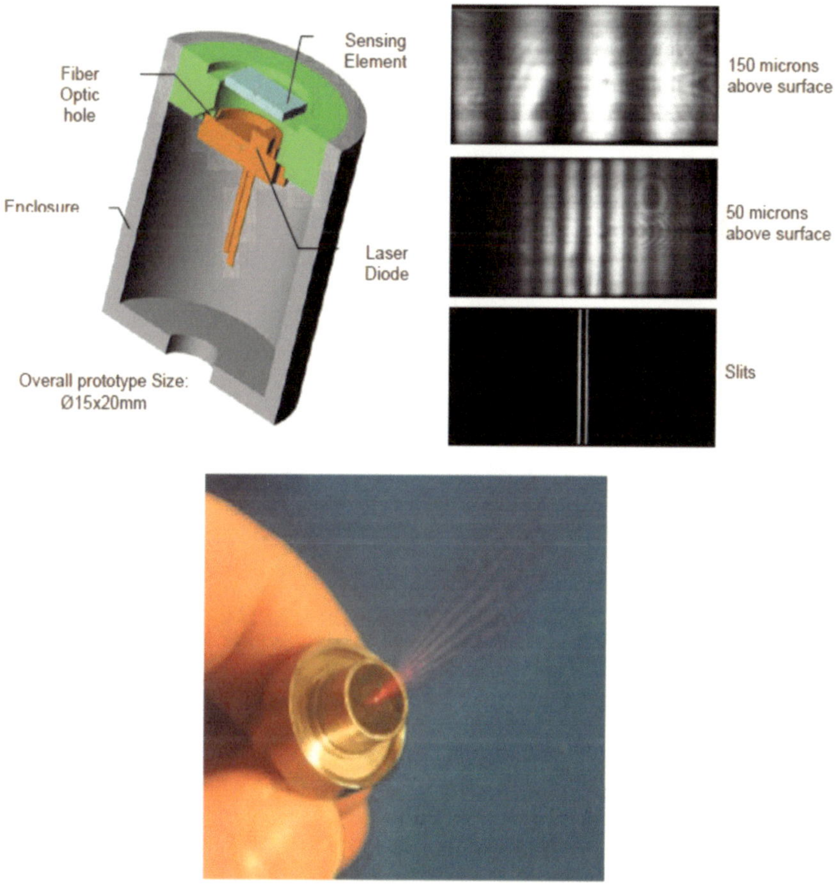

Fig. 10. Shear stress sensor assembly (left) and photographs of the fringes at different heights above the surface (right). Micro-optical shear stress sensor is shown in bottom [8].

4.1 Remote point diffractive optical sensors

The ever-increasing need for sensing in the chemical and physical domain has led to the development of a variety of sensing schemes, including electrochemical and optical sensors. Alternative optical sensing methods use a remotely transmitted light beam to interact with a lossless material exposed to an analyte. A laser beam propagates in free space (or in optical waveguides/fibers), and by interrogating a suitable nanocomposite medium, the "sensor head," it becomes modulated, thus producing the sensor signal. The sensor head is a

powerless (electrically passive) thin-film device of the environmentally sensitive material, not electrically connected to any power or signal processing unit. In a diffractive device, the environmentally induced changes of the optical properties are translated to measureable alterations of the transmitted/reflected diffracted beams. A diffractive NiCl2 /SiO2 nanocomposite photonic sensor for ammonia was demonstrated in [9]. The device has been fabricated on a thin-film structure using direct UV laser microetching techniques. The first-order diffracted beams were found to provide an appreciable sensor response at SNR ~ 18.6, which allowed the detection of quite low, 2 ppm, ammonia levels.
An optimization study of surface relief grating based sensors for use in diffractive optical remote point sensing was presented in [10].

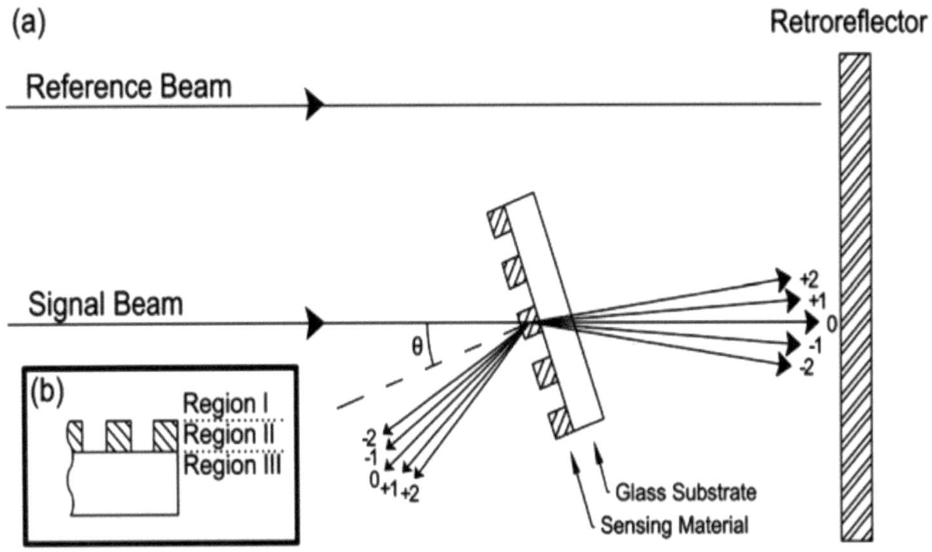

Fig. 11. Remote point sensing by the diffractive device. (a) Schematic overview and (b) SRG geometry used for theory and design [10].

The RPS concept is schematically depicted in figure 11. The passive sensor head consists of a sensing material thin film deposited on a glass substrate. The sensing material is designed to reversibly alter its optical properties upon exposure to a chemical or physical agent. The sensor head is placed in the area to be monitored and it is remotely interrogated in real time by measuring the reflected signal beam. DOEs and photonic crystals offer a unique potential for detecting effective changes of the materials. In the case of DOEs, the intensity and/or the position of the diffraction fringes strongly depend on the refractive index of the patterned material.
The results [10] reveal a strong oscillatory dependence of the diffraction efficiency on the grating depth that can be conveniently tailored by means of direct laser ablation. The dependence on grating period is also oscillatory with decreasing amplitude as the period increases. Duty cycle changes will affect the diffraction efficiencies but in a less pronounced way. It has been shown that an elegant and practical way to enhance the diffraction efficiency and hence the sensor responsivity, inherent to the remote point sensing scheme, is by changing the angle of incidence of the interrogating laser beam.

4.2 DOE sensor for optical computer [1]

Optical computers promise very high speed of computation. The surge in computer speed (as compared with conventional computers of today) will be mostly achieved by greatly parallelized mode of operation. For instance, about 10^6 light beams in a common light beam can interact simultaneously with 10^6 elements of a logic matrix.

In addition to parallelism of data processing, optics has another resource for increasing the speed of optical computers: the polychromatism of radiation. Each of the million of light beams functioning in parallel may contain about a thousand spectrally distinguishable monochromatic components. Each of them is in principle capable of interacting simultaneously and differently with the same logical sell. On a single sell, it is possible to use 16 wavelengths simultaneously and realize a complete set of 16 logical functions. The results of interactions of other radiations with the same cell (repeating the same functions) can be treated as multiplexing, with simple separation of the results using dispersive elements. In this way, optical computers of future generations may have their speed increased by two to three orders of magnitude by using polychromatic nature of the light beam.

The implementation of optical digital computers mostly depends on the creation of optical logic elements (optical analogues of electronic gates) that carry out various logical operations (AND, OR etc) that would go beyond the speed of microelectronic devices and their degree of integration, also reducing cost and power consumption. At the moment, a number of optical switches were created, among which the most promising are optical interference filters, optical etalons (OLE - Optical Logic Etalon), bistable SEED that are devices with their own electrooptic effect (Self ElectroOptic Effect Device) and the so-called QWEST (Quantum Well Envelope State Transition) devices.

Diffractive (dispersive) elements can be used for spectrally selective addressing of signals, can be applied in polychromatic optical processors, serve as a basis for polychromatic logic elements or multiplexer or a focusing element with selectivity in the multimode regime etc.

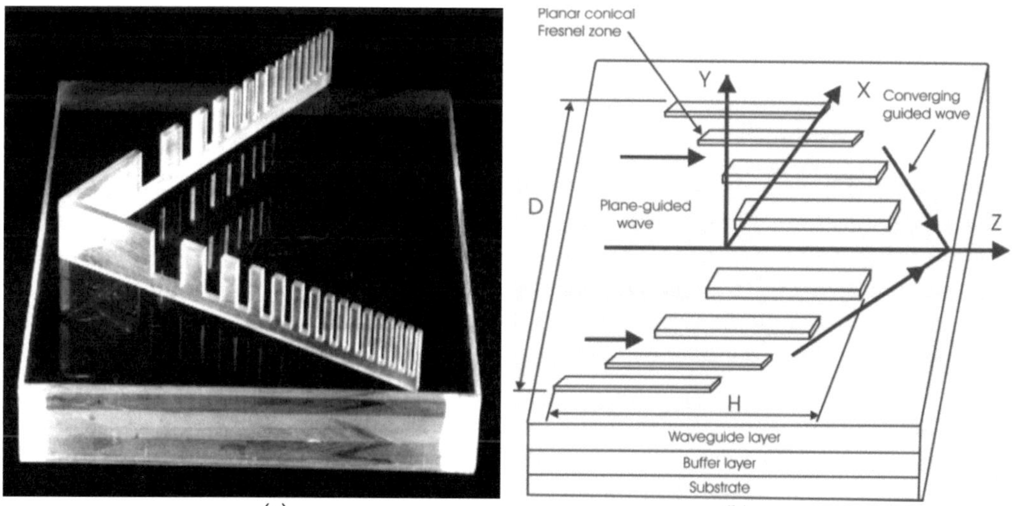

(a) (b)

Fig. 12. (a) Diffractive planar element on a conical surface, (b) principal scheme of novel diffractive integrated optical element.

Diffractive planar elements fabricated on a non-flat surface make it possible to considerably enrich the "pool of devices" of integrated optics of different waveband, including THz, and to design elements with novel properties and potentials. This can be illustrated most clearly using as an example optical elements for optical polychromatic computers. For instance, the "conical" diffractive element discussed above can be used as a nonlinear device for polychromatic radiation or multiplexer or focusing element with selection of multimode regime. Frequency characteristics for such elements are determined by the extent of concavity (convexity) of the surface of the element and by the direction of incidence onto it. Therefore, when working on a wavelength $\lambda \neq \lambda_0$, the position of the focusing area in space (the amount of its displacement) should depend on the direction of incidence of the radiation. Hence, it is possible to distinguish between a signal incident on the "tip" of the element from that falling on its "base" by placing radiation receivers at the corresponding points in space.

It is just as easy to organize logic elements in a similar fashion. Let us consider as an example a diffractive element that focuses radiation emitted from one point onto two points. Such elements are two-dimensional analogues of a three-dimensional element that provides focusing of a point source to a ring. We will use this element "the other way around" - we let it focus radiation from two pointlike sources to a single point. Then, if we change the wavelength of radiation emitted by one of these two pointlike sources, the area of focusing will change its position in space: it will move transversally to the optical axis of the element. This is effectively a polychromatic logic element: if radiation frequencies at its two ends are identical, NO is the output; if input frequencies are not identical, no output signal is generated (or appears depending on the position of radiation receiver). In other words, logical elements "END" and "OR" are realized.

The principle of designing logical "AND", "OR" elements makes it possible to implement a controlled switch. For this purpose, one of the inputs of the diffractive element is a controlling one, while the element itself has several outputs located in the region of focusing of the radiation and corresponding to the value of wavelength at the controlling input [1, 11].

5. Label-free diffractive optical biosensor technology

In the paper [12] describes technology for the detection of small molecule interactions in which a colorimetric resonant diffractive grating is used as a surface binding platform of a microtiter plate and a glass slide. The sensor structure, when illuminated with white light, is designed to reflect only a single wavelength. When molecules are attached to the surface, the reflected wavelength (color) is shifted due to the change of the optical path of light that is coupled into the sensor. By linking receptor molecules to the sensor surface, complementary binding molecules can be detected without the use of any kind of fluorescent probe or radioactive label.

The photonic crystal biosensors, often referred to as optical biosensors, are composed of a periodic arrangement of dielectric material that effectively prevents propagation of light at specific wavelengths and directions [13]. Individual optical biosensors are incorporated within each well of industry standard 96-, 384-, 384lv- and 1536-well microplates or 16-well cartridges. When illuminated with white light, the optical diffractive grating of the photonic crystal reflects a narrow range of wavelengths of light which is measured by the BIND Reader (Figure 13). The wavelength of reflected light shifts upon a change in binding, or adherence, within proximity of the biosensor surface. Real time binding is observed by measuring the shift in peak wavelength value over time, including monitoring the binding of individual components in a multi-component binding complex or upon sequential rounds of cellular stimulation. The shift in peak wavelength is directly proportional to the change in mass.

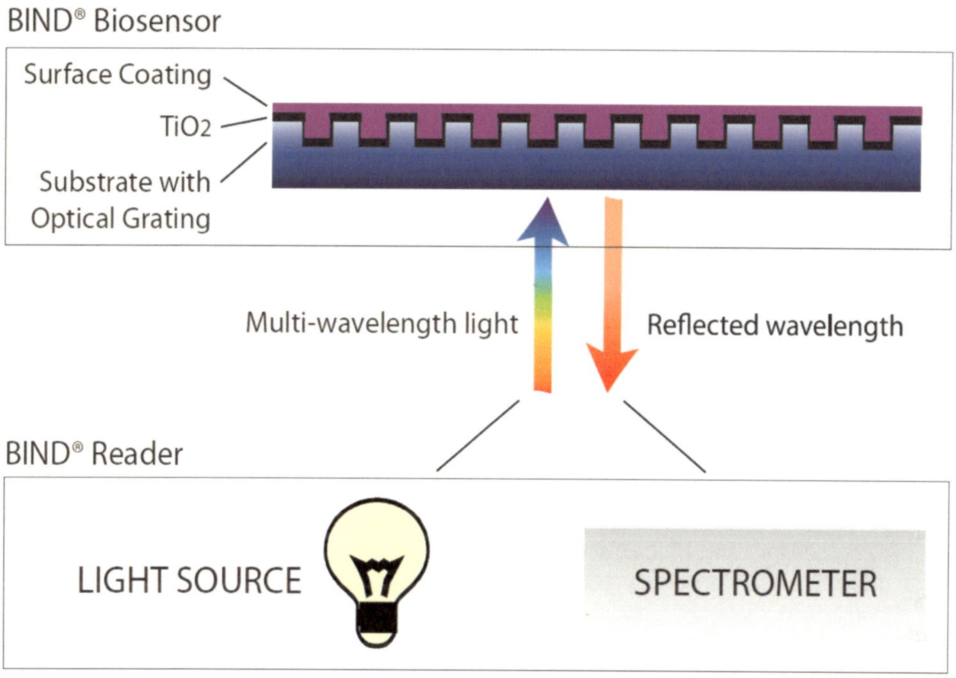

Fig.13. BIND Technology Optical Principles -BIND Biosensors incorporate photonic crystals which reflect a narrow range of wavelengths when illuminated with white light. BIND Biosensors can be coated with a variety of surface chemistries for optimization immobilization and assay performance. The BIND Reader illuminates biosensors and measures the wavelengths of the reflected light. [12-13]

6. References

[1] O.V. Minin, I.V. Minin, *Diffractional Optics of Millimeter Waves*, Institute of Physics Publishing, Bristol, UK, 2004.

[2] Minin I.V., Minin O.V. Concept of near-field millimeter-wave imaging system with a spatial resolution beyond the Abbe barrier. Proc. of the 2008 China-Japan Joint Microwave Conference, September 10-12, Shanghai, CHINA, pp.509-512.

[3] D. Neal et al. One dimensional wavefront sensor development for tomographic flow measurements. / SPIE Vol. 2546, pp. 378-390

[4] Yeonjoon Park et al. Miniaturization of a Fresnel spectrometer. J. Opt. A: Pure Appl. Opt. 10 (2008) 095301

[5] O. Løvhaugen, I. R. Johansen, K. A. H. Bakke, Stephane Nicolas, "Dedicated spectrometers based on diffractive optics: Design, modeling and evaluation", Journal of Modern Optics, Vol. 51, No.14, pp. 2203-2222, (2004)

[6] H. Sagberg, M. Lacolle, I. R. Johansen, O. Løvhaugen, R. Belikov, O. Solgaard, and A. S. Sudbø, "Micromechanical gratings for visible and near-infrared spectroscopy," IEEE Journal of Selected Topics in Quantum Electronics, Vol. 10, No. 3, 604-613 (2004).

[7] H. Sagberg, A. Sudbo, A. Solgaard, K. A. H. Bakke, and I. R. Johansen, "Optical microphone based on a modulated diffractive lens," *IEEE Photonics Technology Letters*, vol. 15, no. 10, pp. 1431-1433, 2003.

[8] Darius Modarress et al. Diffractive optic fluid shear stress sensor. http://measurementsci.com/papers/exp-fluids-s.pdf *See also: US Patent US 6,717,172* http://www.measurementsci.com/papers/6717172_Shear_Stress_Sensor.pdf

[9] M. Vasileiadis et al. Diffractive optic sensor for remote-point detection of ammonia. OPTICS LETTERS / Vol. 35, No. 9 / May 1, 2010, pp.1476-1478.

[10] M. Vasileiadis et al. Optimized design of remote point diffractive optical sensors. J. Opt. 12 (2010) 124016

[11] I.V. Minin, O.V. Minin, S. Shi, C. Chen, J. Mititu, and D. Prather. NOVEL TYPE OF THE ELEMENTS OF INTEGRATED DIFFRACTIVE OPTICS// Terahertz for Military and Security Applications IV, Novel Terahertz Devices and Concepts III Proc. SPIE 6212, *17–21 April 2006 Gaylord Palms Resort and Convention Center • Orlando (Kissimmee), Florida USA*

[12] Bo Lin et al. A label-free optical technique for detecting small molecule interactions //Biosensors and Bioelectronics 17 (2002) 827-834

[13] Label-free Optical Biosensor Technology Case Studies: Direct Binding and Functional Cell-based Assays. http://www.srubiosystems.com/resourceCenter/ApplicationNotes/Case_studies_paper.pdf

Part 4

Microsensors Application

11

Strength Reliability of Micro Polycrystalline Silicon Structure

Shigeru Hamada, Kenji Hasizume,
Hiroyuki Nakaura and Yoshihide Sugimoto
Department of Mechanical Engineering, Faculty of Engineering, Kyushu University
Japan

1. Introduction

Polycrystalline silicon (poly-Si) structure is widely employed in the Micro-Electro-Mechanical Systems (MEMS) [Najafi, 2000; Senturia ,2000]. MEMS devices, which contain mechanical movement, have to maintain their reliability in face of external shock, thermal stress and residual stress from manufacturing processes, and fracture will begin mainly in stress concentration area. Therefore, it is necessary to build up reliability design criterion of the poly-Si structure that has stress concentration [Chen et al., 2002; Greek et al.,1997; Kapels et al., 2000; Muhlstein et al., 2004; Namazu et al., 2000; Sharpe et al., 2001; Tsuchiya et al., 1998]. However, since the size effect is large, the microscopic poly-Si depends for the strength on the effective area caused by the stress concentration of structure. Moreover, as the point peculiar to the microscopic poly-Si at the time of thinking of strength, in order that the techniques of processing the upper surface and the sidewall surface differ, it is mentioned that the surface roughness used as the source of a stress concentration differs. It depends for the strength of the microscopic poly-Si also on surface roughness. Therefore, it is necessary to deal with simultaneously the stress concentration of structure and the stress concentration by surface roughness in the case of strength evaluation. In order to clarify the bending strength and its effective area dependability of poly-Si, bending tests using micro scale cantilever beams with or without notch of several sizes are performed. Moreover, surface roughness measurement using AFM is carried out, it determines for the stress concentration by surface roughness, and a quantitative effective area is defined. Fracture origins are specified by fracture surface observation, and the validity of the effective area are shown. Finally, the static strength design criteria in consideration of scattering in strength which used two parameters, the maximum stress and an effective area, are proposed.

2. Test method

2.1 Specimen

The specimens are illustrated in Fig. 1. Shapes and dimensions of the specimens are shown in Table 1. For bending tests, two types of specimens; Type-A and B are prepared. In the Type-A specimen, the notch of several sizes (1~5 [μm]) is introduced in the root section of micro-cantilever beam. In the Type-B, by the microscopic observation, the 1 [μm] corner radius is

recognized indeed in the root section of micro cantilever beam. Thickness of the specimen (h) is 3.5, 6.4 and 8.3 [μm]. The gap between the cantilever and the substrate is 2 [μm].
The poly-Si is Chemical Vapour Deposited (CVD) on single crystal silicon wafer surface, and the specimens are made from surface micromachining process. The Deep Reactive Ion Etching (DRIE) process were used for processing of the sidewall surface of 6.4, 8.3 [μm] thickness specimens. Therefore, especially in the specimen side surface made by DRIE, microscopic irregularity called "scarop" which is not seen on the upper surface. Figure 2 shows the example of the scarop for the 6.4 [μm] thickness specimen.

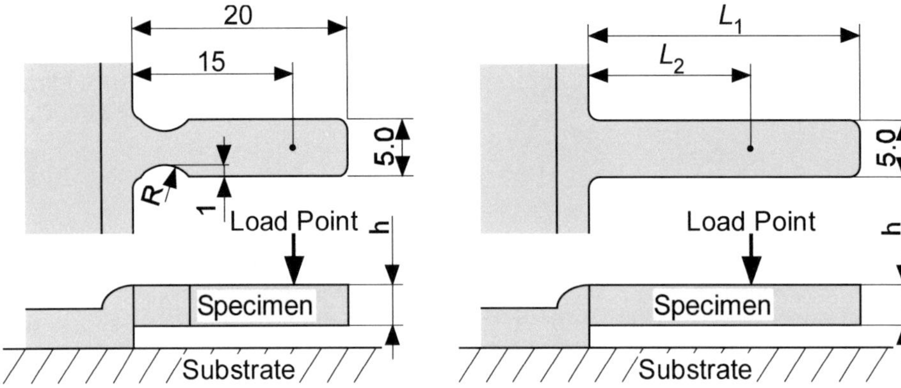

(a) Notched specimen (Type-A) (b) Specimen without notch (Type-B).

Fig. 1. Schematic diagram of the specimens (unit: μm), h = 3.5, 6.4, 8.3 [μm]

Specimen Type		L_1, μm	L_2, μm	R, μm
Type-A	L15R1	20	15	1
	L15R2	20	15	2
	L15R3	20	15	3
	L15R4	20	15	4
	L15R5	20	15	5
Type-B	L10	15	10	–
	L15	25	15	–

Table 1. Shapes and dimensions of the specimen

Fig. 2. Sidewall surface morphology of the h = 6.4 [μm] poly-Si specimen made by DRIE process

2.2 Strength test and stress analysis

A dynamic ultra-micro hardness tester (Shimadzu DUH-W201) with a Berkovich diamond indenter is used for the bending tests. The test machine is for hardness tests but we can obtain the relationship between the load and the displacement with satisfactory accuracy by this machine. The test load speed is 1.421 [mN/sec.]. The bending tests are carried out at room temperature under the atmospheric environment.

In order to quantify the fracture of specimens by the applied stresses in the tests, three-dimensional finite element elastic analyses are performed with Young's modulus: 148[GPa], Poisson's ratio: 0.2. The software used for this analysis is ANSYS 10.0. Figure 3 indicates the examples of FEM models of the specimens (Type-A, L15R1). For the Type-B specimen, the 1μm corner radius is taken into the model based on the microscopic observations mentioned above. The stress concentration factor of Type-B is close to that of Type-A, L15R3. The element sizes of the models are about 1μm in the overall region and are about 0.1μm in the stress concentration area near the notch root.

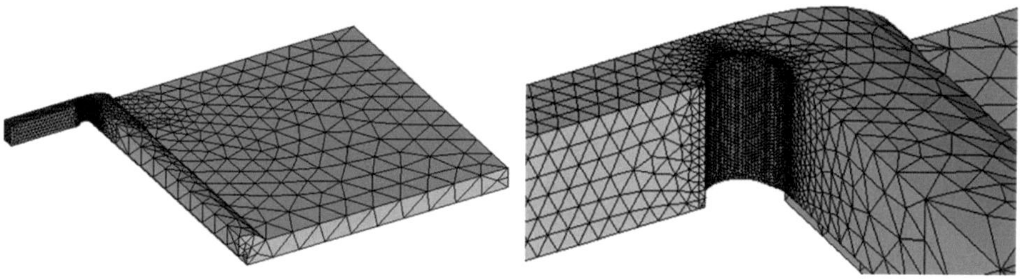

(a) Whole of the FEM model (1/2 model) (b) Detail of the stress concentration (notch) section of the specimen

Fig. 3. Examples of the FEM model (h = 6.4 [μm], L15R1).

2.3 Surface roughness

In order to investigate the stress concentration by the shape of the microscopic surface of a specimen, surface shapes were measured using the atomic force microscope (AFM) (VEECO D-3000). The region to measure was made into 1 [μm] four quarters. The upper surfaces of the specimens were measured by scanning in the specimen longitudinal direction, and the sidewall surface of the specimens scanned and measured the scarop bottom in the specimen longitudinal direction. The maximum stress concentration factor $K_{t\,\max}$ which exists in a specimen was determined using the roughness of the measured surface. It is shown in 5.1 for details. In order to estimate the maximum stress concentration factor which exists in a specimen, the data of stress concentration factor determined by measurement was arranged using statistics of extreme [Gumbel, 1962].

2.4 Effective area

The following equations thought to be able to define an effective area S. The maximum stress $\sigma_{\max}$ of the structure calculated in FEM analysis and the $K_{t\,\max}$ calculated by roughness measurement are used. When the stress concentration shown in the Eq. (1) was taken into consideration, area on which the stress exceeding $\sigma_{\max}$ was made into the effective area.

$$\sigma_{\max} \leq K_{t\,\max}\sigma \tag{1}$$

The S thought to expresses the effective area at the time of evaluating strength here. Within the range of an effective area, it can become fracture origin except the maximum stress working point by the stress concentration of structure by the surface roughness stress concentration.

3. Bending test results

Figure 3 shows examples of the relationship between load and displacement of the bending tests. In this figure, it is known that the polycrystalline silicon deformed elastically until final catastrophic failure in room temperature, showing a brittle nature. The relationship between load and displacement shows a little nonlinear behaviour. This is because of indentation to poly-Si of the indenter.
Figure 5 shows an expression of the test results by use of maximum peak stress σ_{max} in the notch root obtained by FEM analysis. In this figure, the plots of solid mark means the data of fracture specimens and the open mark means the data of non-fracture specimens due to the contact of free edge to the substrate surface before break. The displacement which contact occur changes with specimen, this may be because the residual stress and shape of the specimens has little difference respectively.
In Fig. 5, a tendency can be seen that the plots by the σ_{max} move down with the increase in notch radius because of the effect of the difference on the stress distribution pattern.
In order to investigate the scattering in the fracture strength obtained by a bending strength test and FEM analysis, fracture strength was plotted to Weibull probability paper [Weibull, 1951]. Weibull distribution is used for strength evaluation of a brittle material like the ceramics, and it is thought that Weibull distribution can estimate the strength of the poly-Si which is brittle material.
The function of two population parameters Weibull distribution can be expressed with the following equation.

$$F = 1 - \exp\left\{-A_E\left(\frac{\sigma_B}{\sigma_0}\right)^m\right\} \tag{2}$$

In Eq. (2), F: cumulative probability of failure, σ_0: scale parameter, m: shape parameter, respectively, A_E is the effective area. For Weibull plots, Eq. (2) is changed as follows.

$$\ln\ln\frac{1}{1-F} = m(\ln\sigma_B - \ln\sigma_0) + \ln A_E \tag{3}$$

Figure 6 shows the Weibull plots of bending strength for poly-Si. The non-fracture data are treated statistically [Johnson, 1964]. Figure 6 show the validity of using Weibull analysis for this study.
Figure 7 shows the scale parameters (σ_0) of this study. If shapes of the specimen are different, then effective surface area is different. Then, the scale parameter is thought to be different. Figure 7(a) shows the validity of the effective surface. Figure 7(b) shows the same result. In Fig.7, h = 3.5 [μm] specimen shows different trend. We'll discuss about it later.
Figure 8 shows the result of shape parameter (m). With effective surface area, Weibull parameters are shown as follows.
Equation (3) shows the shape parameter (m) is independent on effective area (A_E). Figure 8 shows the independency but for h = 3.5 [μm] specimen. Then, we found that the scatter in

poly-Si bending strength for h = 3.5 [μm] specimen is smaller than that of h = 6.4, 8.3 [μm] specimen because the m of the h = 3.5 [μm] specimen are large.

Figure 9 shows the fracture surface and around the fracture surface of poly-Si of this study. Fig. 9(a) (h = 3.5 [μm]) are different from Fig. 9 (b,c) (h = 6.4, 8.3 [μm]). This is because the etching processes for these specimens are different. The h = 6.4 [μm] and 8.3 [μm] specimens are made by same DRIE process, but h = 3.5 [μm] is made by another RIE process. Then the side surface condition of these specimen are different and the effect of effective area are thought to be different between h = 3.5 [μm] specimen and h = 6.4, 8.3 [μm] specimen.

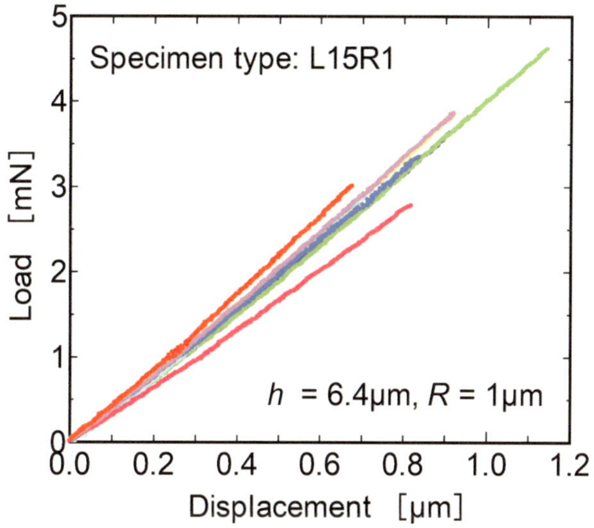

Fig. 4. Example of relationship between load and displacement.

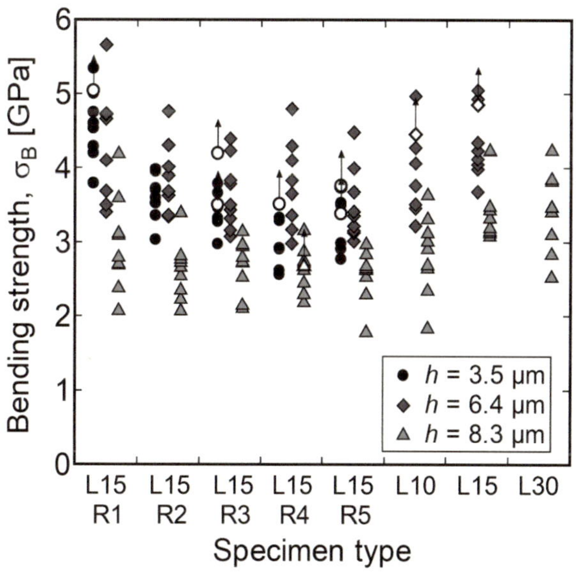

Fig. 5. Weibull plots of bending strength for poly-Si

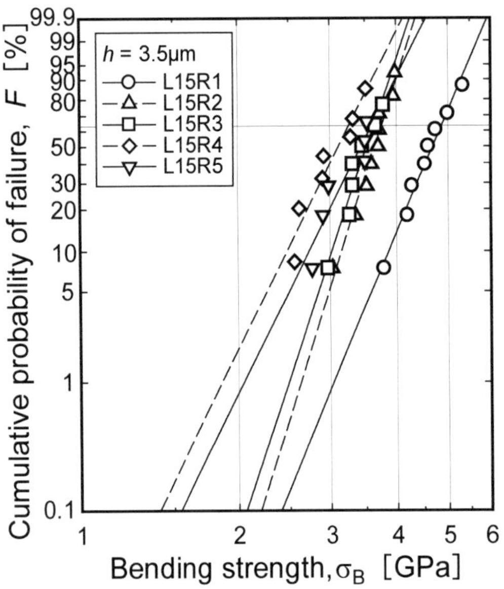

(a) h = 3.5 [μm] specimen

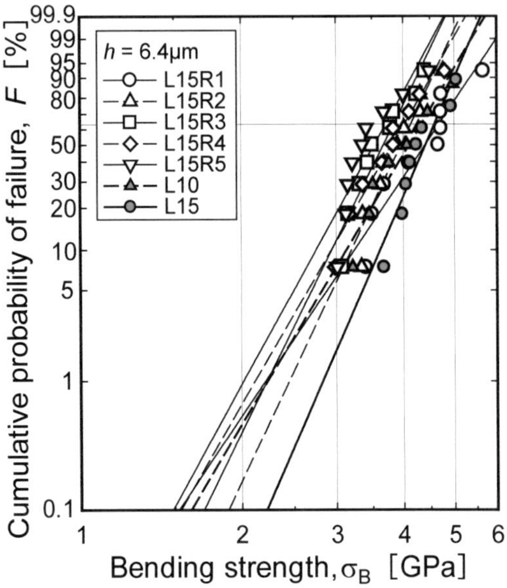

(b) h = 6.4 [μm] specimen

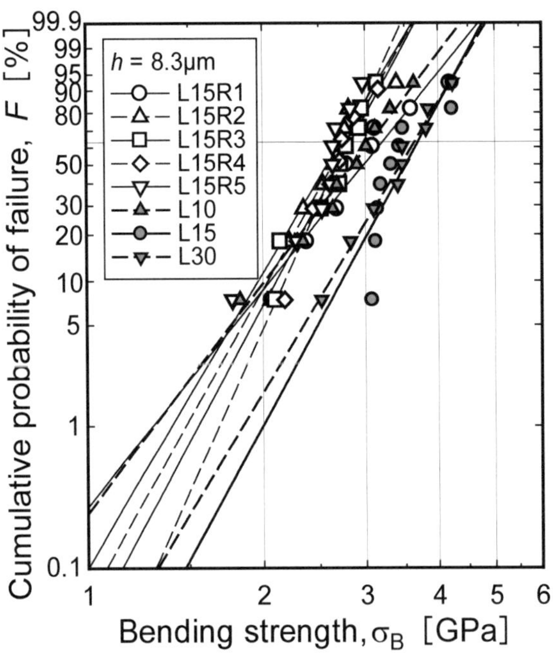

(c) h = 8.3 [μm] specimen

Fig. 6. Weibull plots of bending strength for poly-Si

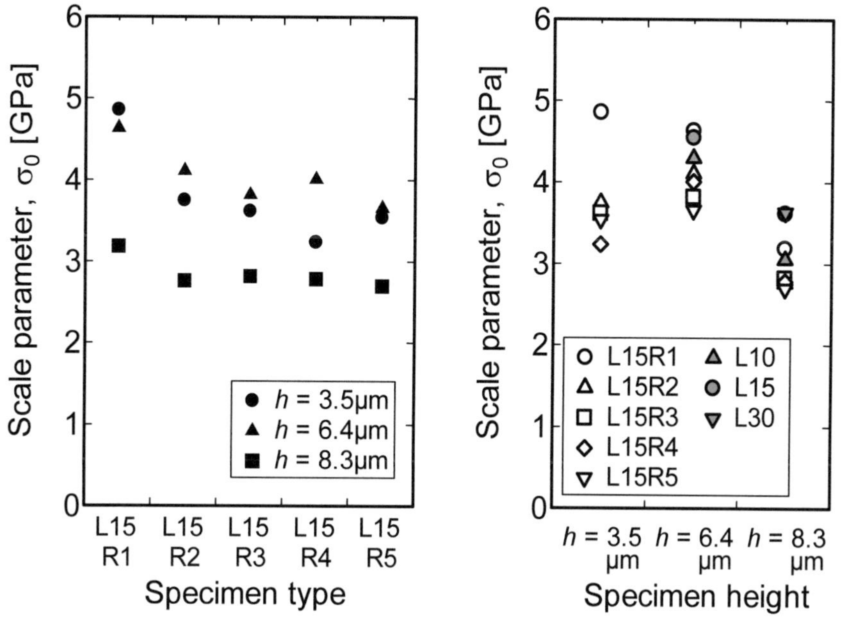

(a) specimen type dependency. (b) specimen height dependency

Fig. 7. Scale parameters of the bending strength for poly-Si.

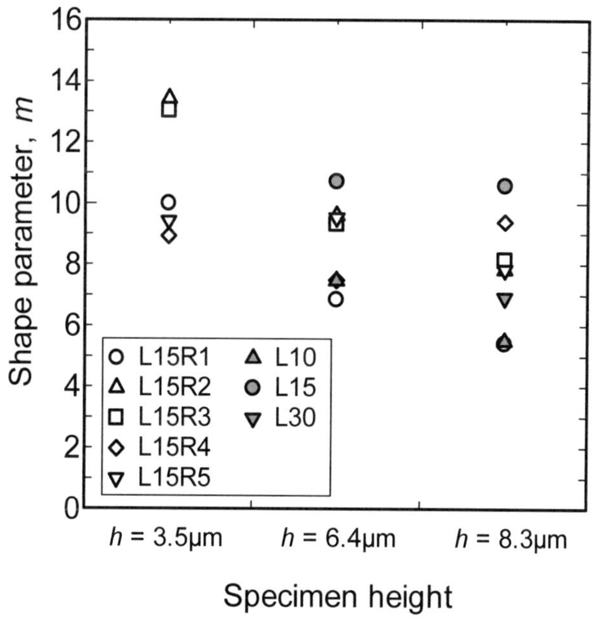

Fig. 8. Shape parameters of the bending strength for poly-Si.

(a) h = 3.5 [μm], L15R3.

(b) h = 6.4 [μm], L30. (c) h = 8.3 [μm], L15.

Fig. 9. Fracture surface and surface image of poly-Si for this study.

4. Fracture toughness and fracture angle

4.1 Fracture toughness

In order to make clear the reason of the strength scattering, analysis of the fracture surface by SEM are performed. If the variation of the strength is dependent on the initial defect size, then calculated fracture toughness is thought to be same. Figure 10 show one example of the analysis. In this figure, in the area of fracture origin, a mirror zone [Hull, 1999] came under observation. Then, we assume that the mirror zone is initial defect, stress intensity factors for the mirror zone are calculated by quarter-elliptical crack in a plate under bending mode stress distribution [Murakami, 1992]. Figure 11 shows the results. The average of the fracture toughness is 2.1 [MPa√m]. Comparing this value with other reports [Hahn et al., 2000; Son et al., 2004], the value found to be about double. Then area of mirror zone is not same as area of initial defect. But Figure 11 shows that the fracture toughness is independent on specimen shape and the value is constant to some extent. Then area of mirror zone assumed to be proportional to the initial defect.

4.2 Fracture angle

More fractographic analysis is performed. Then, we found that the angle of mirror zone is different from the principal stress surface. An example of the observation result and the definition of the angle (θ) are shown in Fig.12. Figure 13 shows the relationship between fracture toughness using stress and mirror zone and the angle. In this figure dependency of the fracture toughness on the angle can be seen. The reasons are thought as follows. (1) The poly-Si crystal on the fracture origin area are different from principal stress direction (2) the poly-Si grain boundary on the fracture initiation area are different from principal stress direction.

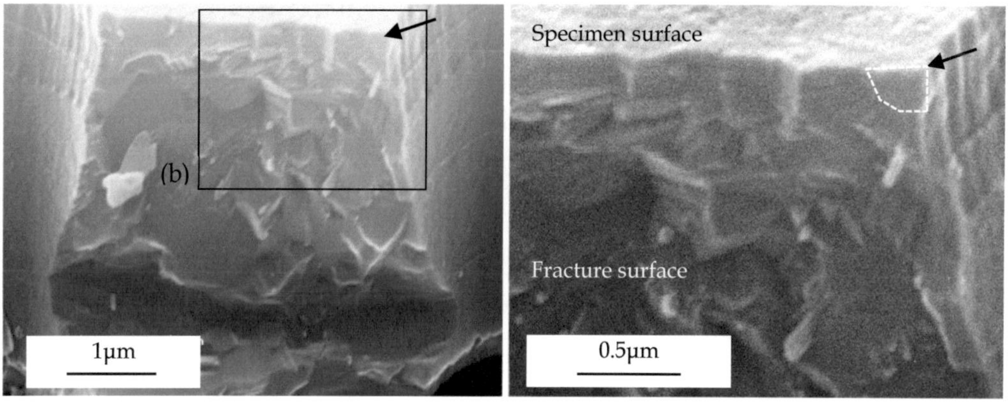

(a) Macroscopic view of fracture surface(arrow shows the fracture origin)

(b) Magnification of fracture origin (dashed line shows the mirror zone)

Fig. 10. Example of fracture surface (h = 3.5 [μm], L15R3)

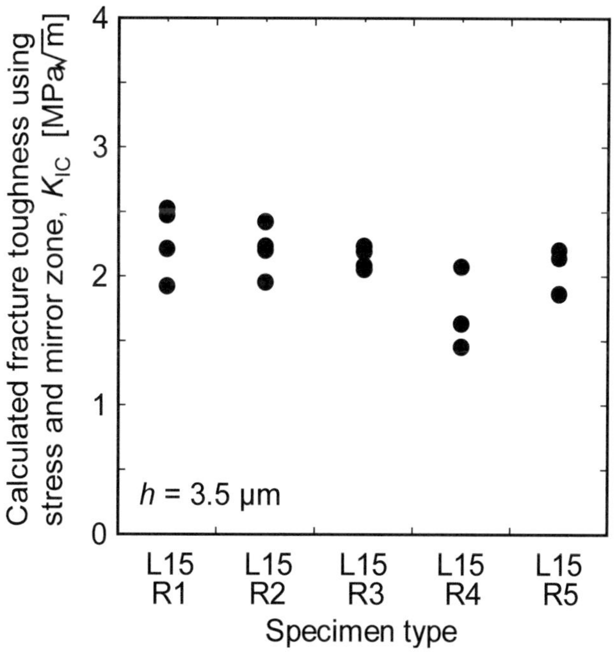

Fig. 11. Scale parameters of the bending strength for poly-Si (h = 3.5 [μm]).

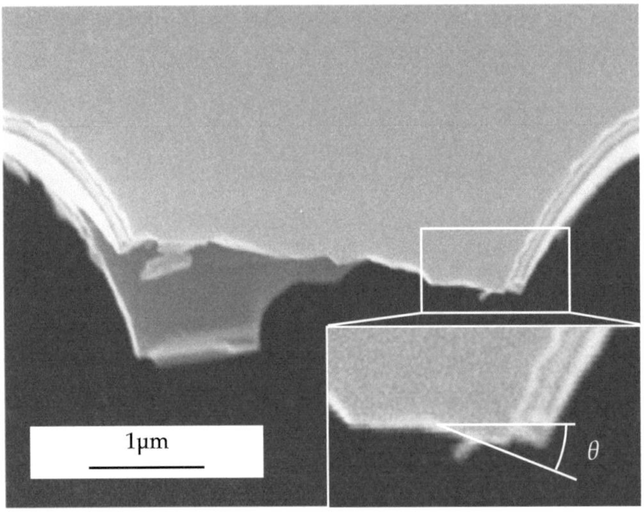

Fig. 12. Definition of fracture surface (mirror zone) angle.

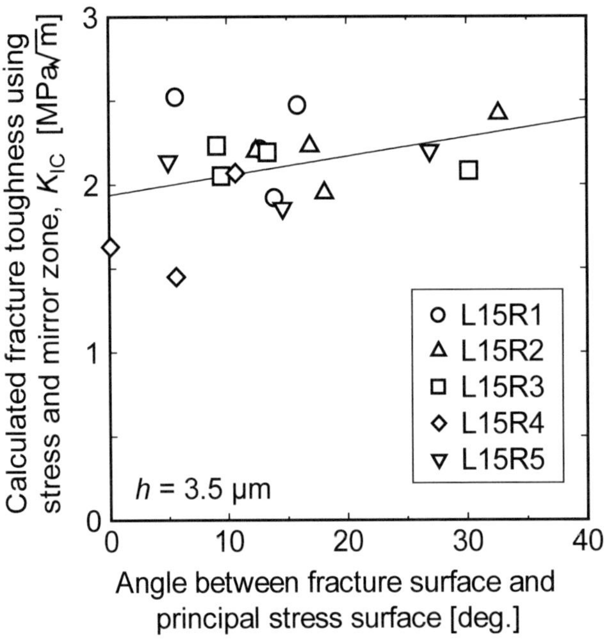

Fig. 13. Angle dependency of fracture toughness (h = 3.5 [μm]).

5. Effective area and application to design

5.1 Effective area definition and calculation results

The stress concentration factors on the specimen surface were determined based on the result of surface roughness measurement by AFM. As shown in Fig. 2, difference occurs in the surface shape of the upper surface and the sidewall surface by the difference in the manufacturing technique. Figure 14 indicates an example of the difference in surface

roughness of the h = 6.4 [μm] specimen obtained by AFM, and Fig. 15 indicates the example of surface section of the scarop bottom on the sidewall surface.
Using the measurement result of the surface roughness, the stress concentration factors K_t of the specimen were calculated. As shown in Fig. 14, the appearance present complicated shapes, therefore FEM analysis is necessary to calculate an accurate stress concentration factors. In this report, in order to simplify, the interference effects by the multiple notches were ignored and the stress concentration factors K_t were determined from width (a) and depth (b) from the roughness measurements using the following equations supposing the equivalent ellipse as shown in Fig. 15.

$$K_t = 1 + \frac{2a}{b} \tag{4}$$

The maximum stress concentration factor $K_{t\,\max}$ which exists in a specimen based on the data of the measured stress concentration factor is estimated using the statistics of extreme. Figure 16 shows the extreme values probability paper. The horizontal axis is the stress concentration factor $K_{t\,j}$ obtained by the Eq. (4). The vertical axis is the reduced variates y_j calculated by the following equation which is a formula of the statistics of extreme.

$$F_j = \frac{j}{n+1}, \quad y_j = -\ln\{-\ln F_j\} \tag{5}$$

(j = 1, 2, 3,…, n n: Number of inspections)

The approximate expression was calculated using the least square method from the obtained distribution. The maximum stress concentration factor which substitutes the return period T for the following equations, and $K_{t\,\max}$ exist in a specimen is estimated.

$$y = -\ln\left\{-\ln\left(\frac{T-1}{T}\right)\right\}, \quad y = \alpha\, K_{t\,\max} + \beta \tag{6}$$

When determining the return period T, evaluation area was made equal to the effective area. The relation between evaluation area and the return period are defined using the following equations. (S_0: inspection area)

$$T_i = \frac{S_i + S_0}{S_0}, \quad S < 10\, S_0 \tag{7}$$

$$T_i = \frac{S_i}{S_0}, \quad S > 10\, S_0 \tag{8}$$

In order to bring evaluation area close to an effective area, calculation performed repeatedly. The computational procedure is as follows. Fig. 17 indicates a computational procedure outline.

1. Define T_0 by the evaluation area to the extent of the whole specimen is included enough
2. Calculate Kt i from defined T_0
3. Calculate assumed effective area S_i from Eq. (1) and FEM
4. Calculate T_i, S_i as evaluation area
5. Calculate $K_{t\,i+1}$ from T_i
6. Compare $K_{t\,i}$ and $K_{t\,i+1}$. If $K_{t\,i}/(K_{t\,i+1}) > 0.99$, then define S_i as effective area
7. If not $K_{t\,i}/(K_{t\,i+1}) > 0.99$, repeat the process after 3).

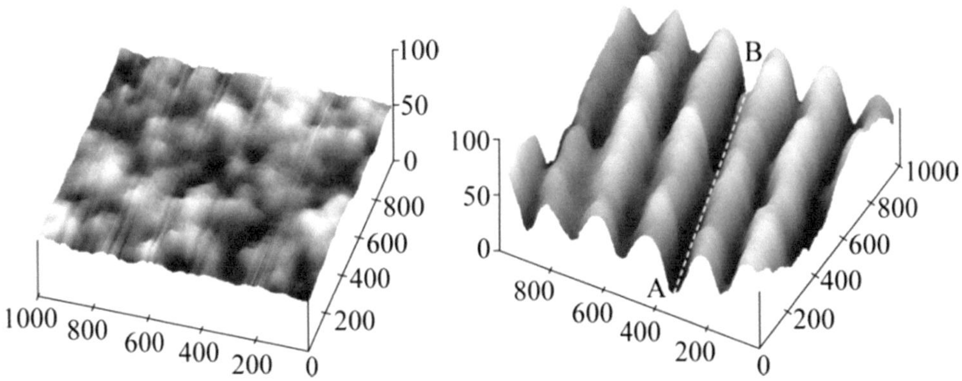

(a) Top surface (b) Sidewall surface

Fig. 14. Surface morphology of top and sidewall (unit: nm) (h = 6.4 [μm]).

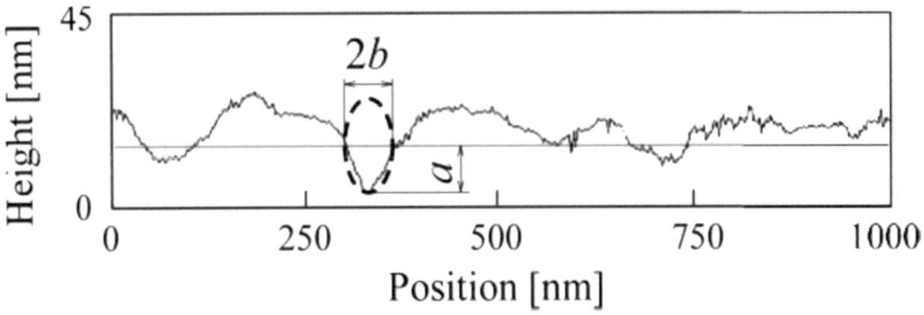

Fig. 15. Surface roughness example of sidewall (Fig. 14 A-B).

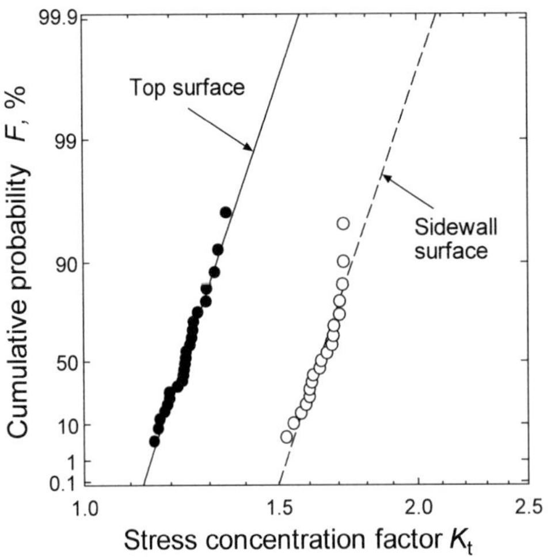

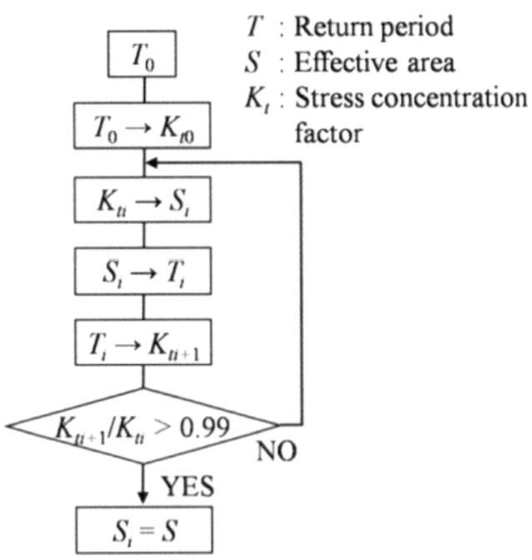

Fig. 16. Variation of stress concentratio factor K_t.

Fig. 17. Schematic diagram of deciding S from T and K_t (h = 6.4 [μm]).

Specimen type	Maximum stress concentration factor $K_{t\,max}$ Top surface	Maximum stress concentration factor $K_{t\,max}$ Sidewall surface	Effective area S [μm²]
L15R1	1.22	1.79	4.02
L15R2	1.28	1.82	8.01
L15R3	1.34	1.84	13.7
L15R4	1.39	1.86	22.7
L15R5	1.39	1.85	22.8
L10	1.20	1.78	3.57
L15	1.21	1.78	3.47

Table 2. Result of calculations, $K_{t\,max}$ and S (h = 6.4 [μm]).

In this study, it calculated as initial return period value $T_0 = 10000$. Table 2 shows the obtained $K_{t\,max}$ and S.

5.2 Effective area and fracture origin

Figure 18 indicates the example in the structure of the effective area. Figure 18 shows that the region of effective area where fracture origin may exist has extended to the specimen sidewall. Fracture surface observation of the specimen was carried out, and the example to which fracture origin exists in the sidewall was observed. Figure 19 shows an example. The scattering in fracture origin is shown in Fig. 20. It turns out that fracture origin varies within an effective area.

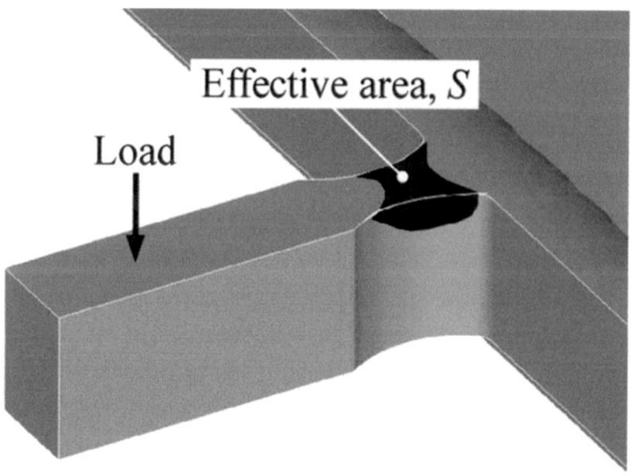

Fig. 18. Calculated effective area S (Specimen type: L15R5, h = 6.4 [μm]).

5.3 Application to design

Bending strength (maximum stress σ_{max} at the time of fracture) σ_B and the maximum stress concentration factor $K_{t\,max}$ were fitted to the Eq. (1) and the effective area was determined. Figure 21 shows the relationship between bending strength and the effective area. Average values of the test data (N = 8) were used for σ_B . The tendency bending strength becomes small as the effective area increased can be seen.

The equation of Weibull distribution which generally took the effective volume V into consideration same as Section 3 is shown as follows.

$$F = 1 - \exp\left\{-V\left(\frac{\sigma_B}{\sigma_0}\right)^m\right\} \tag{9}$$

Eq. (9) can be expressed as follows.

$$\ln \ln \frac{1}{1-F} = m(\ln \sigma_B - \ln \sigma_0) + \ln V \tag{10}$$

It turns out in the Eq. (10) that the effective volume V acts as a value which does not involve at the shape parameter m which shows the level of scattering. Since an effective volume did not participate in scattering, it extrapolated to the reliability needed for a design using the average of the shape parameter determined from the experimental result σ_B. In Fig. 21, an extrapolation example in the case of $F = 0.001$, the relationship between σ_B and S are shown.

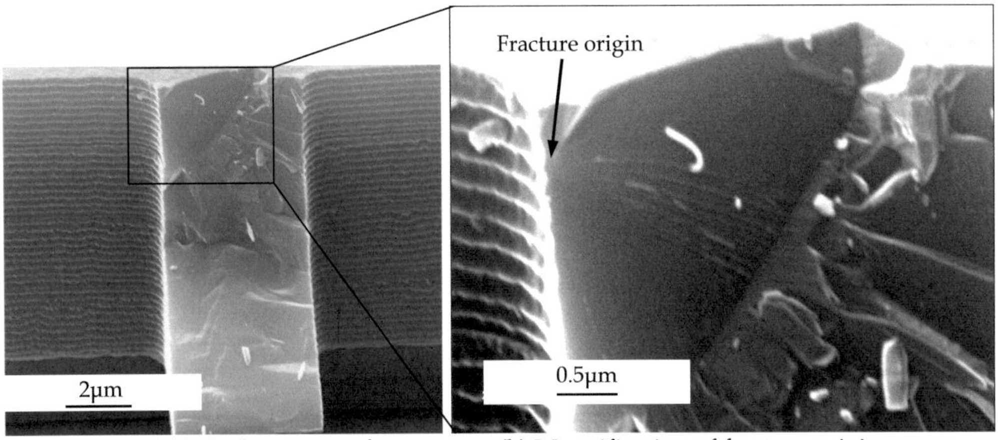

(a) Whole fracture surface (b) Magnification of fracture origin

Fig. 19. Fracture origin on the sidewall surface (h = 6.4 [μm]).

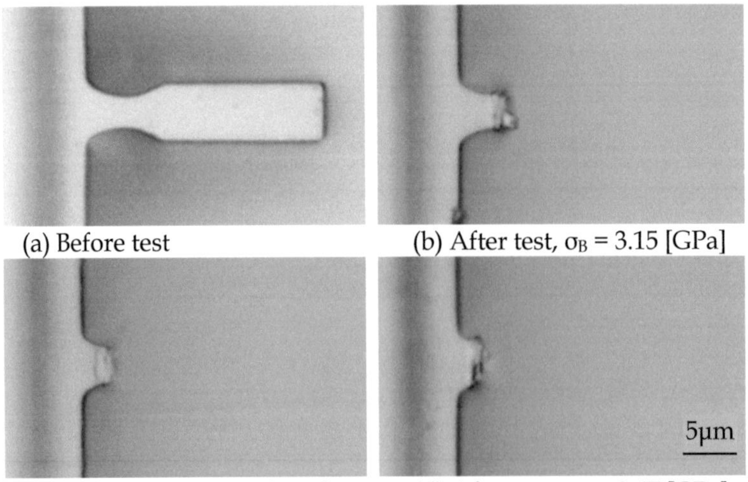

(a) Before test (b) After test, σ_B = 3.15 [GPa]

(c) After test, σ_B = 3.41 [GPa] (d) After test, σ_B = 3.67 [GPa]

Fig. 20. Variation of fracture points, Specimen type: L15R5, h = 6.4 [μm].

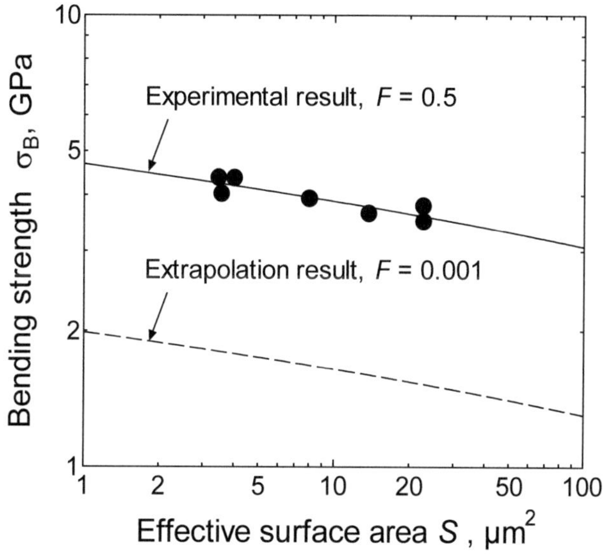

Fig. 21. Relationship between the bending strength and effective area

6. Conclusion

In order to propose the static strength design criteria of the poly-Si structure which has a microscopic dimension, the bending test, surface roughness measurement, FEM analysis, the Weibull statistical analysis, statistics of extreme analysis, and fracture analysis of a cantilever beam were conducted.

The obtained results are as follows.

1. By Weibull analysis, we found that the scatter in poly-Si bending strength made by RIE process is smaller than that of DRIE process.
2. Poly-Si strength is scattered. It depends on surface condition, crystal or grain boundary direction and some other.
3. The definition method of the quantitative effective area in bending cantilever beam was shown to the poly-Si with which the surface roughness on the upper surface and the surface of the sidewall differs.
4. Bending strength depends on the effective area definition are shown.
5. The static strength design criteria in consideration of the scattering in the strength using two parameters, the bending strength (maximum stress at the time of fracture) and the effective area, was proposed.

7. References

Chen, K. S.; Ayón, A. A.; Zhang X. & Spearing, S. M. (2002). Effect of Process Parameters on the Surface Morphology and Mechanical Performance of Silicon Structures After Deep Reactive Ion Etching (DRIE), *IEEE Journal of Microelectromechanical Systems*, Vol. 11, No. 3, pp. 264-275, ISSN 1057-7157

Greek, S.; Ericson, F.; Johansson, S. & Schweitz, J.-Å. (1997). In situ tensile strength measurement and Weibull analysis of thick film and thin film micromachined polysilicon structures, *Thin Solid Films*, Vol. 292, pp. 247-254, ISSN 0040-6090

Gumbel, E. J. (1962). *Statistic of Extremes*, Columbia Univ. Press, New York

Hull, D. (1999). *Fractography*, Cambridge University Press, ISBN 0521640822, Cambridge, pp. 121-129.

Johnson, L. G. (1964). *The Statistical Treatment of Fatigue Experiments*, Elsevier, New York

Kahn, H.; Tayebi, N.; Ballarini, R.; Mullen R. L. & Heuer, A. H. (2000). Fracture toughness of polysilicon MEMS devices, *Sensors and Actuators A*, Vol. 82, pp. 274-280, ISSN 0924-4247

Kapels, H.; Aigner, R. & Binder, J. (2000). Fracture strength and fatigue of polysilicon determined by a novel thermal actuator, *IEEE Transactions on Electron Devices*, Vol. 47, pp. 1522-1528, ISSN 0018-9383

Muhlstein, C. L.; Howe R. T. & Ritchie, R. O. (2004). Fatigue of Polycrystalline Silicon for Microelectromechanical Systems: Crack Growth and Stability under Resonant Loading Conditions, *Mechanics of Materials*, Vol. 36, pp. 13-33, ISSN 0167-6636

Murakami ,Y. Eds. (1992). *Stress Intensity Factors Handbook Vol.3*, Soc. Materials Sci., Japan & Pergamon Press, pp. 591-597

Najafi, K. (2000). Micromachined Micro Systems: Miniaturization Beyond Micro-electronics, *Proc. 2000 Symposium on VLSI Circuits Digest of Technical Papers*, pp. 6-13

Namazu, T.; Isono Y. & Tanaka, T. (2000). Evaluation of Size Effect on Mechanical Properties of Single Crystal Silicon by Nano-Scale Bending Test using AFM, *IEEE Journal of Microelectromechanical Systems*, Vol. 9, pp. 450-459, ISSN 1057-7157

Senturia, S. D. (2000). *Microsystem Design*, Kluwer Academic Publishers, ISBN 0792372468, Dordrecht

Sharpe Jr., W. N.; Jackson, K. M.; Hemker K. J. & Xie, Z. (2001). Effect of Specimen Size on Young's Modulus and Fracture Strength of Polysilicon, *IEEE Journal of Microelectromechanical Systems*, Vol. 10, pp. 317-326, ISSN 1057-7157

Son, D.; Kim, J.; Lim T. W. & Kwon, D. (2004). Evaluation of fracture properties of silicon by combining resonance frequency and microtensile methods, *Thin Solid Films*, Vol. 468, pp. 167-173, ISSN 0040-6090

Tsuchiya, T.; Tabata, O.; Sakata J & Taga, Y. (1998). Specimen size effect on tensile strength of surfacemicromachined polycrystalline silicon thin films, *IEEE Journal of Microelectromechanical Systems*, Vol. 7, pp. 106-113, ISSN 1057-7157

Weibull, W. (1951). A statistical distribution function of wide applicability, *Transactions ASME Journal of Applied Mechanics*, Vol. 18, pp. 293-297

12

MEMS Gyroscopes for Consumer and Industrial Applications

Riccardo Antonello and Roberto Oboe
Dept. of Management and Engineering,
University of Padova - Vicenza branch
Italy

1. Introduction

The advent of MEMS (Micro-Electro-Mechanical-System) technology has enabled the development of miniaturized, low-cost, low-power sensors that are currently replacing their macroscopic scale equivalents in many traditional applications, covering industrial, automotive, biomedical, consumer applications, etc.

Competitiveness of MEMS sensors largely resides in the miniaturization and batch fabrication processes involved in their manufacture, which allow to cut down costs, size and power requirements of the final device. Moreover, miniaturization opens new perspectives and possibilities for the development of completely new class of sensors, where micro-scale phenomena are effectively pursued to achieve results that would be unfeasible at a macro-scale.

Several MEMS sensor typologies are either commercially available or have been presented in technical literature since the beginning of the microsystem technology more than 30 years ago. Pressure and acceleration sensors for the automotive industry have been among the first MEMS devices to be produced in large scale, and they have contributed to foster the further development of MEMS technology. Despite their maturity, these sensors are still dominating, with their sales volume, the market of silicon-based sensors. Recently, another micro-sensor that is becoming relevant in terms of sales volume, especially in the automotive and consumer electronics markets, is the angular rate sensor, or *gyroscope*.

Alongside with accelerometers, micromachined gyroscopes can be used in several applications that require an integrated solution for inertial sensing and motion processing problems (Söderkvist, 1994; Yazdi et al., 1998). In the automotive industry, they can be integrated in advanced safety systems for skid (e.g. ESC - *electronic stability control*) and roll-over (e.g. RSC - *roll-over stability control*) detection and prevention (tilt and yaw control) (Neul et al., 2007; Nonomura et al., 2006; Sparks et al., 1997; Voss et al., 1997), or they can be used in vehicle navigation systems, either as stand-alone or GPS (*Global Positioning System*)-assisted solutions (Brown, 2005; Noureldin et al., 2009). Other automotive applications include *anti-lock braking systems* (ABS), active suspension control for comfort riding and improved vehicle handling, and smart cruise control systems (Fleming, 2001; Seidel et al., 2002).

In consumer electronics, micromachined inertial sensors are used in *Optical Image Stabilization* (OIS) systems for cameras and camcorders (Sachs et al.), in dead reckoning personal navigation systems (either stand-alone or integrated in cell-phones and PDAs, such as the Apple's *iPhone*, *iPad* and *iPod touch* (Apple, 2011)), in 3D-pointing devices such as the Nintendo Wii's game controller (*Wiimote*) with MotionPlus expansion (Nintendo, 2011), the Gyration air mice (Gyration, 2009) or the SCURRY glove-like input device (Kim et al., 2005), and virtual reality headsets. Industrial applications of MEMS inertial sensors may include integrated IMUs (*Inertial Measurement Units*) for AGVs (Automated Guided vehicles) (Microstrain, 2011), distribute sensing and monitoring of industrial plants, motion control of manipulators and machine tools, platform stabilization of heavy machineries, stabilization systems for mobile antennas, integrated inertial platforms for personal transportation systems (Segway, 2011), etc. Some biomedical applications have also been reported, including system for the gait-phase detection (Pappas et al., 2004), the ambulatory analysis of the human posture (Roetenberg et al., 2007), (Kawano et al., 2007) and the estimation of the stride length and walking velocity (Miyazaki, 1997).

To pioneer the commercialization of MEMS gyroscopes has been *Robert Bosch GmbH.*, which introduced the first MEMS gyroscope-based vehicle stability control system in 1998. Since then, many other manufacturers have embarked in the effort of producing and commercializing micromachined angular sensors, attracted by the market opportunities and the predicted sales volumes for MEMS gyros. A list of current manufacturers of MEMS gyros includes the followings (here reported in any specific order): *Analog Devices* (ADI), *Robert Bosch GmbH, Honeywell, Draper Laboratories, Silicon Sensing Systems* - SSS (BAE/*Sumitomo Precision Products*), *Invensense, MEMSense, BAE/AMI Semiconductors, ST Microelectronics, NEC-Tokin, Epson Toyocom, muRata, Systron Donner, Northrop Grumman, Goodrich Corp., Melexis, SensoNor, IMEGO, Gyrooptics, Gladiator Technologies*. Even if most of them are direct competitors, their production is diversified in terms of addressed market segments (i.e. cost and performance of the sensor), technical specifications, fabrication technology, etc.

2. Operating principle of conventional and miniaturized gyroscopes

2.1 Conventional gyroscopes

Gyroscopes are sensors designed for measuring angular rotations about some specific axes with respect to inertial space. *Rate gyroscopes* measure a rotation rate (angular velocity), while *displacement gyroscopes* (also known as *whole-angle gyroscopes*) measure a rotation angle. They have been traditionally designed as either mechanical devices exploiting the conservation of the angular momentum stored in a spinning wheel, or as optical systems exploiting the Sagnac effect experienced by counter-propagating laser beams in a ring cavity or a fiber optic coil. Typical designs include:

- *Momentum wheel gyroscopes*: they consist of a spinning wheel mounted on a gimbal support, whose function is to isolate the wheel spin axis from rotations of the body on which the gyro is attached to. Because of the conservation of the angular momentum, the spin axis remains fixed during the body motion, and therefore the gimbal angles provide a measurement of the body angular displacement with respect to an inertial frame. A key design issue in momentum wheel gyroscope is the reduction of bearing friction, since it strongly affects the measurement drift. Friction reduction can be achieved with innovative suspension designs (e.g. gas bearings, electrostatic and magnetic suspensions, floated gyros, etc.) or by employing a slightly different mechanical design, such as in the *Dynamically Tuned Gyroscope* (DTG). In DTG, a rotor (momentum wheel) is attached to a motor shaft by means of a gimbal mechanism with flexible suspensions. As the rotor spins, driven by the motor, the gimbal rotates without twisting the flexures only if the motor shaft is aligned with the rotor axis, otherwise the rotor starts fluttering and the flexures are twisted by gyroscopic precession torques. As a net effect, such torques produce a reduction of the equivalent stiffness of the flexures (softening effect); therefore, depending on the misalignment of the motor shaft and the rotor axis (i.e. the body attitude with respect to an inertial frame), there is a suitable rotor spin rate at which the equivalent stiffness of the flexures is nulled and the rotor is unrestrained. This properly tuned rotor spin rate provides an indirect measurement of the angular displacement.

- *Ring Laser Gyroscopes* (RLG): they consist of a polygonal closed gas cavity inside which two laser beams are propagated in opposite directions, namely clockwise (CW) and counterclockwise (CCW). After traveling along the whole cavity, the two beams meet at a return point and an interference pattern is obtained. The displacement between adjacent interference fringes is affected by the sensor rotation rate: this is a consequence of the Sagnac effect, namely the difference in the travel path length experienced by the two beams whenever the RLG undergoes a rotation whose axis is normal to the waveguide plane. Therefore, a measurement of the sensor angular velocity can be provided by measuring either the displacement shift of the interference fringes or, more commonly, the phase shift of the two beams (using a photodetector at the return point).
- *Fiber Optic Gyroscopes* (FOG): they are similar to the ROGs, except that a fiber optic is used in place of the gas cavity. Thanks to such replacement, the sensor structure is simplified (no gas or mirrors are necessary) and the manufacturing costs are greatly reduced (no precise mirror alignment is required). The scale factor of the sensor is usually enhanced by arranging the fiber optic as a coil with many turns: in practice, the longer the fiber optic is for a given enclosed area, the larger the sensor scale factor is. A weak point of FOGs is their sensitivity to the strain distribution in the optical fiber, induced by temperature changes and accelerations; minimizing such effects is a major concern in FOGs design.

A comprehensive review of many conventional designs and technologies for gyroscopes can be found in (Lawrence, A., 1993; R.R.Ragan (ed), 1984).

2.2 Micromachined gyroscopes

Traditional designs conceived for macro-scale sensors are usually difficult to reproduce in miniaturized devices because of the limitations imposed by the conventional micro-fabrication techniques. For examples, bearings required to mitigate friction and wear of rotating parts, such as spinning wheels or disks, are difficult to manufacture at the micro-scale; alternative designs for frictionless supports based on levitation of any sort (e.g. mechanical, electrostatic/magnetostatic or based on fluid flotation) are failure-prone and not robust enough for commercial applications. Furthermore, because precision in conventional mechanical gyroscopes is related to the angular momentum impressed to the spinning wheel, which in turns is proportional to the mass and the spin rate, it follows that conventional designs become ineffective when dimensions scale down.

Instead of attempting to replicate existing macro-scale designs at the micro-scale level, miniaturized gyroscopes are more conveniently designed by pursuing a completely different working principle, which is better suited for taking advantage of the miniaturization process. In almost all MEMS gyroscopes, the sensor angular velocity is inferred by detecting, on the sensor frame, the motion deflections experienced by a vibrating member. The deflections can be explained by the onset of an apparent force, called *Coriolis force*, which is observed whenever the observer reference frame rotates with respect to a moving body (vibrating

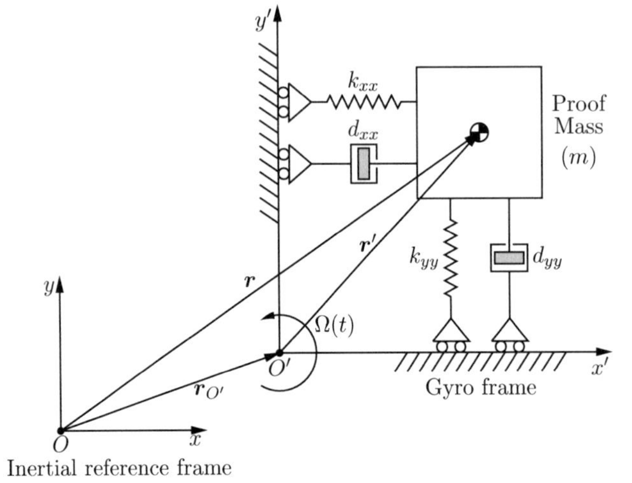

Fig. 1. Simplified lumped-element model of a single axis gyroscope. The gyro frame rotates at an angular rate $\Omega(t)$ about an axis (z'-axis) orthogonal to the $x'O'y'$ plane.

member). A gyroscope adopting this working principle is called *Coriolis Vibrating Gyroscope* (CVG).
The working principle of a CVG can be better understood with the aid of the simplified lumped-element model of a single-axis gyroscope (z-axis gyro) depicted in Fig. 1.
It consists of a proof mass attached to the sensor frame by means of elastic suspensions, in such a way that it can vibrates on the $x'O'y'$ plane with no constraints. The structure has two (ideally orthogonal) modes of vibrations, that are dynamically coupled through the Coriolis force when observed on a sensor-fixed reference frame. In fact, by denoting with $\boldsymbol{r}$ and $\boldsymbol{r}'$ the positions of the proof mass in, respectively, an inertial reference frame and the sensor reference frame, and with $\boldsymbol{r}_{O'}$ and $\boldsymbol{\Omega}$ the position and angular velocity vectors of the sensor frame with respect to the inertial frame, it follows that (Meirovitch, 1970):

$$\dot{\boldsymbol{r}} = \dot{\boldsymbol{r}}_{O'} + \dot{\boldsymbol{r}}' + \boldsymbol{\Omega} \times \boldsymbol{r}' \tag{1}$$

$$\ddot{\boldsymbol{r}} = \ddot{\boldsymbol{r}}_{O'} + \ddot{\boldsymbol{r}}' + 2\boldsymbol{\Omega} \times \dot{\boldsymbol{r}}' + \dot{\boldsymbol{\Omega}} \times \boldsymbol{r}' + \boldsymbol{\Omega} \times (\boldsymbol{\Omega} \times \boldsymbol{r}') \tag{2}$$

Regarding accelerations, $\ddot{\boldsymbol{r}}$ denotes the acceleration of the proof mass relative to the inertial frame, $\ddot{\boldsymbol{r}}_{O'}$ is the acceleration of the origin of the sensor frame and $\ddot{\boldsymbol{r}}'$ is the acceleration of the proof mass as measured in the sensor frame. Rewriting the last equation in terms of $\ddot{\boldsymbol{r}}'$ yields:

$$\ddot{\boldsymbol{r}}' = \ddot{\boldsymbol{r}} - \ddot{\boldsymbol{r}}_{O'} - 2\boldsymbol{\Omega} \times \dot{\boldsymbol{r}}' - \dot{\boldsymbol{\Omega}} \times \boldsymbol{r}' - \boldsymbol{\Omega} \times (\boldsymbol{\Omega} \times \boldsymbol{r}') \tag{3}$$

The contribution $-\dot{\boldsymbol{\Omega}} \times \boldsymbol{r}'$ to the total acceleration in the sensor frame is known as the *Coriolis acceleration*, while $-\boldsymbol{\Omega} \times (\boldsymbol{\Omega} \times \boldsymbol{r}')$ is the centrifugal acceleration; the remaining term $-\dot{\boldsymbol{\Omega}} \times \boldsymbol{r}'$ is due to a non-constant rotation rate of the sensor frame. With the knowledge of the acceleration $\ddot{\boldsymbol{r}}'$, it is possible to derive the following equations of motion for the proof mass in the sensor reference frame:

$$\boldsymbol{M}\,\ddot{\boldsymbol{r}}' + \boldsymbol{D}\,\dot{\boldsymbol{r}}' + \boldsymbol{K}\,\boldsymbol{r}' = \boldsymbol{F} - \boldsymbol{M}\left[\ddot{\boldsymbol{r}}_{O'} + 2\boldsymbol{\Omega} \times \dot{\boldsymbol{r}}' + \dot{\boldsymbol{\Omega}} \times \boldsymbol{r}' + \boldsymbol{\Omega} \times (\boldsymbol{\Omega} \times \boldsymbol{r}')\right] \tag{4}$$

The 3×3 positive definite matrices $\boldsymbol{M}$, $\boldsymbol{D}$ and $\boldsymbol{K}$ are the mass, damping and stiffness matrices; $\boldsymbol{F} = \boldsymbol{M}\ddot{\boldsymbol{r}}$ is the total external force (measured in the inertial frame) applied to the proof mass.
In order to measure the component of the angular velocity around the z'-axis of the sensor reference frame, the proof mass motion can be constrained to lie on the x'-y' plane, by making the spring suspensions along the z'-axis much stiffer than those along the x' and y' directions. Under this assumption, the decomposition of the equations of motion along the two principal directions of vibration yields:

$$m_x\ddot{x}' + d_{xx}\dot{x}' + \left[k_{xx} - m_x(\Omega_y^2 + \Omega_z^2)\right] x' + m_x\left(\Omega_x\Omega_y - \dot{\Omega}_z\right) y' = F_x - m_x\ddot{x}_{O'} + 2m_x\Omega_z\dot{y}' \quad (5)$$

$$m_y\ddot{y}' + d_{yy}\dot{y}' + m_y\left(\Omega_x\Omega_y + \dot{\Omega}_z\right) x' + \left[k_{yy} - m_x(\Omega_x^2 + \Omega_z^2)\right] y' = F_y - m_y\ddot{y}_{O'} - 2m_y\Omega_z\dot{x}' \quad (6)$$

where it has been assumed that the two principal directions of vibrations are aligned with the sensor frame axes, i.e. the $\boldsymbol{D}$ and $\boldsymbol{K}$ matrices are diagonal. With a relatively small sensor bandwidth, i.e. for constant or small varying angular rate inputs, the angular acceleration $\dot{\Omega}_z$ is negligible; moreover, if the angular rate is much smaller than the frequency of the proof mass vibrating motion, then Ω_x^2, Ω_y^2 and $\Omega_x\Omega_y$ can also be neglected. The effect of the linear accelerations $\ddot{x}_{O'}$ and $\ddot{y}_{O'}$ can be compensated either by applying some counterbalancing actuation forces or by adjusting the induced slow trends in the sensor output response, and therefore they can be dropped in the subsequent analysis. Under these circumstances, the equations of motion become:

$$m_x\,\ddot{x}' + d_{xx}\,\dot{x}' + k_{xx}\,x' = F_x + 2m_x\,\Omega_z\,\dot{y}' \quad (7)$$

$$m_y\,\ddot{y}' + d_{yy}\,\dot{y}' + k_{yy}\,y' = F_y - 2m_y\,\Omega_z\,\dot{x}' \quad (8)$$

The conventional mode of operation of a vibration gyroscope consists of driving one mode of vibration (*drive* or *primary mode*) into a controlled-amplitude harmonic oscillation; for an underdamped mode (typical situation in micro-gyroscopes), the oscillation is driven at the vibration mode resonance, in order to take advantage of the mechanical amplification. The remaining modes of vibration (*sense*, *pickoff* or *secondary modes*) are used to sense the angular rate, by measuring the proof mass displacement induced by the Coriolis force. Assuming that the drive and sense modes are directed, respectively, along the x and y axes, the equations of motion for the conventional mode of operation can be written as follows:

$$x' = -X_0 \sin\omega_x t \quad (9)$$

$$\ddot{y}' + \frac{\omega_y}{Q_y}\,\dot{y}' + \omega_{yy}^2\,y' = u_y + 2\,\Omega_z\,\omega_x X_0 \cos\omega_x t \quad (10)$$

where $\omega_x = \sqrt{k_{xx}/m_x}$ and $\omega_y = \sqrt{k_{yy}/m_y}$a are the natural frequencies of, respectively, the drive and sense modes, $Q_y = m_y\omega_y/d_{yy}$ is the quality factor of the second order (under)damped oscillator representing the sense mode dynamics and $u_y = F_y/m_y$ is an external acceleration.
Assuming a cosinusoidal angular rate input $\Omega_z(t) = \Omega\cos\omega_\Omega t$ and no external excitation, i.e. $u_y = 0$, the proof mass displacement produced by the Coriolis acceleration along the sense mode direction is equal to:

$$y(t) := \omega_x X_0\,H_{\Omega+}\,\Omega\cos((\omega_x + \omega_\Omega)t + \phi_{\Omega+}) + \omega_x X_0\,H_{\Omega-}\,\Omega\cos((\omega_x - \omega_\Omega)t + \phi_{\Omega-}) \quad (11)$$

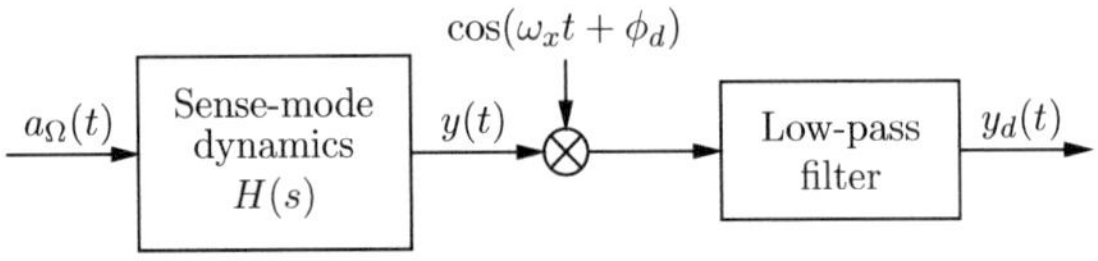

Fig. 2. Synchronous amplitude demodulation scheme used to retrieve the angular rate measurement $\Omega(t)$ from the proof mass displacement $y(t)$. In the block diagram, a_Ω denotes the Coriolis acceleration $a_\Omega = 2\Omega_z \omega_x X_0 \cos \omega_x t$.

where:

$$H_{\Omega\pm} := |H(s)|_{s=j(\omega_x \pm \omega_\Omega)} \qquad \phi_{\Omega\pm} := \angle H(s)|_{s=j(\omega_x \pm \omega_\Omega)} \tag{12}$$

and $H(s)$ is the transfer function of the sense-mode dynamics:

$$H(s) = \frac{1}{s^2 + \dfrac{\omega_y}{Q_y}s + \omega_y^2} \tag{13}$$

The displacement signal is dual-side-band (DSB) modulated signal, with carrier frequency ω_x: hence, an amplitude demodulation scheme (Fig. 2) can be used to retrieve the information about the modulating signal $\Omega_z(t)$. In fact, by using a demodulating carrier equal to $\cos(\omega_x t + \phi_d)$, the output y_d of the demodulator is equal to:

$$\begin{aligned} y_d(t) &= \mathrm{LPF}[y(t)\cos(\omega_x t + \phi_d)] \\ &= \frac{1}{2}\omega_x X_0 H_{\Omega+} \Omega \cos(\omega_\Omega t + \phi_{\Omega+} - \phi_d) + \frac{1}{2}\omega_x X_0 H_{\Omega-} \Omega \cos(-\omega_\Omega t + \phi_{\Omega-} - \phi_d) \end{aligned} \tag{14}$$

where LPF denotes an ideal (brick-wall) low-pass filtering with corner frequency $\omega_{LPF} \ll \omega_x$ (it is understood that the drive-mode vibration frequency ω_x is designed to be much larger than the sensor bandwidth, i.e. $\omega_{\Omega,\mathrm{MAX}} \ll \omega_x$). For a small sensor bandwidth, it can be assumed that $\omega_x \pm \omega_\Omega \approx \omega_x$, so that:

$$H_{\Omega\pm} \approx H_0 := |H(j\omega_x)| = \frac{1}{\sqrt{(\omega_y^2 - \omega_x^2)^2 + \left(\dfrac{\omega_x \omega_y}{Q_y}\right)^2}} \qquad \phi_{\Omega\pm} \approx \phi_0 := \angle H(j\omega_x)$$

and:

$$y_d(t) = \omega_x X_0 H_0 \, \Omega \, \cos(\Delta\phi) \cos(\omega_\Omega t) \tag{15}$$

with $\Delta\phi = \phi_{\Omega\pm} - \phi_d = \phi_0 - \phi_d$. Therefore, the demodulator output y_d is proportional to the angular rate input $\Omega_z(t)$. From the last equation it can be noted that the sensor scale factor (see also Sec. 4) depends on the amplitude X_0 of the drive motion, the drive and sense modes natural frequencies ω_x and ω_y, and the sense-mode quality factor (through H_0). In general, sensitivity can be enhanced by increasing the drive mode amplitude X_0 and by lowering the natural frequencies ω_x and ω_y, namely by employing a bulkier proof mass or by softening the elastic suspensions. For a given drive-mode natural frequency ω_x, the sensitivity is maximized when $\omega_y = \omega_x$, i.e. when the two modes are exactly matched (*mode-matching condition*).

3. MEMS gyroscopes technology

3.1 Mechanical structures

As pointed out in Sec. 2, almost all MEMS gyroscopes operate as Coriolis Vibrating Gyroscopes (CVGs). A CVG comprises a mechanical structure with at least two modes of vibration that are dynamically coupled via the Coriolis force, i.e. an apparent force that arises from the relative motion of the vibrating structure and the sensor frame. The normal operation of a CVG consists of exciting one mode of vibration (*driving mode*) at a prescribed amplitude, and detecting the vibrations induced by the Coriolis force on the remaining modes (*sense* of *pickoff* modes).
Several mechanical designs for micromachined CVGs have been either proposed in technical literature or exploited in commercial products. A classification of the main designs is reported below:

- *vibrating beams*: they consists of tiny clamped-free beams (*cantilevers*) or clamped-clamped beams (*bridges*) that are driven into flexural vibration on a plane. Then, in response to a rotation, the beam starts vibrating along an orthogonal direction, and this motion can be used to infer about the angular rate input - see Fig. 3.

  Several designs involving beam structures have been proposed in literature, especially in connection with the usage of piezoelectric materials (Soderkvist, 1991). A vibrating beam structure has also been chosen for the development of the first silicon integrated micromachined CVG back in 1981 (O'Connor & Shupe, 1983).

  An example of a vibrating beam gyroscope consisting of a silicon-based microcantilever fabricated with bulk-micromachining techniques is reported in (Maenaka et al., 1996). Beam actuation is provided by a piezoelectric element; the Coriolis-induced vibration is electrostatically detected by measuring the capacitance changes between the beam and dedicated sensing electrodes. A combined beam-mass structure with a composite beam is proposed in (Li et al., 1999). The two-section composite beam structure is designed to have a vertical and a lateral highly compliant vibrating modes. The vertical mode is excited by

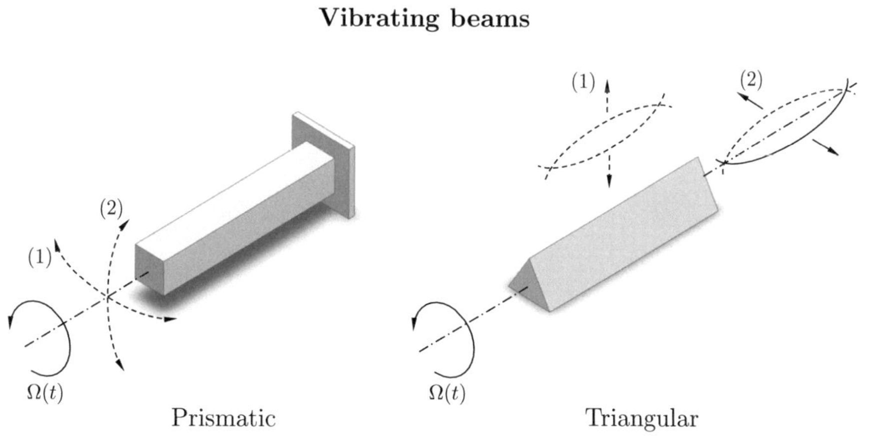

Fig. 3. Most conventional types of vibrating beam CVGs. In figure, $\Omega(t)$ is the input angular rate, (1) is the primary vibration mode, and (2) is the vibration response due to Coriolis forcing.

means of electrostatic actuation, while the Coriolis-induced vibration on the lateral mode is detected by means of embedded piezoresistors.

A commercial product featuring a vibrating beam design is the *Gyrostar*, a piezoelectric CVG produced by (muRata, 2003). It consists of an equilateral prism which is excited into flexural vibration by using piezoelectric elements applied to its sides. The beam is attached to the supporting frame at positions along the beam length that correspond to node points for the free-free flexural modes of vibration: this choice ideally decouples the beam from the supporting structure. The vibrations induced by the Coriolis force on the secondary mode are detected again by using piezoelectric transducers.

A structure resembling a vibrating beam gyroscope can also be found in nature: the *halteres*, a pair of vibrating knob sticks found in many two-winged insects, are indeed a pair of tiny vibrating beam CVGs that are used to stabilize and control the flight attitude (Nalbach, 1993; Nalbach & Hengstenberg, 1994).

- *vibrating forks*: they contain a pair of proof masses that are oscillated with the same amplitude, but in opposite directions. In a traditional fork structure, the tines are excited to resonate in anti-phase in the plane of the fork (drive mode); then, when the sensor rotates, the tines start oscillating along the perpendicular direction to the plane, thus generating a torque that excites the torsional mode around the stem. Forks can be of *single*, *dual* or *multi-tines* types; the latter type is used in order to increase sensitivity and reject common-mode errors (caused by geometrical asymmetries).

  Most of the Quartz Rate Sensors (QRS) that populated the market before the advent of silicon micromachined gyroscopes had a vibrating forks structure. For example, the first

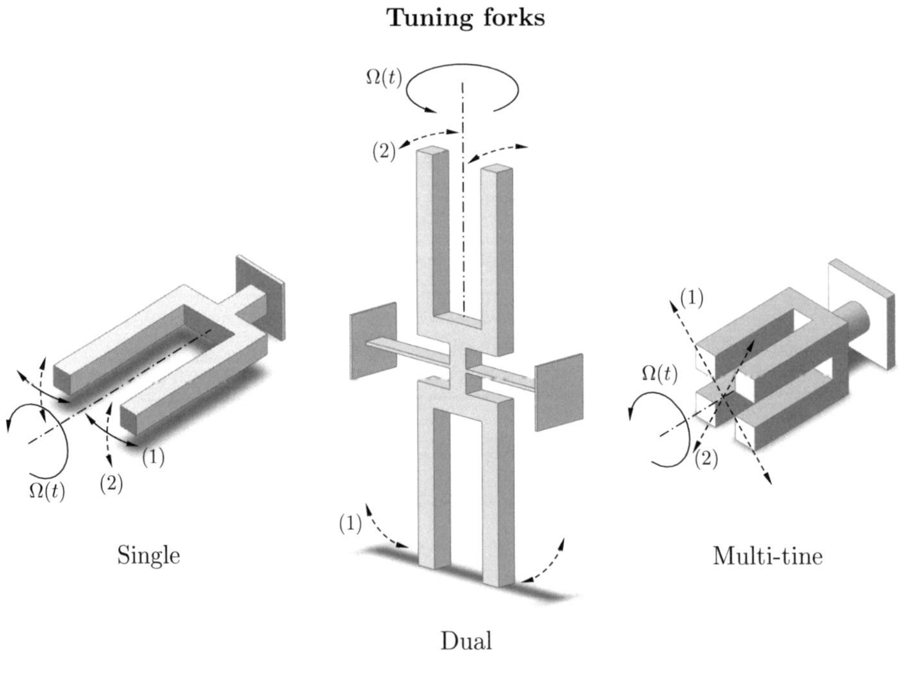

Fig. 4. Most conventional types of tuning fork CVGs. In figure, $\Omega(t)$ is the input angular rate, (1) is the primary vibration mode, and (2) is the vibration response due to Coriolis forcing.

miniaturized gyroscope to have been introduced in the market back in 1980, namely the Systron Donner/BEI QRS, was a H-shaped two-sided tuning fork gyroscope (Madni et al., 2003) (see Fig. 10-a). Epson-Toyocom is another company selling QRS whose structure is a double-T tuning fork with external vibrating tines and central stationary sensing arm.

In silicon micromachined vibrating forks gyroscopes, the fork tines are usually replaced by anti-phase resonating seismic masses vibrating on a common plane (Bernstein et al., 1993). The plane can be either parallel to the substrate, such as in Bosch SMG074 (Lutz et al., 1997), Analog Devices ADXRS150 (Geen et al., 2002) and STMicroelectronics LISY300AL (Oboe et al., 2005), or normal to the substrate, such as in the Invensense IDG family (Nasiri & Flannery Jr., 2007) (see Fig. 9).

A silicon micromachined gyroscope with a conventional vibrating fork structure (i.e. a structure comprising a fork with vibrating tines) is the angular-rate sensor produced by Daimler-Benz (Voss et al., 1997).

- *vibrating plates*: they have a resonant element consisting of a tiny plate, attached to the sensor outer frame by means of linear or torsional elastic suspensions (Tang et al., 1989). Forced vibrations can be induced either along a straight line (*linear plate* configuration (Clark et al., 1996; Tanaka et al., 1995)) or around an axis of rotation (*angular disk* configuration (Geiger et al., 1998; Juneau et al., 1997; Rajendran & Liew, 2004)). Melexis MLX90609-N2 is an example of a commercial MEMS gyroscope based on a vibrating plate structure (actually, a single gimbaled mass with translation drive).The vibrating angular disk structure is exploited in many commercial dual-axis pitch and roll MEMS gyroscopes: examples include the Bosch SMG060 and the STMicroelectronics LPR family.

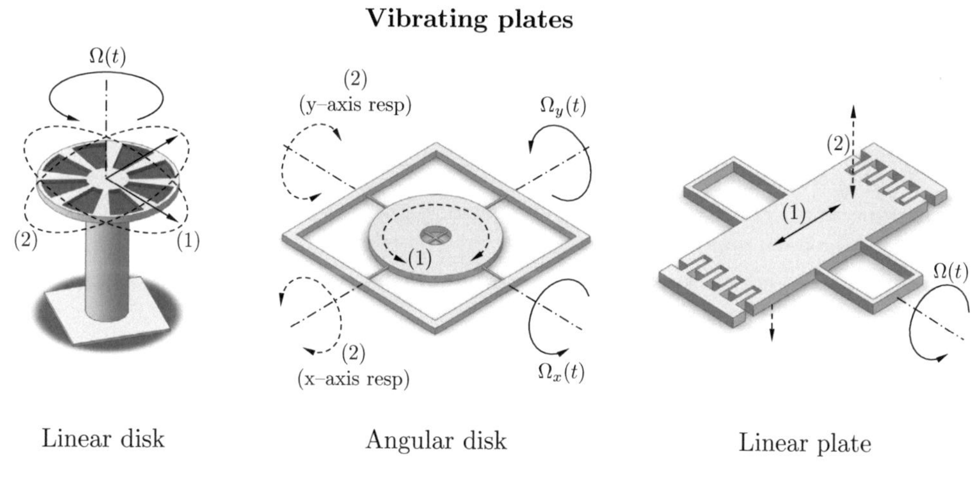

Fig. 5. Most conventional types of vibrating plate CVGs. In figure, $\Omega(t)$, $\Omega_x(t)$ and $\Omega_y(t)$ are the input angular rates, (1) is the primary vibration mode, and (2) is the vibration response due to Coriolis forcing.

- *vibrating shells*: they have circular shapes, such as *rings*, *cylinders* or *hemispheres*, which are set into a standing-wave vibration through external forcing. Whenever the sensor undergoes a rotation around its axis of symmetry, the vibration pattern, consisting of nodes and antinodes of the forced standing-wave, moves with respect to the external case; its

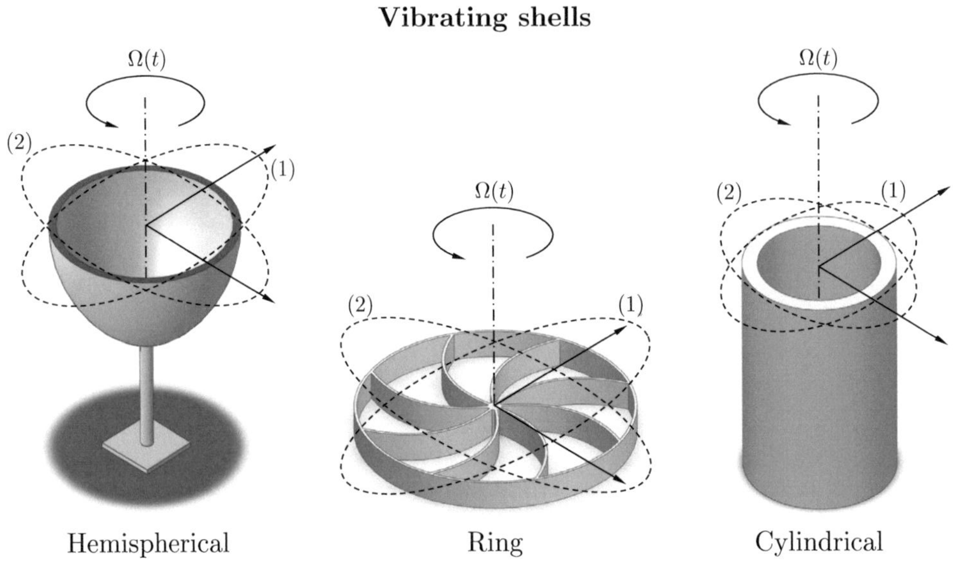

Fig. 6. Most conventional types of vibrating shell CVGs. In figure, $\Omega(t)$ is the input angular rate, (1) is the initial vibration pattern and (2) is the pattern after rotation.

motion can be detected by dedicated displacement sensors and used to infer about the angular rate input.

Most of the MEMS gyroscopes produced by Silicon Sensing Systems (SSS) are based on a vibrating ring structure (Fell, 2006; Hopkin et al., 1999). Delphi Delco Electronics has also reported in (Chang et al., 1998) the design of a vibrating ring gyroscope manufactured by electroplating on a fully processed CMOS wafer.

A vibrating cylinder structure can be found in NEC-Tokin ceramic gyroscopes; the structure actually consists of a cylindrical piezoelectric ceramic oscillator with embedded electrodes for electrostatic detection of Coriolis-induced vibrations (Abe et al., 1992).

Vibrating hemispherical shells have been traditionally used in macro-sized gyroscopes, such as the Delco Hemisperical Resonator Gyro (HRG) (Lawrence, A., 1993). The device consists of a hemispherical shell made of fused silica, which is encased within a sealed vacuum housing. A standing-wave vibration is electrostatically induced on the shell metal-coated rim; wave pattern shifts caused by sensor rotations are detected with capacitive pick-offs. Recently, a micromachined gyroscope with a similar structure has been patented (Stewart, 2009).

- gyroscopes based on the *surface acoustic wave (SAW)* technology.

  In a SAW gyroscope, a set of metallic electrodes (interdigital transducer - IDT) patterned on the surface of a piezoelectric substrate is used to generate a Rayleigh standing-wave. A Rayleigh wave is a mechanical transverse wave whose shear component is normal to the substrate surface, and whose energy is concentrated within one wavelength of the substrate surface (Drafts, 2001; Vellekoop, 1998). The out-of-plane vibration of the particles near the surface is perturbed by the Coriolis force whenever the piezoelectric substrate undergoes a rotation (about an axis vertical to its surface). Such perturbation produces a secondary standing wave polarized parallel to the substrate surface, whose amplitude is proportional to the sensor angular rate: hence, by sensing the amplitude of the secondary wave with an additional IDT, it is possible to retrieve a measurement of the input angular rate.

Some design examples of SAW MEMS gyroscopes are presented in (Jose et al., 2002; Kurosawa et al., 1998; Liu & Wu, 2007).

There have been very few attempts to depart from the conventional designs based on the CVG working principle; the most noticeable examples are:

- gyroscopes based on the conservation of the angular momentum in levitated spinning disks, similarly to conventional (macro-sized) mechanical flywheel gyroscopes. Both the electrostatic (Damrongsak & Kraft, 2005; Ellis & Wilamowski, 2008) and magnetic levitation principles (Dauwalter & Ha, 2005; Shearwood et al., 2000) have been exploited.
- thermal convective gyroscopes.Their working principle is based on the detection of convective heat flow deflections induced by the Coriolis acceleration. The sensor proposed in (Zhu et al., 2006) consists of a hermetically sealed gas chamber obtained by etching a small cavity on a silicon substrate. The cavity contains a suspended central heater that is used to induce a regular gas flow within the chamber, and four suspended thermistor wires placed symmetrically on both sides of the heater for measuring local changes in the gas flow. By measuring the voltage imbalances among the four thermistors readouts (using a Wheastone bridge circuit) it is possible to estimate both angular velocities and linear accelerations.
- gyroscopes using liquid or gas jet flows. In the prototype reported in (Yokota et al., 2008), a jet flow in an electro-conjugate fluid (ECF) is generated by imposing an electric field between two brass electrodes dipped in the liquid. When the sensor is rotated, the jet flow is deflected by the Coriolis acceleration. The deflection, which is an indirect measure of the input angular rate, is sensed as an unbalancing in the electrical resistance of two tungsten hot-wires placed on the sidewalls of the fluid channel.

  A similar design is proposed in (Zhou et al., 2005), except that a gas is used instead of an ECF.
- Micro-Opto-Electro-Mechanical (MOEMS) gyroscopes. This technology is still under development, and no accurate MOEMS gyroscopes exist yet. The goal is the development of a miniaturized optical device that, similarly to a standard interferometric optical gyroscope, relies on the Sagnac effect for measuring a rotation rate. The main design issue for micro-optical gyroscopes is how to create optical path lengths that are large enough to sense useful angular velocities (i.e. greater in strength than the noise inherent in the measurement). In the AFIT MiG prototype reported in (Stringer, 2000), the elongation of the optical path is achieved by creating a spiral path with a set of suitably arranged micro-mirrors placed above the silicon die.

3.2 Fabrication technologies

There are fundamentally two alternative technologies available for the fabrication of micromechanical devices: *bulk micro-machining* and *surface micro-machining* techniques.

- In *bulk micro-machining* (Kovacs et al., 1998) the microstructures are formed by selectively removing (*etching*) parts from a bulk material, which is typically a silicon crystal. The etching process can be performed by either dipping the silicon wafer into an etching solution (*wet etching*) or by exposing the material to vapors or glow-discharge plasmas of chemically reactive gases (*dry etching*). Protective masks are applied on the surface of the bulk material in order to avoid the exposure to etchants: thus, etching takes place only on those portions that are not covered by a layer of protective material. Most wet etching is *isotropic*, meaning that the etching rate does not depend on the orientation of the substrate; nevertheless, for particular etchants, *anisotropic* (i.e. orientation-dependent) wet etching can occur, so that the etching rate along the direction of a certain crystal axis can be hundreds of times greater than others. Larger levels of directionality can be achieved with anisotropic dry etching techniques, such as DRIE (*Deep Reactive Ion Etching*), in which the direction perpendicular to the exposed surface is etched much faster than the direction parallel to the surface. The depth of the etched features can be controlled by either controlling the exposure time to etchants (once the etching rate is known) or by using some kind of etch-stopping technique or material.

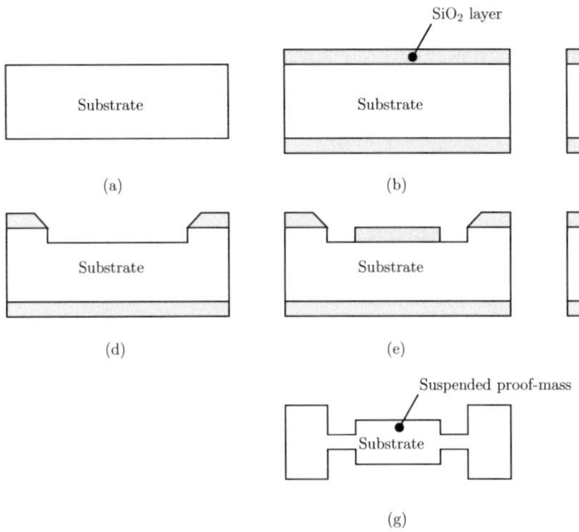

Fig. 7. Typical steps in a bulk micromachining process: (a) substrate preparation - typically, a $500 \div 700\,\mu m$ thick single silicon (Si) crystal; (b) deposition of a silicon dioxide (SiO_2) layer - typical thickness: $1 \div 2\,\mu m$; (c) patterning (photoresist deposition + optical lithography) and etching of the SiO_2 layer; (d) substrate etching; (e) deposition of SiO_2 layer for a selective area (repetition of step (b)); (f) substrate etching for creating deeper trenches (repetition of step (f)); (g) creation of a suspended structure (e.g. a proof mass) after repeating steps (a) $\div$ (f) on the bottom side of the substrate and removing the residual SiO_2 at both sides.

- In *surface micro-machining* (Bustillo et al., 1998; Howe et al., 1996), the microstructures are formed by by depositing, growing and etching different structural layers on top of a substrate. Since the substrate acts only as a supporting structure, it can be made of inexpensive materials such as plastic, glass, quartz, ceramic or other piezoelectric materials (Kotru et al., 2008), instead of the more expensive single-crystal silicon used for IC (integrated circuits) fabrication. On top of the substrate, several layers can be deposited, patterned and released; surface *planarization* is usually required before the deposition of every structural layer, in order to prevent critical issues during photolithography, such as the limited focus depth of high-resolution lithographic tools over non-planar surfaces, and etching - anisotropic etching of non-planar surfaces may leave behind several stringers of unetched material. Apart of structural layers, the fabrication of complex structure with

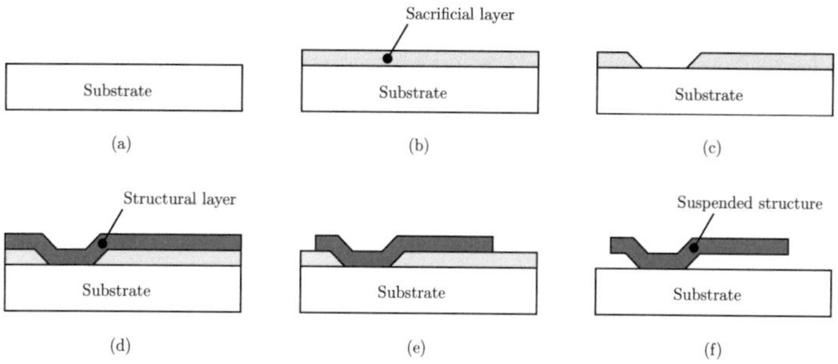

Fig. 8. Typical steps in a surface micromachining process: (a) substrate preparation - typically, a $500 \div 700\,\mu m$ thick single silicon (Si) crystal; (b) deposition of a sacrificial layer - typically, a $1 \div 2\,\mu m$ thick silicon dioxide (SiO_2) layer; (c) creation of a hole by patterning (photoresist deposition + optical lithography) and etching of the sacrificial layer; (d) deposition of a structural layer - typically, a $1 \div 5\,\mu m$ thick polysilicon layer; (e) shape definition by patterning and etching of the structural layer; (f) release of the suspended structure (e.g. a cantilever beam).

suspended or freely moving parts may require the deposition of so-called *sacrificial layers*, i.e. layers that are selectively removed (*release etch* step) after growing one or more thin-film structures above them. Thin-film deposition can be realized with several techniques, such as physical or chemical vapor deposition (PVD or CVD, respectively), electrodeposition, spin coating and Sol-Gel deposition. Thicker structures can be created by either using epi-poly as structural material, or by bonding together two or more silicon wafers, using wafer bonding techniques such as silicon-to-silicon bonding, silicon-on-insulator (SOI) bonding, anodic bonding, adhesive bonding, etc.

3.3 Actuation and sensing mechanisms

Several methods have been exploited so far for generating and detecting vibrating motions inside CVGs. Nevertheless, the foremost methods in industry practice are based on electrostatic and piezoelectric principles, mainly because of the easiness of fabrication, miniaturization and integration with standard manufacturing processes of the IC industry.

In electrostatic actuation, the attractive (repulsive) forces arising on oppositely (similarly) electrically charged objects are used to generate motion; conversely, the capacitance change experienced by electrically charged objects moving apart each other is exploited to detect motion. Typically, an electrostatic actuator or sensor resembles a capacitor with moving plates: indeed, this is the case for the *parallel-plate* and *comb-fingers* structures (Boser, 1997; Senturia, 2001). An example of a MEMS gyroscope exploiting electrostatic actuation and sensing is reported in Fig. 9.

In piezoelectric actuation, the property (*inverse piezoelectric effect*) of certain materials (e.g. quartz, ceramics, special alloys or piezoelectric polymers) to change their shape when subjected to an electric field is effectively exploited to generate a mechanical deformation or displacement. With regards to sensing, either the *direct piezoelectric effect* (Kotru et al., 2008; Soderkvist, 1991) (generation of an electric field in response to a mechanical strain) or the *piezoresistive effect* (Li et al., 1999; Voss et al., 1997) (change of electrical resistance in response to a mechanical stress) are effectively used to sense the Coriolis-induced motion.

Except for electrostatic and piezoelectric methods, rather few alternatives have been investigated and tested; examples of typical solutions presented in literature include designs based on thermal (Shakoor et al., 2009) and magnetic (Paoletti et al., 1996; Tsai et al., 2009) actuation, or optical sensing (Bochobza-Degani et al., 2000; Norgia & Donati, 2001).

3.4 Onboard electronics

The onboard electronics is necessary for:

- *driving and sustaining the oscillations of the vibrating member.* Two requirements must be fulfilled when generating the driving motion: first, the oscillation amplitude must be controlled with a high level of accuracy, since the stability of the sensor scale factor depends on it (see Eqn. 15); second, the oscillation frequency should be as close as possible to the resonant frequency of the vibrating member, in order to maximize the efficiency in motion generation. These two requirements are usually accomplished by employing a dedicated feedback control loop (*driving loop*), which basically excites the vibrating member with a properly gain and phase adjusted version of the driving mode detected motion. The phase is adjusted to meet the *Barkhausen's condition*, thus actually implementing an electromechanical (sinusoidal) oscillator (i.e. an electronic oscillator with a mechanical resonating element); the gain is adjusted to regulate the oscillation amplitude to the desired set-point. Details about the working principle and implementation of a driving loop can be found in (M'Closkey & Vakakis, 1999; Oboe et al., 2005) (conventional design) and (Dong & Avanesian, 2009; Leland et al., 2003) (unconventional designs based on adaptive control schemes).

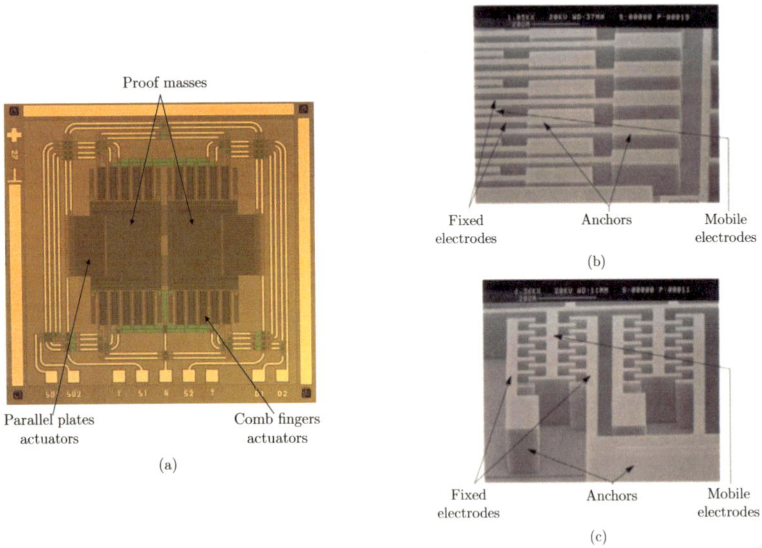

Fig. 9. Example of electrostatic actuation and sensing in a MEMS gyroscope: (a) ST Microelectronics LISY300AL single-axis yaw-gyroscope (die photo); (b) parallel-plate electrodes used for sensing the Coriolis-induced vibration along the sense axis; (c) comb-fingers structures used for actuating/sensing the proof-mass motion along the drive axis.

- *retrieving an angular velocity measurement from the sensing mode vibration.* Several stages are usually involved in retrieving a reliable measure: first, the Coriolis-induced motion must be transduced into an electric signal, possibly ensuring a sufficiently high signal-to-noise ratio. Second, the transduced signal must be demodulated with a carrier that is synchronized with the driving motion, in order to obtain a baseband signal which is proportional to the angular velocity; and finally, the demodulated signal must be conditioned (e.g. scaled, filtered, digitized, etc.).
- *reducing the interaction between the driving and sensing loops.* Differently from the ideal situation described in Sec. 2.2, in the real situation there is always a spurious motion along the sense axis that is directly proportional to the drive vibration. This motion, called *quadrature error*, is mainly due to a lack of orthogonality between the drive and sense axes, which in turn results from structural asymmetries due to fabrication defects and imperfections. The quadrature error requires to be compensated, since it detrimentally affects the measurement. Usual compensation methods consists of either rebalancing the mechanical structure (with mechanical or electrostatic methods - (Painter & Shkel, 2003;

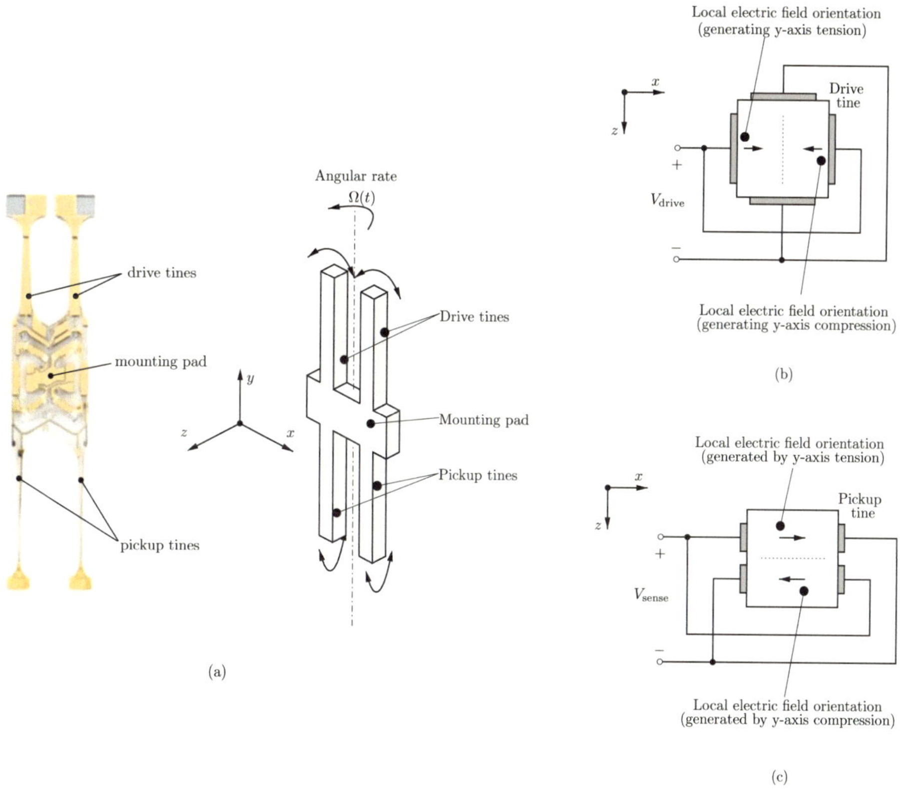

Fig. 10. Example of piezoelectric actuation and sensing in a micro-gyroscope: (a) Systron Donner Quartz Rotation Sensor (QRS) (quartz cut axis orientation ≡ z-axis) (Gupta & Jenson, 1995; Knowles & Moore, 2004); (b) electrodes configuration for generating the drive tine bending vibration; (b) electrodes configuration for sensing the pickup tine bending vibration.

Weinberg & Kourepenis, 2006)) or canceling the error from the measurement (using a feed-forward cancellation scheme - (Antonello et al., 2009; Saukoski et al., 2007)).

- *improving sensor linearity and bandwidth.* This is usually achieved by exploiting a closed-loop sensing interface, in which the sense motion is nulled by employing a control loop. The feedback signal used for nulling the sense motion contains the angular velocity information, which can be extracted with a basic synchronous baseband demodulation circuit. When a digital output must be provided, the feedback signal can be oversampled and quantized with a coarse quantizer: in this case, the closed-loop sensing interface behaves as a (electromechanical) $\Sigma\Delta$ modulator (Dong et al., 2008; Petkov & Boser, 2005).
- *improving scale factor thermal stability.* A temperature compensation loop can be sometimes integrated on-board to reduce the sensitivity of the scale factor to temperature variations (Jiancheng & Jianli, 2009).

Additional electronic functions may include self test and calibration, bias compensation, etc.

4. Industrial requirements

The specifications and test procedures for a single-axis CVG-based angular rate sensor have been standardized in (*IEEE Standard Specification Format Guide and Test Procedure for Coriolis Vibratory Gyros*, 2004). The standard requirements for a CVG are specified in terms of its *performances*, its *mechanical* and *electrical interface* characteristics, the *environmental* conditions, the sensor *life time* and *reliability* (usually measured as *Mean Time Between Failure* - MTBF).
The performance of a CVG is specified according to the following quantities, whose definitions are taken from (*IEEE Standard for Inertial Sensor Terminology*, 2001):

- *Input range*: the interval between the input limits within which a quantity is measured. The *input limits* are defined as the extreme values of the input, generally plus or minus, within which performance is of the specified accuracy. The *full-scale* (FS) input is the maximum magnitude of the two input limits.
- *Accuracy* (or *linearity error*): the deviation of the output from a least-squares linear fit of the input-output data. It is generally expressed as a percentage of the input full-scale, or a percent of output, or both.

  The definition implicitly assumes that the ideal sensor exhibits a *linear* input-output behavior (i.e. the static input-output sensor characteristic is a linear function).
- *Scale factor* [1] : the ratio of a change in output to a change in the input intended to be measured, typically specified in $\lfloor V/^\circ/s \rfloor$. It is evaluated as the slope of the least squares straight line fit to input-output data.

  In the ideal case, the scale-factor is constant over both the entire input range and the whole sensor lifespan. In the real case, the following quantities are used to judge the scale factor quality:
  - *asymmetry error*: the difference between the scale factor measured with positive input and that measured with negative input, specified as a fraction of the scale factor measured over the input range.

[1] Sometimes the term *sensitivity* is used as a synonym for scale factor. However, according to (*IEEE Standard for Inertial Sensor Terminology*, 2001), the term sensitivity is reserved for denoting the ratio of a change in output to a change in an *undesirable* or *secondary input*.

- *scale factor stability*: the variation in scale factor over a specified time of continuous operation. Ambient temperature, power supply and additional factors pertinent to the particular application should be specified.
- *scale factor sensitivities*: the ratio of change in scale factor to a change in an undesirable input, such as the steady state operating temperature (*scale factor temperature sensitivity*) or the constant acceleration along any axis (*scale factor acceleration sensitivity*). Additional sensitivities may be specified such as those due to variations in supply voltage (including frequency, voltage, ripple, starting and operating current), orientation, vibration, magnetic field, radiation, and other environments pertinent to the particular application.

- *Resolution*: the smallest input change, for inputs greater than the noise level, that can be reliably detected. It is usually evaluated as the minimum input change that produces a change in output equal to some specified percentage (at least 50%) of the change in output expected using the nominal scale factor.
- *Drift rate*: the portion of gyro output that is functionally independent of input rotation. The systematic component of the drift rate (*systematic drift rate*) includes:
  1. *Bias* (or *zero rate output* - ZRO): the average over a specified time of gyro output measured at specified operating conditions that has no correlation with input rotation. Bias is typically expressed in $[^\circ/s]$ or $[^\circ/hr]$.
  2. *Environmentally sensitive drift rate*: the components of systematic drift rate that are sensitive to temperature (steady state, gradient, ramp), acceleration, vibration and other quantities.

  The random component of the drift rate (*random drift rate*) includes:
  1. *Angle Random Walk* (ARW): the angular error buildup with time that is due to white noise in angular rate, typically expressed in $[^\circ/hr/\sqrt{hr}]$ or $[^\circ/s/\sqrt{hr}]$.
  2. *Rate Random Walk* (RRW): the drift rate error buildup with time that is due to white noise in angular acceleration, typically expressed in $[^\circ/hr/\sqrt{hr}]$.
  3. *Bias Instability*: the random variation in bias as computed over specified finite sample time and averaging time intervals, characterized by a $1/f$ power spectral density, typically expressed in $[^\circ/hr]$.
- *Bandwidth*: the range of frequency of the angular rate input that the gyroscope can detect. Typically specified as the cutoff frequency coinciding to the $-3\,dB$ point. Alternatively, the frequency response or transfer function could be specified.
- *Activation time*: it includes the *turn-on time*, i.e. the time from the initial application of power until a sensor produces a specified useful output, though not necessarily at the accuracy of full specification performance, and the *warm-up time*, i.e. the time from the initial application of power to reach specified performance under specified operating conditions.

The mechanical and electrical requirements are usually specified in terms of:

- *Outline, mounting dimensions, weight*
- *Gyro axes*: the *input axis* (IA) is the axis about which a rotation of the case causes a maximum output; the *input reference axis* (IRA) is the direction of an axis (ideally parallel to an input axis) as defined by the case mounting surfaces, or external case markings, or both. In case of a multi-axis gyroscope, more than one IA (and, correspondingly, IRA) can be defined.

- *Seal*: CVGs may be sealed using vacuum, gas or ambient environment.
- *Acoustic noise emission*
- *Electrical impedances*: load impedances and impedances of excitation, monitoring, temperature sensing and test circuits.
- *Input power, grounding*
- *Output signals*: the type and characteristics of output signals, such as analog voltage or current, parallel or serial digital, or incremental angle pulses.
- *Electromagnetic interference* and *electrostatic discharge* (ESD) *immunity*.

The environmental requirements specify the environmental conditions (limits) and perturbations during storage, transport and/or operation under which the sensor preserves its functionality. Environmental characteristics account for:

- *Linear* and *angular accelerations*: both the acceleration (axis, direction, intensity) and exposure time should be defined.
- *Linear* and *angular vibrations*: both axes and direction of vibration should be defined. For a deterministic vibration, specifications include the type of vibration (e.g. sinusoidal) and its characteristics (e.g. amplitude, frequency sweep range, exposure time). For a random vibration, its spectral characteristic (e.g. power spectral density - PSD) is usually specified.
- *Mechanical shock*: it is specified in terms of axis, direction, wave shape, intensity (usually measured as a peak value in $[m/s^2]$ or a multiple of the gravity acceleration g) and duration.
- *Temperature range*
- *Others*: may include ambient air pressure, humidity, electromagnetic fields, etc.

Conventionally, gyroscopes are classified into three different categories based on their performance: *inertial-grade*, *tactical-grade*, and *rate-grade* devices (Yazdi et al., 1998). Table 1 summarizes the requirements for each of these categories.
RLGs, together with HLGs (R.R.Ragan (ed), 1984), are currently the angular rate sensors with highest performance available in the market, and exhibit inertial grade performances. They are used in the most demanding applications, especially those requiring extremely high scale factor stability (typically, high precision space applications). FOGs normally achieves tactical grade performances, while typical MEMS CVGs seldom exceed the rate grade performance level, which is however satisfactory for most of the automotive and consumer electronics applications.

Parameter	*Rate grade*	*Tactical grade*	*Inertial grade*
Angle Random Walk $[^\circ/\sqrt{hr}]$	> 0.5	$0.5 \div 0.05$	< 0.001
Bias drift $[^\circ/hr]$	$10 \div 1000$	$0.1 \div 10$	< 0.01
Scale factor Accuracy $[\%]$	$0.1 \div 1$	$0.01 \div 0.1$	< 0.001
Full Scale Range $[^\circ/s]$	$50 \div 1000$	> 500	> 400
Max. Shock in 1 ms $[gs]$	10^3	$10^3 \div 10^4$	10^3
Bandwidth, Hz	> 70	$\simeq 100$	$\simeq 100$

Table 1. Performance requirements for different classes of gyroscopes

5. Benchmark tests for two commercial products

5.1 Invensense IDG-650 Dual-Axis Pitch & Roll MEMS Gyroscope

The Invensense IDG-650 (Invensense, 2010) is a dual-axis MEMS gyroscope consisting of two independent out-of-plane resonating tuning fork CVGs (for pitch and roll axes) integrated on the same silicon die. The MEMS and CMOS integrated circuit (IC) are electrically connected together through wafer level bonding, in such a way that the mechanical structure is covered and hermetically sealed at the wafer level (*Nasiri-Fabrication* - patent protected). The hermetic sealing improves the sensor resistance to humidity, high temperature and electromagnetic/radio frequency interferences (EMI/RFI). The sensor has selectable scale factors and full scale ranges ($FS = \pm 2000^\circ/s$ for faster motions, or $FS = \pm 440^\circ/s$ for slower motions). Both the output measurements are thermally compensated by an internal compensation circuit, comprising an integrated onboard temperature

5.2 STMicroelectronics LPR530AL Dual-Axis Pitch & Roll MEMS Gyroscope

The STMicroelectronics LPR530AL (STMicroelectronics, 2010) is a dual-axis MEMS gyroscope capable of measuring angular rates along pitch and roll axes. The mechanical sensing element is a vibrating disk fabricated using STMicroelectronic's proprietary surface micromachining process called *ThELMA* (*Thick Epipoly Layer for Microactuators and Accelerometers*). The mechanical element is obtained by etching a thicker polysilicon epitaxial layer ($\approx 15\mu m$) grown on top of a sacrificial oxide layer ($\approx 2\mu m$), which is removed at the end of the process (release step) by isotropic etching. A second wafer is bonded to the substrate with the purpose of creating a protecting case for the micromechanical structure. The ThELMA micromachined mechanical element and the CMOS controller are finally assembled together in the same chip (hybrid solution: two chips in a single package) either in a side-by-side or a stacked configuration.

	STM LPR530AL	IS IDG-500	IS IDG-650	
• *Scale factor*	0.83	2	0.5	$mV/^\circ/s$
• *Scale factor* (× 4.5)	3.33	9.1	2.27	$mV/^\circ/s$
• *Scale factor calibration tolerance*	5	6	6	%
• *Scale factor calibration drift over specified temperature* ($T_A = 25^\circ$)	0.034	0.067	0.067	$\%/^\circ C$
• *Full-scale*	1200	500	2000	$^\circ/s$
• *Full-scale* (× 4.5)	300	110	440	$^\circ/s$
• *ZRO level*	1.23	1.35	1.35	V
• *ZRO tolerance*	80	250	150	mV
• *ZRO temperature sensitivity over specified temperature* ($T_A = 25^\circ C$)	0.083	0.417	0.667	$^\circ/s/^\circ C$
• *phase delay* at 10 Hz	4.0	4.5	4.5	$^\circ$
• *Turn-on time* (typ)	200	50	50	ms
• *Turn-on-time* (max)	200	200	200	ms
• *Total RMS noise*	1.4	0.8	0.3	mV
• *Nonlinearity*	1	1	1	% of FS
• *Cross-axis sensitivity*	1	1	1	% of FS
• *Supply voltage*	$2.7 \div 3.6$	$2.7 \div 3.3$	$2.7 \div 3.3$	V
• *Current absorption*	6.8	7	7	mA
• *Temperature range*	$-40 \div 85$	$-20 \div 85$	$-20 \div 85$	$^\circ C$
• *Package size*	$5 \times 5 \times 1.5$	$4 \times 5 \times 1.2$	$4 \times 5 \times 1.2$	mm

Table 2. Specifications comparison for the gyroscopes under test

5.3 Comparative tests

The results of our comparative tests are briefly summarized in the following paragraphs. Parts have been tested on a single-axis precision positioning and rate table (Aerosmith, 2005), providing the desired angular rate profiles. An additional integral thermal chamber has been installed on the table to allow thermal sensitivity/stress analyses.

5.3.1 ZRO temperature sensitivity

The ZRO has been measured by setting the default FS (i.e. full-scale $\times$ 4.5 in Tab. 2) for both STM and IS sensors, and by varying the temperature over the range $-40^{\circ}\,C \div 85^{\circ}\,C$ in steps of $5^{\circ}\,C$. As shown in Fig. 11, the ZRO of the STM sensor appears to be slightly more insensitive to temperature variations than the IS sensor; however, the temperature sensitivities of both sensors are within specifications.

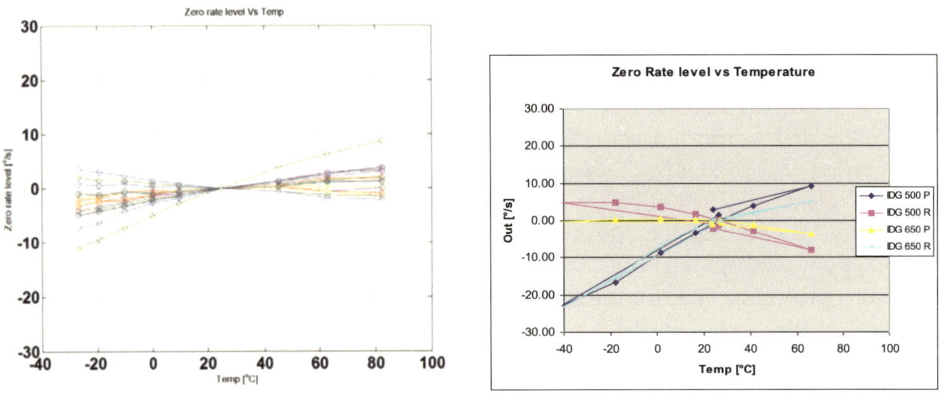

Fig. 11. ZRO temperature sensitivity test results: STM (left); IS (right).

5.3.2 ZRO mechanical stress sensitivity

The ZRO has been measured after applying different calibrated vertical loads on the sensor package, by means of a load cell. It has been noted that only the STM sensor has a ZRO that is immune to the applied mechanical stress; for what concerns the IS sensor, the ZRO variation exceeds the value of $30^{\circ}/s$ as the vertical load varies over the range $0 \div 4\,kg$ (see Fig. 12).

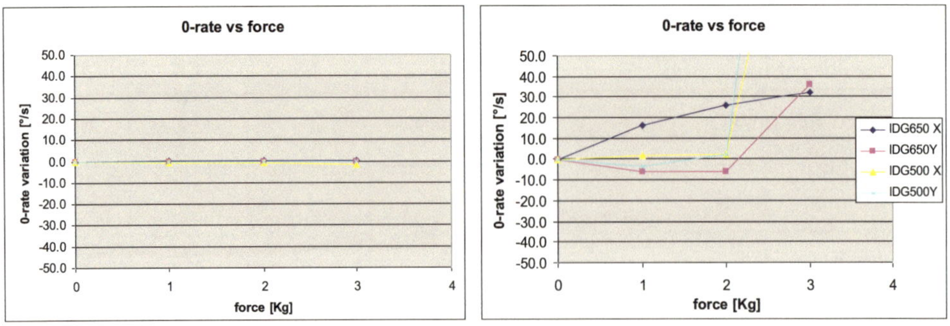

Fig. 12. ZRO mechanical stress sensitivity test results: STM (left), IS (right).

5.3.3 Scale factor temperature sensitivity

Scale factor measurements have been done at a constant angular rate of $150°/s$, and by varying the temperature over the range $-40°\,C \div 85°\,C$ in steps of $5°\,C$. On average, both the STM and IS sensors exhibit a scale factor temperature sensitivity below $0.05\,\%/°C$ and thus within specifications (see Fig. 13) .

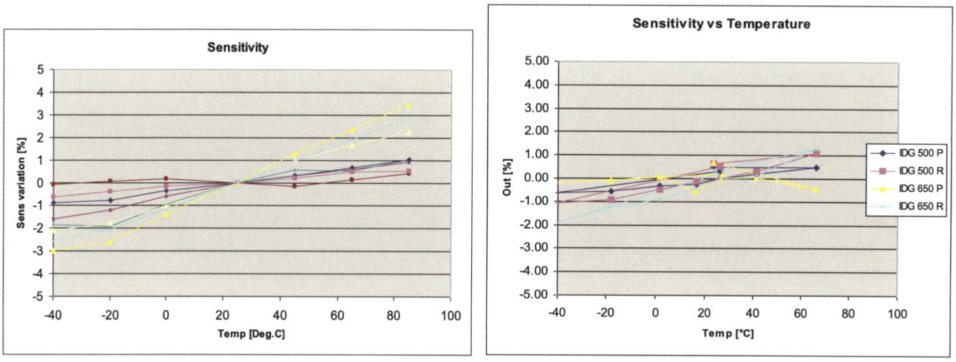

Fig. 13. Scale factor temperature sensitivity test results: STM (left); IS (right).

5.3.4 Frequency response

The sensor frequency response (from the angular rate input to sensor output measurement) has been measured at 16 frequency points, almost regularly spaced in the frequency range $0.1 \div 100\,Hz$. Measurements have been carried out frequency-by-frequency, by evaluating the sensor magnitude and phase responses to an applied sinusoidal angular rate input. As it can be observed in Figs. 14 and 15, the STM and IS sensors have almost identical frequency responses, with a flat magnitude response up to $50\,Hz$ and a phase lag of $45°$ at $80\,Hz$; the only noticeable difference consists of a flatter magnitude response of the STM sensor at higher frequencies.

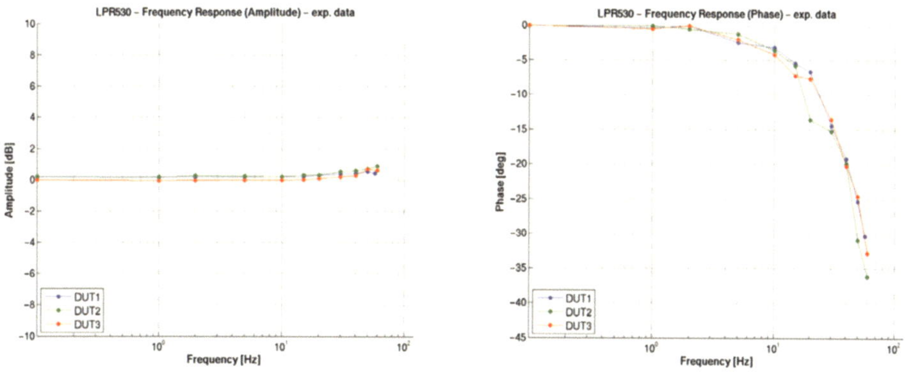

Fig. 14. STM LPR530AL measured frequency response: magnitude (left); phase (right).

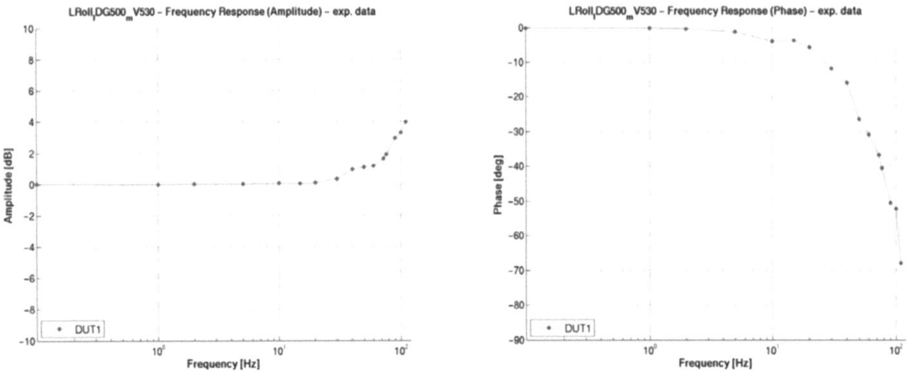

Fig. 15. IS IDG-500/650 measured frequency response: magnitude (left); phase (right).

5.3.5 Noise power spectral density

The output noise Power Spectral Density (PSD) has been measured by using a Digital Signal Analyzer (DSA). For a fair comparison, the full-scales of the two sensors have been selected to be as matched as possible: therefore, the STM sensor FS has been set to $300^\circ/s$ (scale factor $= 3.33\,mV/^\circ/s$), while the IS sensor (IDG-500) FS has been selected as $500^\circ/s$ (scale factor $= 2\,mV/^\circ/s$). With such configuration, the measured output noise floors of the two sensors are almost identical, e.g. $S_n = 0.035\,^\circ/s/\sqrt{Hz}$ for the STM LPR530AL and $S_n = 0.030\,^\circ/s/\sqrt{Hz}$ for the IS IDG-500 (see also Fig. 16).

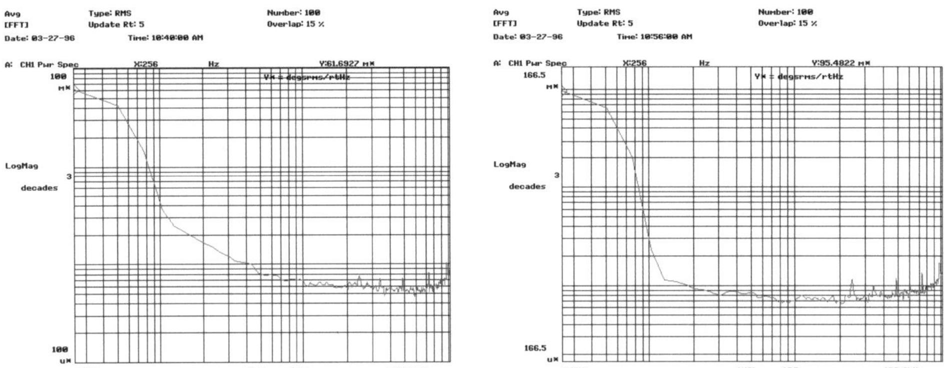

Fig. 16. Output noise power spectral density measurements: STM (left); IS (right).

6. Conclusions

In recent years, the development and commercialization of MEMS gyroscopes have experienced a rapid growth, as a result of performance improvements and cost reductions. Such sensors have begun to be applied in many consumer and industrial applications, either to replace older, bulkier and more expensive angular rate sensors, or to become essential parts in completely new applications requiring small and inexpensive devices.

This paper has provided a brief introduction to the design, technology and industrial requirements aspects behind the recent commercialization of many MEMS gyroscopes for consumer and industrial applications.
In order to assess the performance levels currently achieved by many sensors available in the market, two commercially available sensors, e.g. the STMicroelectronics LPR530AL and the Invensense IDG-500/650 dual-axis pitch & roll MEMS gyroscopes, have been compared by running several benchmark tests. The tests have shown similar results for the two devices, except for the ZRO immunity to mechanical stress, for which the STMicroelectronics sensor has exhibited better performances.
In general, the average performance levels achieved by current MEMS gyroscopes available in the market are sufficient for most of the consumer and industrial applications; nevertheless, it is perhaps only a matter of time before they will become adequate also for the most demanding inertial navigation applications.

7. References

Abe, H., Yoshida, T. & Turuga, K. (1992). Piezoelectric-ceramic cylinder vibratory gyroscope, *Japanese Journal of Applied Physics* 31(Part 1, No. 9B): 3061–3063.

Aerosmith, I. (2005).
URL: *http://www.ideal-aerosmith.com/*

Antonello, R., Oboe, R., Prandi, L., Caminada, C. & Biganzoli, F. (2009). Open loop compensation of the quadrature error in mems vibrating gyroscopes, *Industrial Electronics, 2009. IECON '09. 35th Annual Conference of IEEE*, pp. 4034 –4039.

Apple (2011).
URL: *www.apple.com*

Bernstein, J., Cho, S., King, A., Kourepenis, A., Maciel, P. & Weinberg, M. (1993). A micromachined comb-drive tuning fork rate gyroscope, *Micro Electro Mechanical Systems, 1993, MEMS '93, Proceedings An Investigation of Micro Structures, Sensors, Actuators, Machines and Systems. IEEE.*, pp. 143–148.

Bochobza-Degani, O., Seter, D. J., Socher, E. & Nemirovsky, Y. (2000). A novel micromachined vibrating rate-gyroscope with optical sensing and electrostatic actuation, *Sensors and Actuators A: Physical* 83(1-3): 54 – 60.

Boser, B. (1997). Electronics for micromachined inertial sensors, *Solid State Sensors and Actuators, 1997. TRANSDUCERS '97 Chicago., 1997 International Conference on*, Vol. 2, pp. 1169–1172 vol.2.

Brown, A. (2005). Gps/ins uses low-cost mems imu, *Aerospace and Electronic Systems Magazine, IEEE* 20(9): 3–10.

Bustillo, J., Howe, R. & Muller, R. (1998). Surface micromachining for microelectromechanical systems, *Proceedings of the IEEE* 86(8): 1552–1574.

Chang, S., Chia, M., Castillo-Borelley, P., Higdon, W., Jiang, Q., Johnson, J., Obedier, L., Putty, M., Shi, Q., Sparks, D. & Zarabadi, S. (1998). An electroformed CMOS integrated angular rate sensor, *Sensors and Actuators A: Physical* 66(1-3): 138 – 143.

Clark, W. A., Howe, R. T. & Horowitz, R. (1996). Surface micromachined Z-axis vibratory rate gyroscope, *Tech. Dig. Solid-State Sens. Actuator Workshop, Hilton Head Island*, pp. 283–287.

Damrongsak, B. & Kraft, M. (2005). A micromachined electrostatically suspended gyroscope with digital force feedback, *Sensors, 2005 IEEE.*

Dauwalter, C. & Ha, J. (2005). Magnetically suspended mems spinning wheel gyro, *Aerospace and Electronic Systems Magazine, IEEE* 20(2): 21–26.
Dong, L. & Avanesian, D. (2009). Drive-mode control for vibrational mems gyroscopes, *Industrial Electronics, IEEE Transactions on* 56(4): 956–963.
Dong, Y., Kraft, M., Hedenstierna, N. & Redman-White, W. (2008). Microgyroscope control system using a high-order band-pass continuous-time sigma-delta modulator, *Sensors and Actuators A: Physical* 145-146: 299 – 305.
Drafts, B. (2001). Acoustic wave technology sensors, *Microwave Theory and Techniques, IEEE Transactions on* 49(4): 795–802.
Ellis, C. D. & Wilamowski, B. M. (2008). Fabrication and control of an electrostatically levitated rotating gyro, *in* T. George & Z. Cheng (eds), *Micro (MEMS) and Nanotechnologies for Space, Defense, and Security II*, Vol. 6959, SPIE, p. 69590Q.
Fell, C. (2006). Development of a second generation low cost mems gyroscope: Design for manufacture, *MEMS Sensors and Actuators, 2006. The Institution of Engineering and Technology Seminar on*, pp. 75–82.
Fleming, W. (2001). Overview of automotive sensors, *Sensors Journal, IEEE* 1(4): 296–308.
Geen, J., Sherman, S., Chang, J. & Lewis, S. (2002). Single-chip surface micromachined integrated gyroscope with 50 °/h allan deviation, *Solid-State Circuits, IEEE Journal of* 37(12): 1860–1866.
Geiger, W., Folkmer, B., Sobe, U., Sandmaier, H. & Lang, W. (1998). New designs of micromachined vibrating rate gyroscopes with decoupled oscillation modes, *Sensors and Actuators A: Physical* 66(1-3): 118 – 124.
Gupta, P. K. & Jenson, C. E. (1995). Rotation rate sensor with center mounted tuning fork, U.S. Patent 5396144.
Gyration (2009).
URL: *http://www.gyration.com*
Hopkin, I. D., Fell, C. P., Townsend, K. & Mason, T. R. (1999). Vibrating structure gyroscope, U.S. Patent 5932804.
Howe, R. T., Boser, B. E. & Pisano, A. P. (1996). Polysilicon integrated microsystems: technologies and applications, *Sensors and Actuators A: Physical* 56(1-2): 167 – 177.
IEEE Standard for Inertial Sensor Terminology (2001). *IEEE Std 528-2001* .
IEEE Standard Specification Format Guide and Test Procedure for Coriolis Vibratory Gyros (2004). *IEEE Std 1431-2004* .
Invensense (2010).
URL: *http://www.invensense.com/products/idg_650.html*
Jiancheng, F. & Jianli, L. (2009). Integrated model and compensation of thermal errors of silicon microelectromechanical gyroscope, *Instrumentation and Measurement, IEEE Transactions on* 58(9): 2923–2930.
Jose, K. A., Suh, W. D., Xavier, P. B., Varadan, V. K. & Varadan, V. V. (2002). Surface acoustic wave MEMS gyroscope, *Wave Motion* 36(4): 367 – 381.
Juneau, T., Pisano, A. & Smith, J. (1997). Dual axis operation of a micromachined rate gyroscope, *Solid State Sensors and Actuators, 1997. TRANSDUCERS '97 Chicago., 1997 International Conference on*, Vol. 2, pp. 883–886.
Kawano, K., Kobashi, S., Yagi, M., Kondo, K., Yoshiya, S. & Hata, Y. (2007). Analyzing 3d knee kinematics using accelerometers, gyroscopes and magnetometers, *System of Systems Engineering, 2007. SoSE '07. IEEE International Conference on*, pp. 1–6.

Kim, Y. S., Soh, B. S. & Lee, S.-G. (2005). A new wearable input device: Scurry, *Industrial Electronics, IEEE Transactions on* 52(6): 1490–1499.

Knowles, S. J. & Moore, R. H. (2004). Tuning fork with symmetrical mass balancing and reduced quadrature error, U.S. Patent 6701785.

Kotru, S., Zhong, J., Highsmith, A. & Jackson, J. (2008). Design and fabrication of a meso-scale gyroscope, *Microelectronics and Electron Devices, 2008. WMED 2008. IEEE Workshop on*, pp. 5–8.

Kovacs, G., Maluf, N. & Petersen, K. (1998). Bulk micromachining of silicon, *Proceedings of the IEEE* 86(8): 1536–1551.

Kurosawa, M., Fukuda, Y., Takasaki, M. & Higuchi, T. (1998). A surface-acoustic-wave gyro sensor, *Sensors and Actuators A: Physical* 66(1-3): 33 – 39.

Lawrence, A. (1993). *Modern inertial technology - Navigation, guidance, and control*, Springer-Verlag.

Leland, R., Lipkin, Y. & Highsmith, A. (2003). Adaptive oscillator control for a vibrational gyroscope, *American Control Conference, 2003. Proceedings of the 2003*, Vol. 4, pp. 3347 – 3352 vol.4.

Li, X., Bao, M., Yang, H., Shen, S. & Lu, D. (1999). A micromachined piezoresistive angular rate sensor with a composite beam structure, *Sensors and Actuators A: Physical* 72(3): 217 – 223.

Liu, Q. H. & Wu, X. Z. (2007). Design of a Novel MEMS IDT Dual Axes Surface Acoustic Wave Gyroscope, *Nano/Micro Engineered and Molecular Systems, 2007. NEMS '07. 2nd IEEE International Conference on*, pp. 817–820.

Lutz, M., Golderer, W., Gerstenmeier, J., Marek, J., Maihofer, B., Mahler, S., Munzel, H. & Bischof, U. (1997). A precision yaw rate sensor in silicon micromachining, *Solid State Sensors and Actuators, 1997. TRANSDUCERS '97 Chicago., 1997 International Conference on*, Vol. 2, pp. 847–850.

Madni, A., Costlow, L. & Knowles, S. (2003). Common design techniques for BEI GyroChip quartz rate sensors for both automotive and aerospace/defense markets, *Sensors Journal, IEEE* 3(5): 569–578.

Maenaka, K., Fujita, T., Konishi, Y. & Maeda, M. (1996). Analysis of a highly sensitive silicon gyroscope with cantilever beam as vibrating mass, *Sensors and Actuators A: Physical* 54(1-3): 568 – 573.

M'Closkey, R. & Vakakis, A. (1999). Analysis of a microsensor automatic gain control loop, *American Control Conference, 1999. Proceedings of the 1999*, Vol. 5, pp. 3307 –3311 vol.5.

Meirovitch, L. (1970). *Methods of Analytical Dynamics*, McGraw-Hill, New York.

Microstrain (2011).
URL: *http://www.microstrain.com*

Miyazaki, S. (1997). Long-term unrestrained measurement of stride length and walking velocity utilizing a piezoelectric gyroscope, *Biomedical Engineering, IEEE Transactions on* 44(8): 753–759.

muRata (2003).
URL: *http://www.murata.com*

Nalbach, G. (1993). The halteres of the blowfly Calliphora - Part I: kinematics and dynamics, *Journal of Comparative Physiology A: Neuroethology, Sensory, Neural, and Behavioral Physiology* 173(3): 293–300.

Nalbach, G. & Hengstenberg, R. (1994). The halteres of the blowfly Calliphora - Part II: three-dimensional organization of compensatory reactions, *Journal of Comparative Physiology A: Neuroethology, Sensory, Neural, and Behavioral Physiology* 175(6): 695–708.

Nasiri, S. & Flannery Jr., A. F. (2007). Method of making an x-y axis dual-mass tuning fork gyroscope with vertically integrated electronics and wafer-scale hermetic packaging, U.S. Patent 7250112.

Neul, R., Gomez, U.-M., Kehr, K., Bauer, W., Classen, J., Doring, C., Esch, E., Gotz, S., Hauer, J., Kuhlmann, B., Lang, C., Veith, M. & Willig, R. (2007). Micromachined angular rate sensors for automotive applications, *Sensors Journal, IEEE* 7(2): 302–309.

Nintendo (2011).
URL: *http://www.nintendo.com/wii/what/accessories/wiimotionplus*

Nonomura, Y., Fujiyoshi, M., Omura, Y., Fujitsuka, N., Mizuno, K. & Tsukada, K. (2006). Soi rate gyro sensor for automotive control, *Sensors and Actuators A: Physical* 132(1): 42 – 46.

Norgia, M. & Donati, S. (2001). Hybrid opto-mechanical gyroscope with injection-interferometer readout, *Electronics Letters* 37(12): 756–758.

Noureldin, A., Karamat, T., Eberts, M. & El-Shafie, A. (2009). Performance enhancement of mems-based ins/gps integration for low-cost navigation applications, *Vehicular Technology, IEEE Transactions on* 58(3): 1077–1096.

Oboe, R., Antonello, R., Lasalandra, E., Durante, G. & Prandi, L. (2005). Control of a Z-axis MEMS vibrational gyroscope, *Mechatronics, IEEE/ASME Transactions on* 10(4): 364–370.

O'Connor, J. M. & Shupe, D. M. (1983). Vibrating beam rotation sensor, U.S. Patent 4381672.

Painter, C. & Shkel, A. (2003). Active structural error suppression in mems vibratory rate integrating gyroscopes, *Sensors Journal, IEEE* 3(5): 595 – 606.

Paoletti, F., Gretillat, M.-A. & de Rooij, N. (1996). A silicon micromachined vibrating gyroscope with piezoresistive detection and electromagnetic excitation, *Micro Electro Mechanical Systems, 1996, MEMS '96, Proceedings. 'An Investigation of Micro Structures, Sensors, Actuators, Machines and Systems'. IEEE, The Ninth Annual International Workshop on*, pp. 162–167.

Pappas, I., Keller, T., Mangold, S., Popovic, M., Dietz, V. & Morari, M. (2004). A reliable gyroscope-based gait-phase detection sensor embedded in a shoe insole, *Sensors Journal, IEEE* 4(2): 268–274.

Petkov, V. & Boser, B. (2005). A fourth-order ΣΔ interface for micromachined inertial sensors, *Solid-State Circuits, IEEE Journal of* 40(8): 1602 – 1609.

Rajendran, S. & Liew, K. M. (2004). Design and simulation of an angular-rate vibrating microgyroscope, *Sensors and Actuators A: Physical* 116(2): 241 – 256.

Roetenberg, D., Slycke, P. & Veltink, P. (2007). Ambulatory position and orientation tracking fusing magnetic and inertial sensing, *Biomedical Engineering, IEEE Transactions on* 54(5): 883–890.

R.R.Ragan (ed) (1984). Inertial technology for the future, *Aerospace and Electronic Systems, IEEE Transactions on* AES-20(4): 414–444.

Sachs, D., Nasiri, S. & Goehl, D. (n.d.). Image stabilization technology overview. Available at: http://www.invensense.com/cn/shared/pdf/ImageStabilizationWhitepaper_051606.pdf.

Saukoski, M., Aaltonen, L. & Halonen, K. (2007). Zero-rate output and quadrature compensation in vibratory mems gyroscopes, *Sensors Journal, IEEE* 7(12): 1639–1652.

Segway (2011).
URL: *http://www.segway.com*

Seidel, H., Aikele, M., Rose, M. & Toelg, S. (2002). Safety relevant microsystems for automotive applications, *Microsystem Technologies* 7(5): 244–248.

Senturia, S. D. (2001). *Microsystem Design*, Kluwer Academic Publishers.

Shakoor, R. I., Bazaz, S. A., Kraft, M., Lai, Y. & ul Hassan, M. M. (2009). Thermal Actuation Based 3-DoF Non-Resonant Microgyroscope Using MetalMUMPs, *Sensors* 9(4): 2389–2414.

Shearwood, C., Ho, K. Y., Williams, C. B. & Gong, H. (2000). Development of a levitated micromotor for application as a gyroscope, *Sensors and Actuators A: Physical* 83(1-3): 85 – 92.

Soderkvist, J. (1991). Piezoelectric beams and vibrating angular rate sensors, *Ultrasonics, Ferroelectrics and Frequency Control, IEEE Transactions on* 38(3): 271–280.

Söderkvist, J. (1994). Micromachined gyroscopes, *Sensors and Actuators A: Physical* 43(1-3): 65 – 71.

Sparks, D., Zarabadi, S., Johnson, J., Jiang, Q., Chia, M., Larsen, O., Higdon, W. & Castillo-Borelley, P. (1997). A CMOS integrated surface micromachined angular rate sensor: it's automotive applications, *Solid State Sensors and Actuators, 1997. TRANSDUCERS '97 Chicago., 1997 International Conference on*, Vol. 2, pp. 851–854 vol.2.

Stewart, R. E. (2009). Micro hemispheric resonator gyro, U.S. Patent 2009/0031831.

STMicroelectronics (2010).
URL: *http://www.st.com/stonline/products/literature/ds/15812/lpr530al.pdf*

Stringer, J. (2000). *The Air Force Institute of Technology (AFIT) Micro Electro-Mechanical Systems (MEMS) Interferometric Gyroscope (MiG)*, Master's thesis, Air Force Institute of Technology (AFIT).

Tanaka, K., Mochida, Y., Sugimoto, M., Moriya, K., Hasegawa, T., Atsuchi, K. & Ohwada, K. (1995). A micromachined vibrating gyroscope, *Sensors and Actuators A: Physical* 50(1-2): 111 – 115.

Tang, W., Nguyen, T.-C. & Howe, R. (1989). Laterally driven polysilicon resonant microstructures, *Micro Electro Mechanical Systems, 1989, Proceedings, An Investigation of Micro Structures, Sensors, Actuators, Machines and Robots. IEEE*, pp. 53–59.

Tsai, N.-C., Huang, W.-M. & Chiang, C.-W. (2009). Magnetic actuator design for single-axis micro-gyroscopes, *Microsystem Technologies* 15(4): 493–503.

Vellekoop, M. J. (1998). Acoustic wave sensors and their technology, *Ultrasonics* 36(1-5): 7 – 14. Ultrasonics International 1997.

Voss, R., Bauer, K., Ficker, W., Gleissner, T., Kupke, W., Rose, M., Sassen, S., Schalk, J., Seidel, H. & Stenzel, E. (1997). Silicon angular rate sensor for automotive applications with piezoelectric drive and piezoresistive read-out, *Solid State Sensors and Actuators, 1997. TRANSDUCERS '97 Chicago., 1997 International Conference on*, Vol. 2, pp. 879–882 vol.2.

Weinberg, M. & Kourepenis, A. (2006). Error sources in in-plane silicon tuning-fork MEMS gyroscopes, *Microelectromechanical Systems, Journal of* 15(3): 479–491.

Yazdi, N., Ayazi, F. & Najafi, K. (1998). Micromachined inertial sensors, *Proceedings of the IEEE* 86(8): 1640–1659.

Yokota, S., Imamura, T., Takemura, K., Edamura, K. & Kumagai, H. (2008). A liquid rate gyroscope using electro-conjugate fluid, *Intelligent Sensors, Sensor Networks and Information Processing, 2008. ISSNIP 2008. International Conference on*, pp. 459–464.

Zhou, J., Yan, G., Zhu, Y., Xiao, Z. & Fan, J. (2005). Design and fabrication of a microfluid angular rate sensor, *Micro Electro Mechanical Systems, 2005. MEMS 2005. 18th IEEE International Conference on*, pp. 363–366.

Zhu, R., Ding, H., Su, Y. & Zhou, Z. (2006). Micromachined gas inertial sensor based on convection heat transfer, *Sensors and Actuators A: Physical* 130-131: 68 – 74.

13

Planar Oxygen Sensors for Non Invasive Imaging in Experimental Biology

Henning Tschiersch[1], Gregor Liebsch[2], Achim Stangelmayer[2], Ljudmilla Borisjuk[1] and Hardy Rolletschek[1]
[1]*Institut für Pflanzengenetik und Kulturpflanzenforschung (IPK), Gatersleben,*
[2]*PreSens Precision Sensing GmbH, Regensburg,*
Germany

1. Introduction

The presence of molecular oxygen is a *sine qua non* for aerobic metabolism. In both plant and animal mitochondria, it acts as the terminal electron acceptor for oxidative phosphorylation occurring during cellular respiration, and is necessary for the generation of ATP, the common energy currency within the living cell (Atkinson, 1977; Cooper, 2000). It is the major by-product of photosynthesis in which plant biomass is accumulated by the conversion of carbon dioxide into polymeric compounds. Since respiration and photosynthesis are so fundamental to life on earth, an understanding of the mechanisms underlying oxygen consumption, production and homoeostasis has become a significant field of both biological and biotechnological research (Volkmer et al., 2009).

Oxygen (micro-) sensors, which are widely used in the life sciences, are designed to provide a precise measurement of the concentration of oxygen within a localized region of a tissue or an organ (Borisjuk & Rolletschek, 2009). Most of these devices have been based on miniaturized Clark-type electrodes (Revsbech & Jørgensen, 1986), in which oxygen diffuses into the sensor via a permeable membrane, following which its reduction at the cathode generates a measurable electrical current. This approach can deliver a spatial resolution at the low μm scale. Increasingly this technology is being replaced by optical oxygen microsensors (micro-optodes) based on fibre optic materials (Klimant et al., 1995; Rolletschek et al., 2009), in which the concentration is assessed in tapered glass fibres of tip size ~50μm via the dynamic quenching of a luminophore. This approach enjoys several advantages over the electrochemical detection system, as detailed elsewhere (Kühl & Polerecky, 2008; Rolletschek et al., 2009).

Importantly, microsensor-based approaches are invasive, which means that a given biological sample cannot be readily studied over a prolonged time period. Furthermore, the internal structure of most biological samples is far from homogeneous, with complex compartmentation being the norm. As a result, whole tissue measurements can only reflect the mean performance of a tissue, and cannot report variation between distinct compartments. This loss of richness compromises the value of such data for elucidating the biology of the tissue as a whole. At best, conventional sensor systems assess oxygen

concentrations across a transect, leaving its two dimensional distribution unknown. Lifting this limitation requires the development of a planar sensor.
Here, we present a novel oxygen sensing approach, in which image processing has been combined with optical sensor technology. The optical sensor foil (i.e. the planar optode) attached to the surface of the sample translates the oxygen signal into a light signal, which is then captured and interpreted pixel by pixel by a digital camera. Since a single image captures an array of sensor points, the system permits an instantaneous two-dimensional mapping of oxygen distribution. While some analagous approaches have already been described in the literature (Liebsch et al., 2000; Glud et al., 2005; Kühl & Polerecky, 2008), the system we describe here represents a significant improvement with respect to spatial resolution, handling and image processing, and eventually ease of use. Two applications of the system are described in some detail: the first involves a respiring (oxygen consuming) root of oilseed rape (*Brassica napus*), and the second a photosynthetically active (oxygen generating) leaf of *Cabomba caroliniana*, an aquatic perennial herbaceous plant. In both, marked oxygen gradients were detected across both time and space. In combination with the use of specific inhibitors, the planar sensor system can be expected to permit a spatially well resolved analysis of respiration or photosynthesis. We conclude that the new planar sensor setup provides fascinating opportunities for research in all areas of life sciences.

2. Planar oxygen sensors – design, calibration and applications

The following chapters provide an overview on (i) the technical features of the novel planar sensor setup, and (ii) the possibilities for its use in plant biology, in particular to study respiration (oxygen consumption) and photosynthesis (oxygen production).

2.1 Experimental design for life time imaging of oxygen

Digital revolution in photography induced a giant trend towards capturing images and creating movies of nearly everything one can think of. Beside scientific and industrial cameras the market of consumer imaging devices is constantly growing and continually new products are launched showing increased resolution while being miniaturized. The enhancement of image quality and downsizing affects all market segments of consumer cameras, high-quality SLR cameras as well as low-tech webcams and mobile phone cameras. As a result, the use of such consumer devices is also of increasing interest in the field of opto-chemical sensing where the response of a fluorescent sensor is recorded in order to measure chemical analytes. Typically, for this application fairly bulky and sophisticated camera systems (Holst et al., 1998; Schröder et al., 2007; Kühl & Polerecky, 2008) are used which support time resolved measurement. Measuring a lifetime dependent parameter is generally preferred because of the favourable accuracy due to suppressing common interferences including heterogeneous lightfield or sample coloration and auto-fluorescence allowing even transparent sensor foils (Holst et al., 2001). This is not possible if using even high-tech standard consumer cameras which allow ratiometric calibration schemes at best. However, beside the restriction of transparent sensors ratiometric imaging has proved to be also an excellent solution for measuring analyte contents of a sample quantitatively and two dimensionally (Wang et al., 2008). Then, it depends on the sample target and analytical problem if the possibility of miniaturisation and mobility overcompensates the restriction. Especially in biological application fields of imaging with fluorescent optical sensors it is desirable to use compact devices which are close to pocket size and can easily be taken to

the place of measurement. As a result, complex biological systems are not disturbed and can be measured "as is" in their natural environment or green house. New reports address the topic of applying portable consumer technology by using SLR cameras (Wang et al., 2010) or even mobile phones (Filippini & Lundstrom, 2006) with the side-effect of substantially reducing the costs for imaging devices at the same time. However, in these solutions the question about suitable optical filters, macro lense and light source combinations is not sufficiently solved. Especially a micrometer resolution is indispensable if investigating biological processes of plant seeds, embryos, collenchyma or rhizospheres. Also in other medical and biotechnical applications a high resolution is needed including monitoring phase transitions in aquatic biology, mini bioreactors, tissue engineering and skin microcirculation. Therefore, we developed the idea of using consumer camera technology further and identified a type which fits perfectly to the demands of fluorescent optical imaging: the USB microscope. The results presented here were measured with a prototype of new imaging product series "VisiSens" (PreSens GmbH, Regensburg, Germany).
The market of USB microscopes is allocated to many types showing huge differences in image quality. We based our development on a current high-end USB microscope with good sensitivity and image quality and improved it for imaging fluorescence-optical sensors by integrating an optic block with high quality LED PCB and optical filters. Figure 1 shows a solid (left panel) and a transparent (middle panel) technical drawing of the measurement head, showing its compactness and the arrangement of the respective components. The three images beside show millimeter paper measured with different magnification settings. Maximum magnification is approximately 200-fold where the field of view is ~ 2.5 x 2.0 mm.

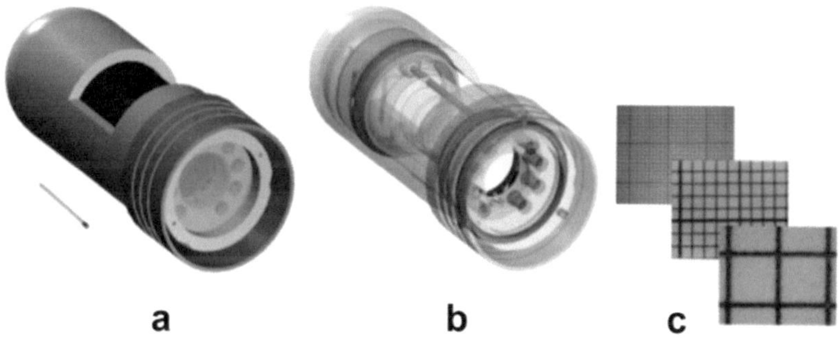

Fig. 1. (a) Solid and (b) transparent technical drawing of the compact measurement head incorporating a USB-microscope for imaging fluorescent sensor foils. Camera and light source are powered via standard USB connector. (c) Images of millimeter paper demonstrating the magnification up to 200-fold.

Figure 2 shows an explosion drawing of the USB microscope where the components are addressed in detail. The all-aluminium detector head (a) integrates a color RGB CMOS chip (b), a microscope lense (c) with manual focus, 8 blue emitting LEDs (i) which are driven by a printed circuit board (PCB) (g) and aligned in an aluminium block (h) and optical filters for light diffusion (j), excitation (k) and emission (l). The up to 200-fold magnified images are recorded with a 1.3 megapixel (1280 x 1024) color chip which results in more than 300,000 independent sensing points (= pixel) for the respective sensor response (i.e. color channel of the RGB chip). Maximum spatial image resolution is ~ 2.5 mm per 1280 pixel (~ 2 μm per pixel). Maximum spatial sensor resolution depends on the sensor used and is typically ~ 25 to 100 µm. Power supply of the camera and the LED light source is provided via the standard USB connector which makes the system laptop compatible and a portable device. The dimensions of the detection head are 10 cm in length and 3.8 cm in diameter, the working distance is typically from 1 to 5 cm. The camera can be used free hand or mounted to a stand.

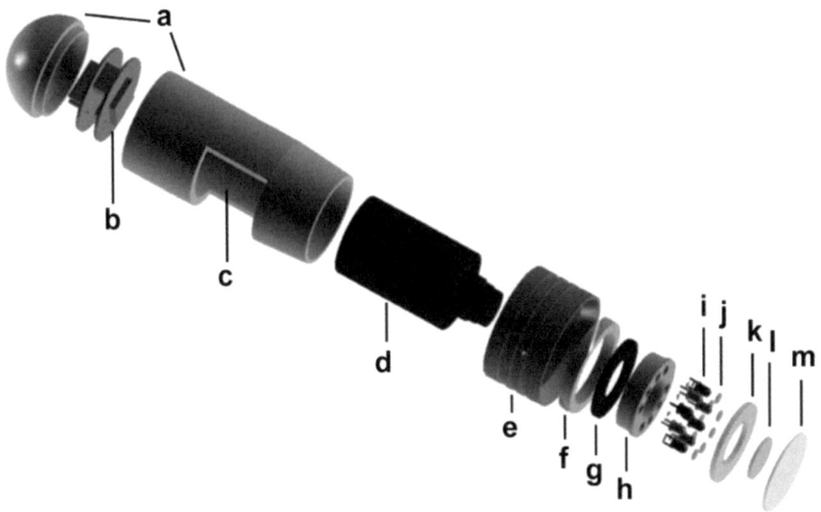

Fig. 2. Explosion drawing of the PreSens USB-microscope "VisiSens" optimized for imaging of fluorescencent sensor foils: (a) aluminium housing, (b) RGB CMOS-chip, (c) opening for manual focus, (d) microscope lense, (e) filter tube, (f) mounting and adjustment ring (g) transmitter PCB, (h) LED reflector block, (i) LEDs, (j) diffusor, (k) short pass excitation filter, (l) long pass emission filter and (m) sapphire glass.

For the measurement the imaging system uses flexible sensor foils which allow two dimensional recording of oxygen distributions in aqueous phase over an area typically ranging from 5 x 5 mm² up to 40 x 40 mm². The planar sensors consist of an oxygen sensitive dye and a reference dye which are immobilized in an oxygen permeable polymer matrix and fixed on a transparent polyester support and overlaid with a white oxygen transparent layer for optical isolation. We used a PreSens sensor foil which is not described here but similar to that described in detail by Wang et al., (2010).

During the measurement the sensitive layer is in contact with the sample and the fluorescence is measured from the backside. Every single indicator dye molecule is interacting independently with oxygen in the form that the red fluorescence of the sensitive dye is dynamically quenched. This means that the energy of the excited dye is transferred to the oxygen molecule by collision (see Fig. 3) and consequently the intensity of the sensitive dye is reduced with increasing oxygen content of the sample.

The reference dye, however, is not affected by oxygen and shows constantly a green light signal. The working range of the oxygen sensor covers the typical biological range from 0 to 100% air saturation (corresponding to 6.04 mL of oxygen per liter freshwater at 25 °C and 101.3 kPa standard atmosphere). The sensors can be shaped to any desired geometry using a scissor.

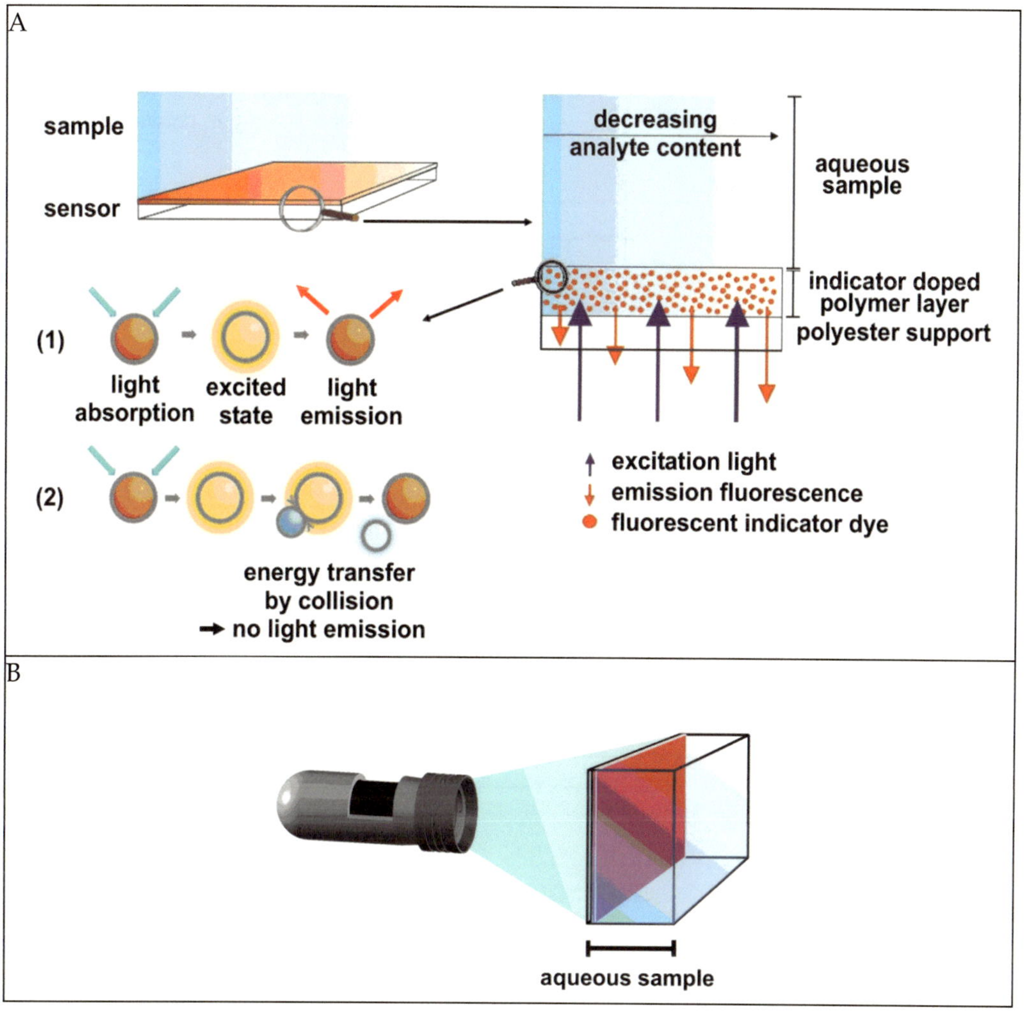

Fig. 3. (A) Schematic of the sensor foil and principle of dynamic fluorescence quenching. The sensor responds to oxygen by passing the energy of an excited state dye via collision to the oxygen molecule. This results in fluorescence quenching with increasing oxygen content of the sample. (B) Measurement set-up. The sensor consists of an indicator doped polymer layer which is fixed on a transparent polyester support. During the measurement the sensor has to be in direct contact with the sample. The sensor response is measured from outside through a transparent window of the respective vessel.

For quantitative evaluation of the sensor response we applied a ratiometric calibration scheme. The sensitive dye and reference dye are excited with the identical light source at the same time but emit at different bands of wavelength. In our case, both dyes are excited with blue light and while the sensitive dye emits red light, the reference emits green light. We selected dyes whose emission bands of wavelength meet exactly the red and green channel sensitivity of our color RGB chip. This enables to obtain the green reference information independently from the red sensor information within a single image at the same time. A quantitative evaluation is done by rationing the red and green channel of the RGB image in order to reference out the main interferences of intensity based measurements namely inhomogeneous light field and dye concentration including varying sensor layer thickness. The respective oxygen content is computed from the ratio applying a calibration function which was derived from measuring the sensor response at known oxygen concentrations in a chamber (Figure 4).

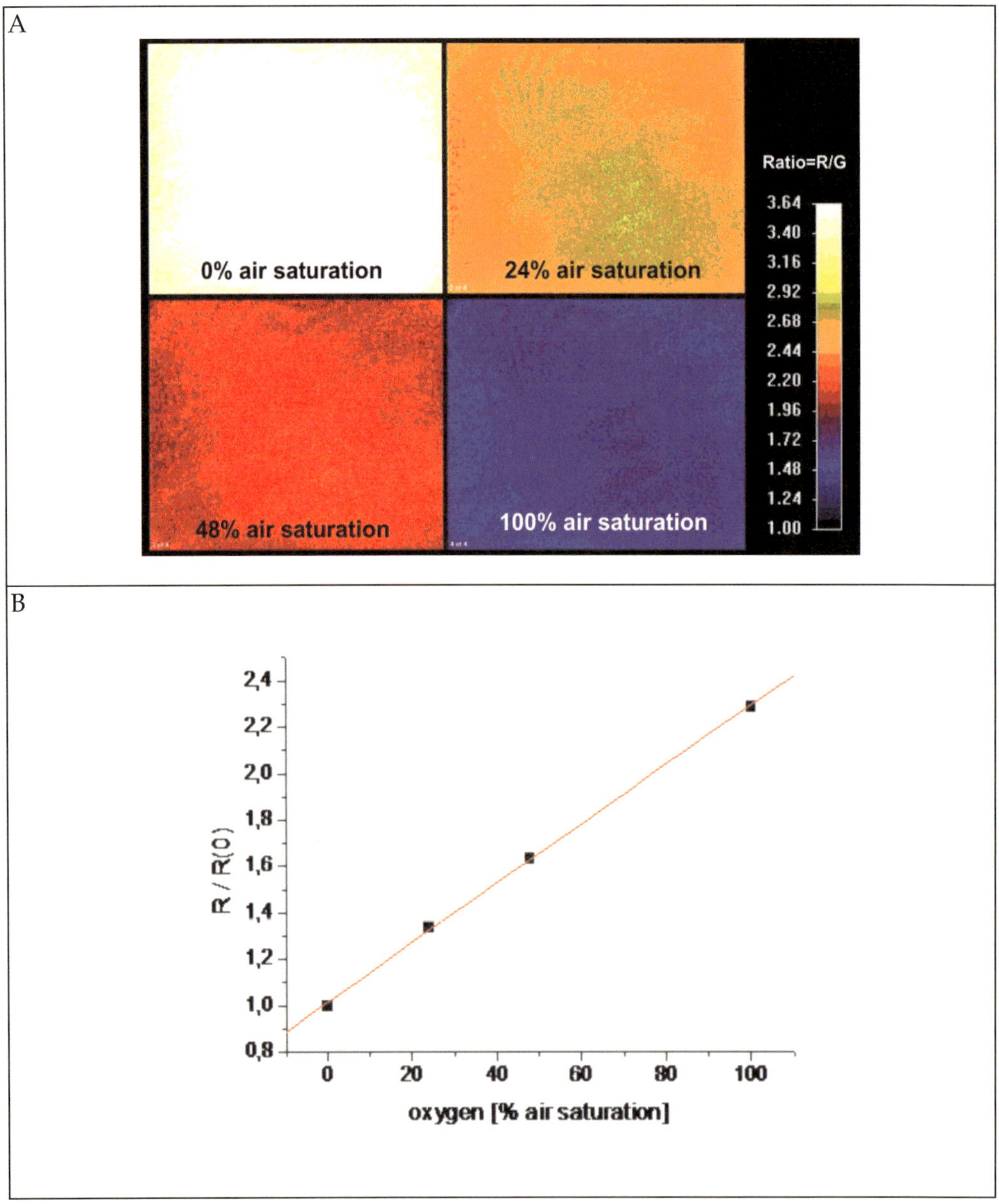

Fig. 4. (A) Calibration Images and data. The sensor was exposed to different oxygen gas mixtures and the response was imaged. (B) The Stern-Vollmer plot leads to a linear relation which was used for calculating the respective oxygen contents of the measurement images.

2.2 Application of planar optical sensors for oxygen measurements on plants

Unlike animals, plants are both producers and consumers of oxygen via photosynthesis and cellular respiration, respectively. Plant leaves, stems and seed tissues generally possess chloroplasts, producing oxygen under illumination ((Borisjuk and Rolletschek, 2009; Tschiersch et al., 2011) whereas roots are a typical example of non-green tissues. Their oxygen homoeostasis and exchange capabilities depend on the developmental state (age), several tissue characteristics (e.g. cuticula) as well as environmental conditions (e.g. temperature) (Armstrong et al., 1994, 2009). In particular, the complex anatomy of tissues often hampers oxygen diffusion, thereby causing steep diffusion gradients and local oxygen deficiencies.

Planar oxygen sensors are an alternative to conventional microsensor-based techniques, and allow to viualize the oxygen dynamics between tissues and surrounding media. Here, we demonstrate the applicability of the planar sensor system to study oxygen dynamics in two plant models: (i) the respirating (O_2-consuming) root system of the crop plant oilseed rape (*Brassica napus*), and (ii) the photosynthetically active (O_2-producing) leaf of the water plant *Cabomba caroliniana.*

2.2.1 Imaging of oxygen consumption in living plant roots

Several contrasting plant species, that differ in their relative growth rates (herbs, grasses, shrubs and trees), possess respiration rates in a relative narrow range between 20 and 52 nmol oxygen (g DW)$^{-1}$ s^{-1} (Loveys et al., 2003). Similar values were reported for roots of crop plant seedlings using clark-type electrodes (Tab. 1). For roots of oilseed rape seedlings we measured mean respiration rates of ~ 79 nmol oxygen (g DW)$^{-1}$ s^{-1} (own unpublished data).

Species	Root respiration rate [nmol O_2 (gDW)$^{-1}$ s^{-1}]	References
Oilseed rape (*Brassica napus*)	79.3	this study
Barley (*Hordeum vulgare*)	16.2	Bloom et al., 1992
Wheat (*Triticum aestivum*)	64.0	Kurimoto et al., 2004
Rice (*Oryza sativa*)	39.1	Kurimoto et al., 2004
Corn (*Zea mays*)	56.7	Hejl & Koster, 2004
Potato (*Solanum tuberosum*)	14.8	Bouma et al., 1996
Tomato (*Solanum lycopersicum*)	15.8	Hadas & Okon, 1987
Pea (*Pisum sativum*)	69.2	DeVisser et al., 1986
Soybean (*Glycine max.*)	67.5	Millar et al., 1998

Table 1. Root respiration rates of various crop plants

Using the planar oxygen sensor we here aimed to visualize quantitatively the oxygen consumption of intact roots. For this purpose oilseed rape was grown on 0.9% Difco-agar for 14 days. For the measurement the root segments of seedlings were covered with the transparent sensor foil. The use of *transparent* foils is of special importance because it allows the alignment between the sample structure and the measured oxygen distribution. The sample was enclosed by an incubation chamber to limit oxygen diffusion from the outside. This custom-made system allowed the combined imaging of the root (Fig. 5a) and the oxygen concentration (Fig. 5b). In our experiment, the oxygen distribution in the sample was followed over six hours with measuring points every 15 min. Based on the decline in oxygen concentration over time, the respiration rate of the central root zone was calculated as 0.015 % air saturation min^{-1}, which corresponds to ~ 12.5 µmol oxygen h^{-1}. The final image of oxygen distribution in Fig. 5b demonstrates that oxygen consumption of the root system can be mapped for distinct root regions with sub-millimetre resolution using the planar sensor setup. It should be noted that the use of thin sensor foils and incubation chambers is preferable, because this reduces diffusive smearing of the spatial oxygen distribution, which otherwise can mask the true oxygen dynamics as well as its gradients. Thus, it allows maximizing the spatial resolution in oxygen mapping.

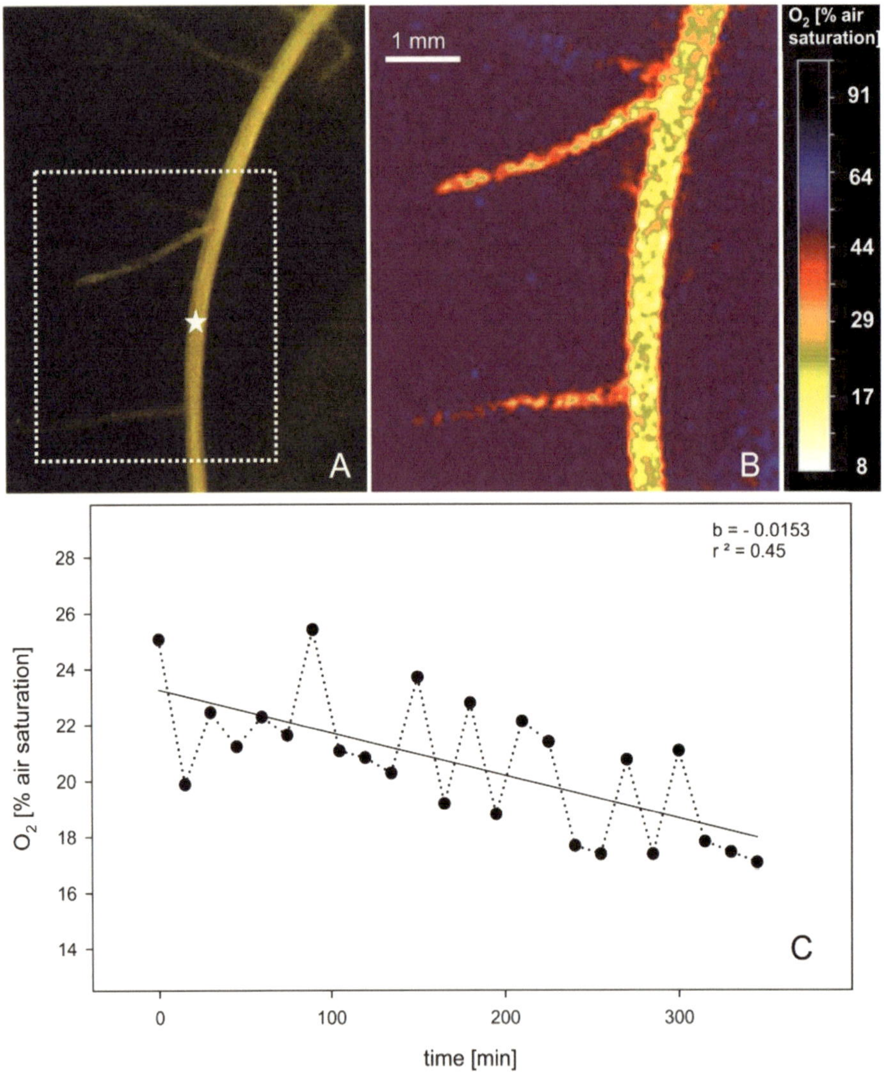

Fig. 5. Imaging of oxygen consumption for roots of oilseed rape. (A) Picture of the root system; the square indicates the area where oxygen imaging was applied. (B) Image of oxygen distribution after 6 h of dark incubation. (C) Time course of oxygen concentration measured within the main root area (region indicated by the star in (A))

2.2.2 Monitoring of oxygen dynamics in photosynthetically-active leaves

The water plant *Cabomba caroliniana* is known as a classical object in photosynthesis research (Smith, 1936; Kok, 1949). Traditionally, fronds of *Cabomba* were used in the classroom to demonstrate bubble formation (photosynthetic oxygen evolution) upon illumination. *Cabomba* leaves freshly collected from surface waters show a photosynthetic rate at light saturation of about 2.5 µmol CO_2 consumed (mg Chl)$^{-1}$ h^{-1} and a respiratory rate of 2-3 µmol CO_2 evolved (mg Chl)$^{-1}$ h^{-1} (Van et al., 1976). In a more recent study, respiration rates of up to 30 µmol O_2 (g DW)$^{-1}$ h^{-1} were reported (Penuelas et al., 1988).

Here, we aimed to visualize both consumption and production of oxygen in the *Cabomba* leaf. For this purpose, one *Cabomba* frond was covered with the sensor foil, and enclosed in the incubation chamber (restricting oxygen diffusion from the exterior). This custom-made system allowed the combined imaging of leaf structure and oxygen concentration. Figure 6a shows a picture of the *Cabomba* leaf. Its incubation under non-lit conditions for only 30 min led to a remarkable decrease in oxygen concentration, which was attributable to cellular respiration. The overlay of the color image and the sensor response image in Fig. 6b demonstrates that this decrease co-localised with the characteristic leaf shape of the plant. The oxygen level declined to ~13% of air saturation after 60 min under non-lit conditions (Fig. 6c). This indicates a substantial respiratory activity of *Cabomba* leaves. Upon illumination (15 µmol photons m^{-2} s^{-1}), a quick increase in oxygen concentration became apparent (Fig. 6c), indicating a strong oxygen production by the leaf. This activity could again be clearly aligned to the leaf shape (Fig. 6d versus 6e).

We suggest that the increase in oxygen concentration under lit conditions was caused by leaf photosynthesis. To prove this suggestion, we performed a similar light/dark experiment with *Cabomba* leaves in the presence of the photosynthesis inhibitor 3-(3,4-Dichlorophenyl)-1,1-dimethylurea (DCMU). The addition of 20µM DCMU completely abolished the light-dependent oxygen production of the leaf (Fig. 6c). This clearly demonstrates that the light-induced increase in oxygen levels is due to photosynthesis.

The differences between the oxygen exchange rates in darkness and light were remarkable, and demonstrate the high photosynthetic capacity of *Cabomba caroliniana* even under low light supply. The imaging experiment revealed that (i) the oxygen production under low light (15 µmol photons m^{-2} s^{-1}) fully compensated dark respiration, and (ii) the oxygen evolution rate in the light was five times higher as compared to the oxygen consumption rate in the dark. This result was confirmed by conventional microsensor-based analysis (optical oxygen sensor SP-PSt3 connected to Fibox3 oxygen meter; PreSens Sensing GmbH, Germany). Under non-lit conditions, a respiration rate of 26.1 µmol O_2 (g DW)$^{-1}$ h^{-1} was measured for *Cabomba* leaves. Under lit conditions (< 10 µmol photons m^{-2} s^{-1}), the photosynthetic oxygen production was sufficient to fully compensate oxygen consumption by respiration (own unpublished data).

The planar sensor system can be used to monitor fast responses of leaves to dark/light switches. Figure 7 displays in detail the differential effect of light versus darkness on the oxygen exchange of the *Cabomba* leaf. Within a few minutes clear changes in oxygen concentration could be monitored in darkness. Extended incubation under dark resulted on the one hand in a stronger decrease in oxygen concentration and on the other hand in a smearing of the sensor signal due to the diffusion of oxygen (Fig. 7, left panels). Upon start of illumination (15 µmol photons m^{-2} s^{-1}), the fluorescence signal declines steadily, which indicates the photosynthetic oxygen release. We conclude that the planar sensor system can perform both spatial and temporal oxygen monitoring.

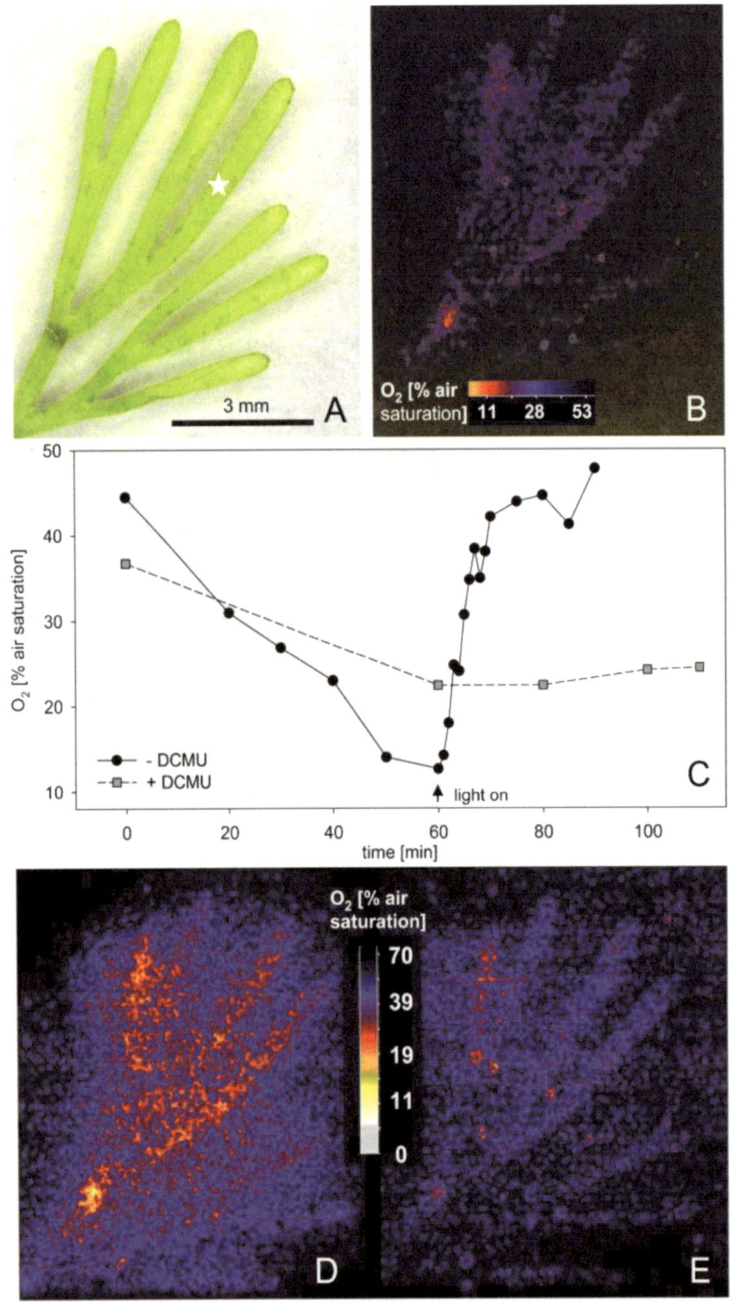

Fig. 6. Imaging of oxygen exchange by the leaf of *Cabomba caroliniana*. (A) Color image of the leaf. (B) Superimposed images of leaf and oxygen distribution after 20 min dark incubation. (C) Time course of oxygen dynamics of leaf with and without addition of DCMU; analysed region corresponds to 1.4 mm^2 leaf area marked with a star in (A). (D) Image of oxygen distribution after 30 min dark incubation. (E) Image of oxygen distribution after 30 min light incubation (15 µmol photons m^{-2} s^{-1}) subsequent to dark incubation.

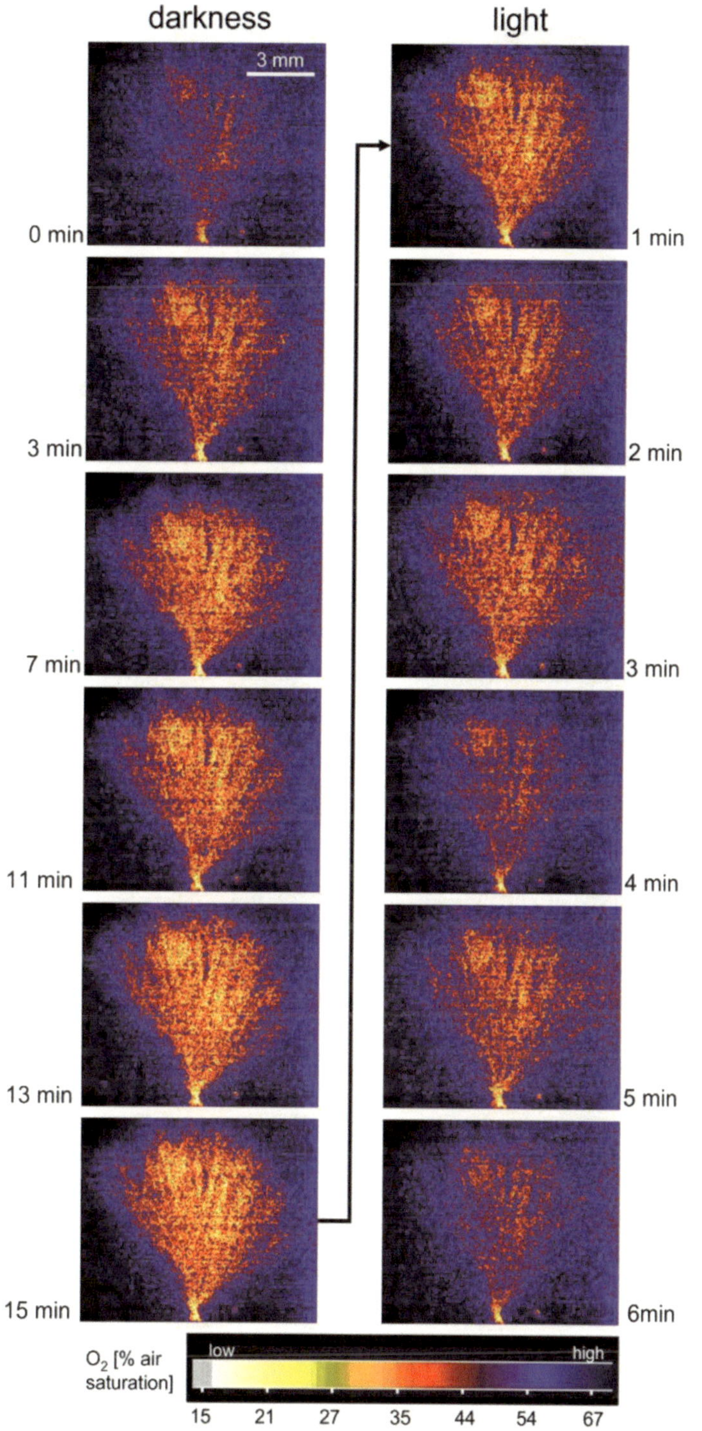

Fig. 7. Monitoring of oxygen dynamics of a *Cabomba caroliniana* leaf during 15 min of dark incubation and the subsequent light phase (15 µmol photons m^{-2} s^{-1}).

3. Prospects for planar oxygen sensing

Above data clearly illustrate the major advantage of planar oxygen sensing as a non-invasive imaging technique. For the first time, rates of oxygen production and consumption could be spatially resolved and visualized. The acquired color-coded oxygen maps are quantitative and have a resolution in the sub-millimetre range. In this way, dynamic changes in oxygen concentration within the complex root (leaf) system of a plant and its surrounding media can be studied. This non-invasive approach will allow investigating mechanisms of cellular growth and interactions among organisms and their environment. Although not studied here, the planar oxygen sensor system will be of high significance in other research areas like biotechnology or medicine. For example, documenting the oxygen dynamics during cell infection and cancerogenesis could help identify specific drug targets to slow or stop the uncontrolled growth of cancer cells (Babilas et al., 2005).
Currently, planar sensors have been developed to specifically detect oxygen, carbon dioxide or pH. It is also conceivable that planar sensor foils have multi-analyte properties. These sensors will combine an oxygen sensitive dye (and its reference dye) with dyes specific for other analytes. In this way, several analytes can be quantitatively visualized in single experiments.

4. Acknowledgment

We acknowledge funding by the Bundesministerium für Wirtschaft und Technologie within the framework of Zentrales Innovationsprogramm Mittelstand (ZIM). We also wish to thank Steffen Wagner for excellent technical assistance.

5. References

Armstrong, W., Strange, M.E., Cringle, S. & Beckett, P.M. (1994). Microelectrode and modelling study of oxygen distribution in roots. Annals of Botany, Vol.74, No.3, (September 1994), pp. 287-299, ISSN 0305-7364

Armstrong, W., Webb, T., Darwent, M. & Beckett, P.M. (2009). Measuring and interpreting respiratory critical oxygen pressures in roots. Annals of Botany, Vol.103, No.2, (January 1994), pp. 281-293, ISSN 0305-7364

Atkinson, D.E. (ed.). (1977). Cellular energy metabolism and its regulation. Academic Press, ISBN 0120661500 , New York, USA

Babilas, P., Liebsch, G., Schacht, V., Klimant, I., Wolfbeis, O.S., Szeimies, R.M. & Abels, C. (2005). In vivo phosphorescence imaging of pO_2 using planar oxygen sensors. Microcirculation, Vol.12, No.6, (September 2005), pp. 477-487, ISSN 1073-9688

Bloom, A.J., Sukrapanna, S.S. & Warner, R.L. (1992). Root respiration associated with ammonium ore nitrate absorption and assimilation by barley. Plant Physiology, Vol.99, No.4, (August 1992), pp. 1294-1301, ISSN 0032-0889

Borisjuk, L. & Rolletschek, H. (2009). The oxygen status of the developing seed. New Phytologist, Vol.182, No.1, (April 2009), pp. 17-30, ISSN 0028-646X

Bouma, T.J., Broekhuysen, A.G.M. & Veen, B.W. (1996). Analysis of root respiration of Solanum tuberosum as related to growth, ion uptake and maintenance of biomass. Plant Physiology and Biochemistry, Vol. 34, No.6, (June 1996), pp. 795–806, ISSN 0981-9428

Cooper, G.M. (2000). The Cell - A Molecular Approach (2nd edition), Sinauer Associates, ISBN-10: 0-87893-106-6, Sunderland (MA),USA

DeVisser, R., Brouwer, K.S. & Posthuma, F. (1986). Alternative path mediated ATP-synthesis in roots of Pisum sativum upon nitrogen supply. Plant Physiology, Vol.80, No.2, (February 1986), pp. 295-300, ISSN 0032-0889

Filippini, D. & Lundstrom, I. (2006). Method and system for chemical or biochemical analysis of a target analyte in a target environment. US Pat. 7,092,089.

Glud, R.N., Wenzhöfer, F., Tengberg, A., Middelboe, M., Oguri, K. & Kitazato, H. (2005). Distribution of oxygen in surface sediments from central Sagami Bay, Japan: In situ measurements by microelectrodes and planar optodes. Deep Sea Research Part I: Oceanographic Research Papers, Vol.52, No.10, (October 2005), pp. 1974–1987, ISSN 0967-0637

Hadas, R. & Okon, Y. (1987). Effect of Azospirillum brasilense inoculation on root morphology and respiration in tomato seedlings. Biology and Fertility of Soils, Vol.5, No.3, (December 1987), pp. 241-247, ISSN 0178-2762

Hejl, A.M. & Koster, K.L. (2004). Juglone disrupts root plasma membrane H^+-ATPase activity and impairs water uptake, root respiration and growth in soybean (Glycine max.) and corn (Zea mays). Journal of Chemical Ecology, Vol.30, No.2, (February 2004), pp. 453-471, ISSN 0098-0331

Holst, G. & Grunwald, B. (2001). Luminescence lifetime imaging with transparent oxygen optodes. Sensors and Actuators B, Vol.74, No.1-3, (April 2001), pp. 78–90, ISSN 0925-4005

Holst, G., Kohls, O., Klimant, I., König, B., Kühl, M. & Richter, T. (1998). A modular luminescence lifetime imaging system for mapping oxygen distribution in biological samples. Sens. Actuators B, Vol.51, No.1-3, (August 1998), pp. 163-170, ISSN 0925-4005

Klimant, I., Meyer, V. & Kühl, M. (1995). Fiber-optic oxygen microsensors, a new tool in aquatic biology. Limnology and Oceanography, Vol.40, No.6, (May 1995), pp. 1159-1165, ISSN 1541-5856

Kok, B. (1949). On the interrelation of respiration and photosynthesis in green plants. Biochimica et Biophysica Acta, Vol.3, No.1, (January 1949), pp. 625-631, ISSN 0005-2728

Kühl, M. & Polerecky, L. (2008). Functional and structural imaging of phototrophic microbial communities and symbioses. Aquatic Microbial Ecology, Vol.53, No.1, (September 2008), pp. 99-118, ISSN 0948-3055

Kurimoto, K., Day, D.A., Lambers, H. & Noguchi, K. (2004). Effect of respiratory homeostasis on plant growth in cultivars of wheat and rice. Plant Cell and Environment, Vol.27, No.7, (July 2004), pp. 853-862, ISSN 0140-7791

Liebsch, G., Klimant, I., Frank, B., Holst, G. & Wolfbeis, O.S. (2000). Luminescence lifetime imaging of oxygen, pH, and carbon dioxide distribution using optical sensors. Applied Spectroscopy, Vol. 54, No.4, (April 2000), pp. 548–559, ISSN 0003-7028

Loveys, B.R., Atkinson, L.J., Sherlock, D.J., Roberts, R.L., Fitter, A.H. & Atkin, O.K. (2003). Thermal acclimation of leaf and root respiration: an investigation comparing inherently fast- and slow- growing plant species. Global Change Biology, Vol.9, No.6, (June 2003), pp. 895-910, ISSN 1365-2486

Millar, A.H., Atkin, O.K., Menz, R.I., Henry, B., Faquhar, G. & Day, D.A. (1998). Analysis of respiratory chain regulation in roots of soybean seedlings. Plant Physiology, Vol.117, No.3, (July 1998), pp. 1083-1093, ISSN 0032-0889

Penuelas, J., Murillo, J. & Azcon-Bieto, J. (1988). Actual and potential dark respiration rates and different electron transport pathways in freshwater aquatic plants. Aquatic Botany, Vol.30, No.4, (May 1988), pp. 353-362, ISSN 0304-3770

Revsbech, N.P. & Jørgensen, B.B. (1986). Microelectrodes: their use in microbial ecology. In: Advances in Microbial Ecology, Vol.9, K.C. Marshall (ed.), pp. 293-352, Springer , ISBN 0306421844 , New York, USA

Rolletschek, H., Stangelmayer, A. & Borisjuk, L. (2009). The methodology and significance of microsensor-based oxygen mapping in plant seeds - an overview. Sensors, Vol.9, No.5, (April 2009), pp. 3218-3227, ISSN 1424-8220

Schröder, C. R., Neurauter, G. & Klimant, I. (2007). Luminescent dual sensor for time-resolved imaging of p CO_2 and p O_2 in aquatic systems. Microchimica Acta, Vol.158, No.3-4, (May 2007), pp. 205-218, ISSN 0026-3672

Smith, E.L. (1937). The influence of light and carbon dioxide on photosynthesis. The Journal of General Physiology., Vol.20, No.6, (July 1937), pp. 807-830, ISSN 0022-1295

Tschiersch, H., Borisjuk, L., Rutten, T. & Rolletschek, H. (2011). Gradients of seed photosynthesis and its role for oxygen balancing. Biosystems, Vol.103, No.2, (February 2011), pp. 302-308, ISSN 0303-2647

Van, T.K., Haller, W.T. & Bowes, G. (1976). Comparison of photosynthetic characteristics of three submersed aquatic plants. Plant Physiology, Vol.58, No.6, (December 1976), pp. 761-768, ISSN 0032-0889

Volkmer, E., Drosse, I., Otto, S., Stangelmayer, A., Stengele, M., Cherian Kallukalam, B. & Mutschler, W. (2008). Hypoxia in static and dynamic 3D culture systems for tissue engineering of bone. Tissue Engineering Part A, Vol.14, No.8, (August 2008), pp. 1331-1340, ISSN 2152-4947

Wang, X.D., Chen, X., Xie, Z.X. & Wang X.R. (2008). Reversible optical sensor strip for oxygen. Angewandte Chemie, Vol.120, No.39, (September 2008), pp. 7560-7563, ISSN 1521-3757

Wang, X.D., Meier, R.J., Link, M. & Wolfbeis, O.S. (2010). Photographing oxygen distribution. Angewandte Chemie International Edition, Vol.49, No.29, (July 2010), pp. 4907-4909, ISSN 1433-7851